RADIATIVE PROPERTIES OF HOT DENSE MATTER III

PROCEEDINGS OF THE
3RD INTERNATIONAL CONFERENCE ON

RADIATIVE PROPERTIES OF HOT DENSE MATTER III

Williamsburg, Virginia, Oct 14-18, 1985

Editors
Balazs Rozsnyai
Charles Hooper
Robert Cauble
Richard Lee
Jack Davis

World Scientific

Published by
World Scientific Publishing Co Pte Ltd
P. O. Box 128, Farrer Road, Singapore 9128.

Library of Congress Cataloging-in-Publication data is available.

RADIATIVE PROPERTIES OF HOT DENSE MATTER III

Copyright © 1987 by World Scientific Publishing Co Pte Ltd.

All rights reserved. This book, or parts theoreof, may not be reproduced in any form or by any means, electronic or mechanical, including photocopying, recording or any information storage and retrieval system now known or to be invented, without written permission from the Publisher.

ISBN 9971-50-235-6

Printed in Singapore by Kim Hup Lee Printing Co. Pte. Ltd.

FOREWORD

This volume contains the proceedings of the Third International Conference on Radiative Properties of Hot Dense Matter, held in Williamsburg, Virginia October 14–18, 1985. Accordingly, the volume bears the label III for distinction. The proceedings of the first and second conferences of the same title were published in the March 1982 edition of JQSRT and by World Scientific 1985, respectively. As was the case in the previous conferences, the organizers wish to share with the scientific community the papers presented at this conference; hence we are publishing the proceedings in this volume.

Balazs Rozsnyai
Charles Hooper
Robert Cauble
Richard Lee
Jack Davis

Editors

CONTENTS

Foreword   v

I. FREE-FREE RADIATION AND SCATTERING

The Inverse Bremsstrahlung Absorption Coefficient in Strongly Coupled Plasmas
R. Cauble, W. Rozmus & J. Davis   3

Can Simple Methods and Simple Potentials be Used in Obtaining Free-Free Gaunt Factors for Strongly Coupled Aluminum Plasmas?
M. Lamoureux, R. Cauble, L. Kim, F. Perrot & R. H. Pratt   15

Continuum Emission (both Isotropic and Polarized) in a Laser-Plasma Described by a Fokker Planck Simulation
M. Lamoureux, J. P. Matte, C. Möller & R. Y. Yin   33

Comparison of Models for the Free-Free Gaunt Factor at Low Temperatures and Frequencies
L. A. Collins & A. L. Merts   57

Computation of Angle Averaged Cross Sections in a Degenerate Compton Scattering Medium
J. M. Pochkowski, G. H. Nickel & W. F. Bailey   77

Optical Potential Approach for Scattering of Electrons by Ions
D. H. Oza & J. Callaway   84

II. X-RAY SPECTROSCOPY, THEORY AND EXPERIMENT

Non-Thermal Effects in a Hot Dense Plasma
L. A. Jones   107

X-Ray Spectroscopy and Determining Plasma Dynamics
R. L. Kauffman, K. G. Estabrook & R. W. Lee   128

Sub-KeV X-Ray Emission from Laser-Produced Plasmas
K. Eidmann, T. Kishimoto, G. D. Tsakiris, R. F. Schmalz, R. Sigel & S. Witkowski   136

X-Ray Spectroscopy of Ne-Like and Na-Like Strontium Ions
J. C. Gauthier, J. P. Geindre, P. Monier, C. Chenais-Popovics, J. F. Wyart & E. Luc-Koenig   154

Radiative Emission near Critical Density in Laser-Produced Plasma
 P. G. Burkhalter, J. P. Apruzese & D. Duston   170

Argon Puff Gas Soft X-Ray Laser
 J. Davis, J. P. Apruzese, C. Agritellis & P. Kepple   197

Collective Vector Method for Calculation of E1 Moments in Atomic
Transition Arrays
 S. D. Bloom & A. Goldberg   230

Photoionization and Photorecombination Cross Sections of Non-
Hydrogenic States in Plasmas
 B. F. Rozsnyai & V. L. Jacobs   264

III. RADIATIVE TRANSFER AND OPACITY

Radiation Transport Phenomena affecting X-Ray Laser Design
 J. P. Apruzese   293

Studies of the Solution of a Radiation Transfer Benchmark
 L. Yobs, J. M. Salter & R. W. Lee   324

Physics of Partial Redistribution and Radiative Transfer in
Stellar Atmospheres
 J. L. Linsky   333

A Fast Accurate Code for Hydrogen and Helium Opacities
 T. R. Carson   374

IV. EQUATION OF STATE, RATE EQUATIONS AND ELECTRODYNAMICS

Electrical Conductivity of Dense Plasmas at Temperatures of
.5 to 1.0 eV
 S. W. Daniels, H. R. Griem & A. D. Krumbein   389

Density Fluctuations and Channel Mixing Effects in the
Absorption of Light by Dense Plasmas
 F. Grimaldi, A. Grimaldi-Lecourt & M. W. C. Dharma-wardana   403

Atomic Process Calculations in Hot Dense Plasmas using Average
Atom Models
 G. Velarde, J. M. Aragones, L. Gamez, J. J. Honrubia, J. M. Martinez-Val,
 E. Minguez, J. L. Ocaña, J. M. Perlado & J. F. Serrano   433

Ion Microfields in Plasmas with Strong Electron-Ion Interactions
Density-Functional Theory
   M. W. C. Dharma-wardana & F. Perrot   462

Solid Density, Low Temperature Plasma Formation in a Capillary
Discharge
   D. R. Kania, L. A. Jones, M. D. Maestas & R. L. Shepherd   483

A Time-Dependent Ionization Balance Model for Non-LTE Plasma
   Y. T. Lee, G. B. Zimmerman, D. S. Bailey, D. Dickson & D. Kim   495

The Ponderomotive Force: Derivation of the Stress-Tensor for High-
Frequency Radiation in Isotropic Dispersive Matter by Generalisation
of the Helmholtz Method
   B. J. B. Crowley   534

I. FREE-FREE RADIATION AND SCATTERING

The Inverse Bremsstrahlung Absorption Coefficient in Strongly Coupled Plasmas

R. Cauble, Berkeley Research Associates
Springfield, VA 22151

W. Rozmus, University of Alberta
Edmonton, Alberta CANADA

and

J. Davis
Naval Research Laboratory, Plasma Physics Division
Washington, D. C. 20375

The calculation of the inverse bremsstrahlung coefficient, which describes the absorption of laser light in plasma, reduces classically to a derivation of the frequency dependent electron-ion collision frequency, $\nu(\omega)$. In laser-plasma interactions at densities and temperatures typical of discussion at this conference, the usual Dawson-Oberman formulation of the collision frequency can be very much in error, since $\nu(\omega)$ in that case depends on the use of binary collisions and the concomitant assumption of weak particle correlations. A prescription of $\nu(\omega)$ has been constructed from classical kinetic theory using a semiclassical two-particle interaction thought to be valid when the plasma is moderately to strongly coupled. The prescription calls for static interparticle correlations functions which are supplied by the solution to the hypernetted chain (HNC) equations utilizing the semiclassical potential. These results are contrasted with the Dawson-Oberman result and results using employing Debye-Hückel correlations. Since the latter provide simple analytic forms for $\nu(\omega)$, we compare these results with those from HNC in hydrogen when the plasma coupling paramter Γ is $\lesssim 1$.

I. Introduction

In the simplest notion of laser light absorption in dense matter, laser radiation interacts with the resulting plasma mainly via inverse bremsstrahlung. This is an environment in which many things are happening, including the production of a dense cool plasma which is said to be strongly coupled (a situation in which usual plasma perturbation theories break down). When the usual theories break down, so does the classical description of inverse bremsstrahlung since this description is based on an electron-ion collision frequency valid only to the kinetic level of the Vlasov equation.

In order to estimate inverse bremsstrahlung in strongly coupled plasmas, a kinetic description of the electron-ion collision frequency appropriate in this regime must be used. This quantity must be frequency dependent and thus calls for a non-Markovian kinetic collision operator. Another application of this collision frequency is the AC electrical conductivity. The DC conductivity for strongly coupled plasmas has been treated elsewhere.[1,2] The collision frequency finds ready application in plasmas heated from low temperatures such as in inertial confinement fusion experiments. We will examine fully ionized hydrogen plasmas with densities between 10^{20} and 10^{24} cm^{-3}.

II. Kinetic Description of the Collision Frequency

The inverse bremsstrahlung absorption coefficient is given by[3]

$$\kappa = \frac{\nu(\omega_L)}{c} \frac{n_e}{n_{cr}} (1-n_e/n_{cr})^{-1/2} , \qquad (1)$$

where n_e is the mean plasma electron density, n_{cr} is the critical density given by

$$n_{cr} = \omega_L^2 \frac{m_e}{4\pi e^2} ,$$

and ω_L is the laser light frequency. The quantity $\nu(\omega)$ is the frequency dependent electron-ion collision frequency. It is this quantity valid in strongly coupled hydrogen that we will calculate below. Expression (1) is strictly only valid for $\omega > \omega_{pe}$, the electron plasma frequency. For $\omega \lesssim \omega_{pe}$ collective effects will substantially affect the plasma dispersion relation from which (1) was derived.

The starting point for the solution of $\nu(\omega)$ is a plasma kinetic equation especially suitable for the description of collisionally dominated plasmas. Here $\nu(\omega)$ is found in terms of the collision operator in the equation for the time evolution of the phase-space density equilibrium correlation function,

$$C(1,2;t) = \langle [\sum_{j=1}^{N_{\alpha_1}} \delta(p_1 - p_j(t)) \delta(r_1 - r_j(t)) - n_{\alpha_1} M_{\alpha_1}(p_1)]$$

$$\times [\sum_{j=1}^{N_{\alpha_2}} \delta(p_2 - p_j(0)) \delta(r_2 - r_j(0)) - n_{\alpha_2} M_{\alpha_2}(p_2)] \rangle \qquad (2)$$

Here, $1 = (r_1, p_1, \alpha_1)$ and $2 = (r_2, p_2, \alpha_2)$, where $\alpha_1, \alpha_2 = i, e$, and M_{α_1} and M_{α_2} are the Maxwell-Boltzmann distribution functions.

The kinetic equation for $C(t)$ reads:

$$\frac{\partial}{\partial t} C^{\alpha_1 \alpha_2}(k, p, p_2; t) - \sum_{\alpha_3 = e,i} \int_0^\infty d\bar{t} \int dp_3 \Phi_{\alpha_1 \alpha_3}(k, p_1, p_3; \bar{t})$$

$$\times C^{\alpha_3 \alpha_2}(k, p_3, p_2; t - \bar{t}) = 0 , \qquad (3)$$

where

$$C^{\alpha_1 \alpha_2}(k, p_1, p_2; t) = \int d(r_1 - r_2) e^{-ik \cdot (r_1 - r_2)} C^{\alpha_1 \alpha_2}(1, 2; t)$$

and $\Phi_{\alpha_1 \alpha_3}(k, p_1, p_3; t)$ is the complete collision operator. $\Phi_{\alpha_1 \alpha_3}$ consists of two static terms and a time-dependent part,[4]

$$\Phi_{\alpha_1 \alpha_3}(k, p_1, p_3; t) = \Phi^{FP}_{\alpha_1 \alpha_3}(k, p_1, p_3) + \Phi^{MF}_{\alpha_1 \alpha_3}(k, p_1, p_3) + \Phi^{C}_{\alpha_1 \alpha_3}(k, p_1, p_3; t)$$

where

$$\phi^{FP}_{\alpha_1\alpha_3}(k,p_1,p_3) = i\,\frac{k\cdot p_1}{m_{\alpha_1}}\,n_{\alpha_1}M_{\alpha_1}(p_1)c_D^{\alpha_1\alpha_3}(k)\delta(t)$$

and   (4)

$$\phi^{MF}_{\alpha_1\alpha_3}(k,p_1,p_3) = -i\,\frac{k\cdot p_1}{m_{\alpha_1}}\,n_{\alpha_1}M_{\alpha_1}(p_1)c_D^{\alpha_1\alpha_3}(k)\delta(t)\ .$$

In $\phi^{MF}_{\alpha_1\alpha_3}$, $c_D^{\alpha_1\alpha_3}(k)$ is the Fourier transformed direct correlation function for pair $\alpha_1\alpha_3$. For simplicity, we subsequently will often delete the α-subscripts and superscripts; their meaning as species identifiers will be understood.

The first term in (4) represents free streaming of the α_1-particles. The second term contains the mean force (via c_D) seen by the particles streaming in the plasma. The time-dependent third term contains the effects of particle collisions. The solution for $C(t)$ in (3) depends on the approximations made for $\phi^c(t)$.

The frequency-dependent electron-ion collision frequency can be identified with an integral of the collisional part of the memory function,[5]

$$\nu(\omega) = \frac{1}{\dfrac{n_i Z^2 e^2 k_B T}{m_i} + \dfrac{n_e e^2 k_B T}{m_e}}\ \mathrm{Re}\Big\{\sum_{\alpha_1\alpha_2}\frac{Z_{\alpha_1}Z_{\alpha_2}}{m_{\alpha_1}m_{\alpha_2}}$$

$$\times \int dp_1 dp_2 (\hat{k}\cdot p_1)\,i\phi^c_{\alpha_1\alpha_2}(k=0,p_1,p_2,z=\omega) n_{\alpha_2} M_{\alpha_2}(p_2)(\hat{k}\cdot p_2)\Big\}\ ,\qquad (5)$$

where

$$\phi^c_{\alpha_1\alpha_2}(k,p_1,p_2,z) = \int_0^\infty dt\, e^{izt}\,\phi^c_{\alpha_1\alpha_2}(k,p_1,p_2;t)$$

and

$$\hat{k} = k\,/|k|\ .$$

We assume for $\phi^c(t)$ the two component generalization of the collision term given by Wallenborn and Baus[6] in the approximate localized form,

$$i\phi^c_{\alpha_1\alpha_2}(k=0,\mathbf{p}_1,\mathbf{p}_2;t)n_{\alpha_2}M_{\alpha_2}(\mathbf{p}_2) = \frac{-k_BT}{2}\int d\mathbf{p}_3 d\mathbf{p}_4$$

$$\times \sum_{\alpha_3\alpha_4}\int\frac{d\boldsymbol{\ell}}{8\pi^3}\{c_D^{\alpha_1\alpha_4}(\ell)V^{\alpha_2\alpha_3}(\ell)(\boldsymbol{\ell}\cdot\frac{\partial}{\partial\mathbf{p}_1})(\boldsymbol{\ell}\cdot\frac{\partial}{\partial\mathbf{p}_2})\times \qquad (6)$$

$$[C^{\alpha_1\alpha_2}(-\ell,\mathbf{p}_1,\mathbf{p}_2;t)C^{\alpha_3\alpha_4}(\ell,\mathbf{p}_3,\mathbf{p}_4;t)-C^{\alpha_1\alpha_3}(-\ell,\mathbf{p}_1,\mathbf{p}_3;t)C^{\alpha_4\alpha_2}(\boldsymbol{\ell},\mathbf{p}_4,\mathbf{p}_2;t)]$$

$$+ [1 <\text{-}> 2]\},$$

where $V^{\alpha_2\alpha_3}(\ell)$ is the transformed pair potential.

In order to find $\phi^c(t)$, we require a knowledge of $C(t)$, but Eq. (6) implicitly requires the full solution for $C(t)$ in terms of $\phi^c(t)$. Therefore, Eq. (3) must be solved in some "lower order" approximation so that the solutions can be introduced in (6). The simplest form for $C(t)$ is the solutin of Eq. (3) with $\phi^{MF} = \phi^c(t) = 0$. This free particle (FP) form is a gaussian in time, but includes known statics via initial time correlations.

Eq. (3) for $C(t)$ with $\phi^c(t)=0$ yields a form identical to the linearized Vlasov equation, except for the appearance of $c_D(k)$.[7] This mean field (MF) solutin of $C(t)$ is more complete than FP, but more complicated. It has been speculated[8] that the presence of correct static correlations is more significant than the exact dynamical evolution of $C(t)$. We can compare FP and MF evaluations of $\nu(\omega)$ to see if this is the case, at least for the collision frequency.

The forms for $\nu(\omega)$ are found by solving (3) for $C(t)$ in either the FP or MF approximation, substituting $C(t)$ into Eq. (6), and performing the integrations dictated by (5). Details can be found elsewhere.[9]

For example, the FP form reads

$$\nu^*(\omega) = \frac{\nu_{FP}(\omega)}{\omega_{pe}} = -\sqrt{\frac{2}{\pi}} \frac{1}{9\bar{k}_{De}} \int_0^\infty dx\, x^3 c_D^{ei}(x)\, V^{ei}(x)\, L(x)\, \exp[-\bar{\omega}^2 \bar{k}_{De}^2/2x^2]\,, \qquad (7)$$

where $\bar{k}_{De} = k_{De}a$ (k_{De} is the electron inverse Debye length; a is the ion sphere radius), $\bar{\omega} = \omega/\omega_{pe}$, $c_D^{ei}(x)$ is the electron-ion direct correlation function, $V^{ei}(x)$ is the transformed pair potential, and $L(x)$ is the combination of static structure factors

$$L(x) = S^{ee}(x)S^{ii}(x) - S^{ei}(x)S^{ie}(x)\,.$$

In the limit of weak coupling, employing the Coulomb potential, Eq. (7) rigorously reduces to

$$\nu_{FP}^*(\omega)\Big|_{\Gamma \to 0} = \sqrt{\frac{2}{3\pi}}\, \Gamma^{3/2}\, \frac{1}{2}\, E_1\left\{ \frac{\bar{\omega}^2 k_{De}^2}{2k_{max}^2} \right\}\,, \qquad (8)$$

where $\Gamma = Ze^2/k_B Ta$ is the usual plasma coupling parameter and E_1 is an exponential integral. The quantity k_{max} is a large k-vector cutoff. When the argument of E_1 is small ($n_e < n_{cr}$, $k_{De} < k_{max}$), Eq. (8) becomes

$$\nu_{DO}^*(\omega) = \sqrt{\frac{2}{3\pi}}\, \Gamma^{3/2}\, \ln(k_{max}/k_{De}\bar{\omega})\,, \qquad (9)$$

which is the classical Dawson-Oberman[10] result if the cutoff is chosen by

$$k_{max} = \text{MIN}\left\{ \frac{k_B T}{Ze^2},\, \frac{(2\pi m_e k_e T)^{1/2}}{\hbar} \right\}\,.$$

The static correlations, $c_D^{ei}(k)$ and $S^{\alpha\beta}(k)$, can be estimated by their Debye-Hückel (DH) forms, but it is known that DH correlations are far from adequate in strongly coupled hydrogen plasmas.[11] An alternative method of generating these correlations is the hypernetted chain (HNC) approximation.[12] HNC is an approximation to the hierarchy of integral equations defining static correlation functions. Solved iteratively with other relations, HNC provides correlations that are in excellent agreement with those from computer simulations[11] for plasmas where Γ is of order one.

HNC is a classical statistical mechanical procedure, so in order to prevent the system from collapsing, as well as to introduce short range quantum effects, a model pair potential is assumed. Such a model potential is[13]

$$V^{\alpha_1 \alpha_2}(r) = \frac{Z^{\alpha_1} Z^{\alpha_2} e^2}{r} [1 - \exp(-r/\lambda_{\alpha_1 \alpha_2})]$$

$$+ \delta_{\alpha_1 e} \delta_{\alpha_2 e} \ln(2) k_B T \exp\{-r^2/[\pi \ln(2) \lambda_{ee}^2]\}$$

(10)

where $\alpha_1, \alpha_2 = e, i$ and $\lambda_{\alpha_1 \alpha_2} = \hbar/(2\pi \mu_{\alpha_1 \alpha_2} k_B T)^{1/2}$ is the thermal de Broglie wavelength for the pair α_1, α_2 of reduced mass $\mu_{\alpha_1 \alpha_2}$. The non-Coulombic part of the first term takes approximate account of short range diffraction effects. The second term reflects electron symmetry due to the Pauli exclusion principle. This term is repulsive and non-negligible when T is of the order of the Fermi temperature. This model has been used in the computer simulations.[11]

III. Results

Using Eq. (10) as the pair potential, static correlations from DH and HNC approximations and correlation dynamics from the FP and MF approximations provide four alternatives to the Dawson-Oberman (DO) collision frequency.

Figure 1 compares the forms for $\nu(\omega)$. The hydrogen plasma has a density of 10^{20} cm^{-3} and a temperature of 13.6 eV. At this density the difference between DH and HNC statics is insignificant. Near the plasma frequency, the FP/DH form is 20% larger than the DO result; this difference increases with frequency. Near $\omega = \omega_{pe}$, MF/DH is over 45% larger than DO. At higher frequencies than shown here, the MF form reduces to the FP result. For this case, the inclusion of the more complete MF dynamic representation as opposed to FP represents as much a difference from DO as the use of the kinetic theory presented here. For $\nu(\omega)$, then it is important to employ the better dynamics if a resolution of about 20% is needed.

The case of 10^{21} cm^{-3} is displayed in Fig. 2. The use of HNC correlations here leads to corrections of about 20% over use of DH correlations. The additional forces represented by the mean field term in the MF solution for $C(t)$ lead to a larger $\nu(\omega)$ throughout the frequency spectrum. The DO form becomes negative at $\omega/\omega_{pe} \approx 5.3$; its validity is questionable, however, at much lower frequencies.

Much the same observations can be made from Fig. 3 where $n_e = 10^{22}$ cm^{-3}. The KrF laser critical frequency is a few times 10^{22} cm^{-3}, so we present curves for 10^{23} cm^{-3} in Fig. 4. Here the DO result is negative at all ω above ω_{pe}. The reduction of MF to FP at high frequencies can easily be seen. Also at very high frequencies, spatial variations are less important, so that the HNC result is no different from the DH result.

Overall, all corrections to Dawson-Oberman lead to larger collision frequencies and thus larger inverse bremsstrahlung coefficients (and smaller AC conductivities). Use of HNC statics instead of DH correlations increase $\nu(\omega)$ by about 20% at larger densities. The difference decreases at higher frequency. Employing MF dynamics instead of FP dynamics leads to an increase of the same order at most densities, also diminishing at higher frequency.

An improved dynamical result might be obtained by accepting the MF result for $C(t)$ - the seed for the MF solution for $\nu(\omega)$ via (5) and (6) - finding $\phi^C(t)$ by (6), solving (3) for $C(t)$ in the new collision-inclusive approximation. This new $C(t)$ could then be substituted directly in (6) for use in (5) to get $\nu(\omega)$ or the process iterated to a converged $C(t)$ before using (6). It is not known whether this solution would provide significant differences in the MF result found here. The inclusion of collective effects near the plasma frequency should also be included.

Of primary importance, however, is the introduction of non-classical absorption coefficient. $\nu(\omega)$ from Eq. (5) rigorously reduces to the work of Dawson and Oberman[10] (if cut-offs are assumed), but itself introduces no artificial cut-offs. Correct screening, through numerical correlations can be included. Even when the plasma is only moderately coupled, the corrections to the classical $\nu(\omega)$ can be substantial (factors of two or more). For strong coupling or higher absorption frequency, the classical form is not useful while even the Debye form here is finite and positive. Since coupling increases linearly with Z, deviations from classical results will occur at lower densities and higher temperatures for higher-Z plasmas.

ACKNOWLEDGEMENTS

This work was supported by the Office of Naval Research Laboratory and Natural Science and Engineering Research Council of Canada.

References

1. D. B. Boercker, F. Rogers, and H. DeWitt, Phys. Rev. A**25**,1623 (1982).

2. R. Cauble and W. Rozmus, to appear in Phys. Lett A (1986).

3. C. E. Max, in *Laser-Plasma Interactions*, ed. by R. Balian and J. C. Adam (North-Holland, Amsterdam, 1982).

4. G. F. Mazenko and S. Yip, in *Statistical Mechanics, Part B*, ed. by B. J. Berne (Plenum, New York, 1977).

5. D. B. Boercker, Phys. Rev. A **23**,1969 (1981).

6. J. Wallenborn and M. Baus, Phys. Rev. A **18**,1737 (1978).

7. J. L. Lebowitz, J. K. Percus, and J. Sykes, Phys. Rev. **188**,487 (1969).

8. R. Cauble and D. B. Boercker, Phys. Rev. A **28**,944 (1983).

9. R. Cauble and W. Rozmus, Phys. Fluids **28**,3387 (1985).

10. J. M. Dawson and C. Oberman, Phys. Fluids **5**,517 (1962).

11. J. P. Hansen and I. R. McDonald, Phys. Rev. A **23**,2041 (1981).

12. J. F. Springer, M. A. Pokrant, and F. A. Stevens, Jr., J. Chem. Phys. **58**,4863 (1973).

13. C. Deutsch, Phys. Lett. A **60**,317 (1977); C. Deutsch, M. M. Gombert, and H. Minoo, Phys. Lett. A **66**,381 (1978) and **72**,481 (1979).

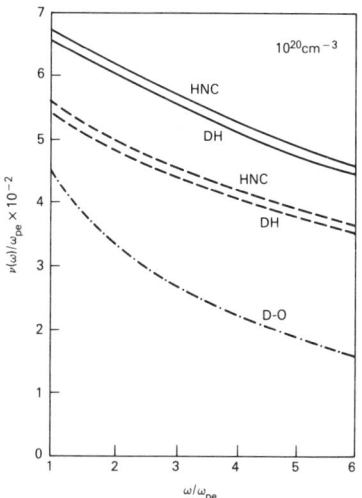

Fig. 1. $\nu(\omega)$ vs. ω for hydrogen plasma at 13.6 eV and 10^{20}cm^{-3}. Solid lines are MF results; dashed are FP results. HNC employs hypernetted chain correlations; DH employs Debye-Hückel. D-O is the Dawson-Oberman form.

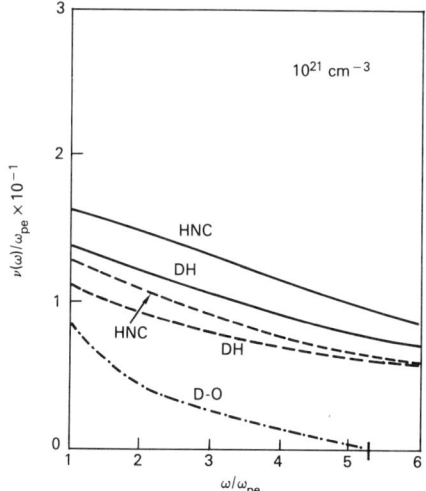

Fig. 2. Same as Fig. 1 except the density is 10^{21}cm^{-3}.

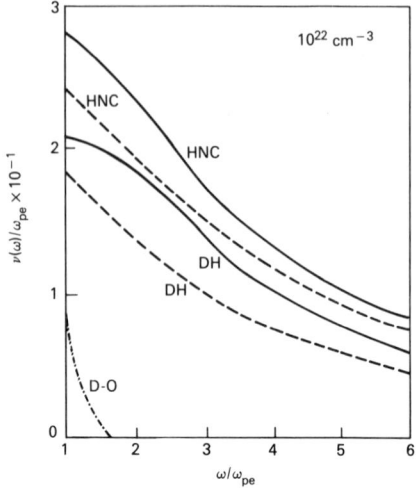

Fig. 3. Same as Fig. 1 except the density is 10^{22}cm^{-3}.

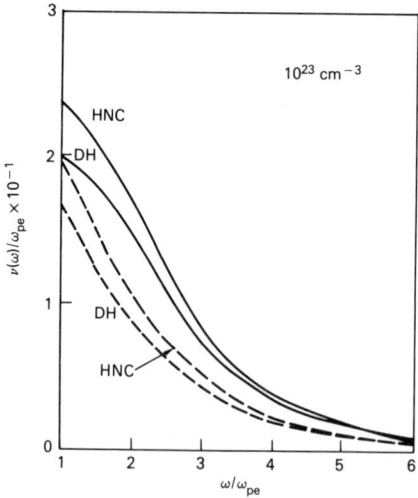

Fig. 4. Same as Fig. 1 except the density is 10^{23}cm^{-3}.

CAN SIMPLE METHODS AND SIMPLE POTENTIALS BE USED IN OBTAINING FREE-FREE GAUNT FACTORS FOR STRONGLY COUPLED ALUMINUM PLASMAS?

M. Lamoureux, Universite Paris-Sud, Orsay, FRANCE

R. Cauble, Berkeley Research Associates, Springfield, VA*

L. Kim, University of Pittsburgh, Pittsburgh, PA

F. Perrot, Centre d'etudes de Limeil-Valenton, Villeneuve-St. Georges, FRANCE

R.H. Pratt, University of Pittsburgh, Pittsburgh, PA

A detailed comparison is made of three methods of obtaining bremsstrahlung Gaunt factors in strongly coupled aluminum plasmas. Each method -- the usual Born-Elwert form, a fully relativistic multipole partial wave expansion, and a "classical" form derived from classical mechanics-- requires an effective interparticle potential as input. We examine the three atomic methods employing five different electron-ion potentials: the ion-sphere potential, the Debye-Hückel potential, the Thomas-Fermi form, a hypernetted chain atomic potential derived from the solution of the Poisson equation using charge densities obtained from the hypernetted chain equations, as well as a density functional theory potential. By comparing the results, the regions of validity of the simpler methods and atomic models can be estimated. Results indicate that even for very dense plasmas, these simpler potentials and methods can be very useful. Thus, large amounts of computation time can be saved.

I. INTRODUCTION

Electron bremsstrahlung is a primary source of radiative cooling as well as an important diagnostic in laboratory plasmas. The bremsstrahlung process as treated in a static potential is fairly well understood and in many situations, accurate theoretical descriptions of the resulting radiation can be obtained. However, it is often necessary to predict the evolution of the plasma using large computer codes in which bremsstrahlung is but one, though costly, process. It is, therefore, very useful to know how accurately simple methods of finding bremsstrahlung cross sections work relative to well-known but time-intensive methods. The problem is especially difficult in the case of a strongly coupled plasma (SCP) since the density is high and in SCP's the calculation of the interatomic potential is very involved.

Here we will examine three distinct methods of finding free-free Gaunt factors, each of which explicitly incorporates the electron-ion potential. The relativistic partial wave method, known to provide experimentally verified results in neutral atom cross section calculations, is used as the standard. Five different potentials, again with one designated as a standard, are examined within the methods. All potentials are "average atom" (AA) potentials as opposed to those from "distinct configuration" models. AA can be an appropriate model for plasmas, either when correlation times are short in comparison to reaction times or when interior regions of the potential dominate. As a testbed, we examine strongly coupled aluminum plasmas at temperatures of 0.5 and 1.0 keV and electron densities, n_e, between 10^{23} and $10^{24} cm^{-3}$.

II. ELECTRON ION POTENTIALS

Two important analytical potentials are the Debye-Hückel (DH) and ion-sphere (IS) potentials. The DH form is appropriate for weakly coupled plasmas and for spatial separations greater than a Debye length. In terms of the unitless screening parameter $V(r)$, which satifies $V(r=0)=1$, the DH form is:

$$V_{DH}(r) = \frac{rU_{DH}(r)}{eZ} = e^{-r/\lambda_D}, \qquad (1)$$

where $U(r)$ is the potential and λ_D is the usual Debye length in which the ionic charge has been taken to be $\bar{Z}$, the mean charge per ion. The IS potential is thought to be valid when the density is very high. The form,

$$V_{IS}(r) = \left[1 - \frac{r}{2r_o}\left[3 - \left(\frac{r}{r_o}\right)^2\right]\right]\frac{\bar{Z}}{Z} \qquad (2)$$

is a solution of the Poisson equation for a uniform electron density inside a sphere of radius, r_o, the ion-sphere radius, containing an ion of charge $\bar{Z}$. Note that $U_{IS}(r=0) \neq 1$. This model includes no electron correlations and includes ion correlations only by the exclusion of other ions from the sphere. In addition, the model contains no temperature dependence and is purely classical.

The potential we use as a standard is derived from density functional theory (DFT).[1] The fundamental quantity of this theory is the free energy of the two component (electron and ion) system, which is a unique functional of the electron an ion density profiles. In the approach utilized here, electron-electron and ion-ion correlations have been included and minimization of the free energy leads to one electron Schrödinger equations in an effective self-consistent potential and an ion equation of hypernetted chain form. The effective potential, as with the two described below, is obtained numerically and cannot be approximated quantitatively in a simple manner.

Fig. 1 is a comparison of these screening functions in 0.5 keV aluminum at $n_e = 10^{24}$ and $10^{25} cm^{-3}$. At these densities the IS potential might be expected to more closely mirror the DFT form than the less appropriate DH potential. This is indeed the case away from the nucleus (beyond r=0.2au for $n_e = 10^{24} cm^{-3}$ and r=0.1au for $n_e = 10^{25} cm^{-3}$).

The log scale of the figure reveals the large deviation of the IS form close to the nucleus. The DH potential is of course more accurate closer to the nucleus. This can be significant for the problem considered here, since for high energy scattering, the most important region of the potential can be close to the nucleus. Higher energy electrons will penetrate more of the potential. In the cases considered here, this important region extends in as far as 0.1au, so it is expected that the IS form can give erroneous results for these plasmas.

We also consider the finite temperature, finite density Thomas-Fermi (TF) potential.[2] The system of equations required to find the TF potential (Poisson equation with an integral over the Fermi-Dirac distribution) contains no ion-ion or electron-ion correlations other than the exclusion of other ions from the sphere of electrons which neutralize the charge of a given ion.

The remaining potential results from the solution of the Poisson equation with particle densities derived from a solution of the two-component hypernetted chain equations (HNC).[3] The HNC equations with a semi-classical pair interaction[4] provide particle pair correlations (charge densities) that have been shown to be especially accurate for strongly coupled plasmas.[5] In this case, involved quantum mechanical considerations are sacrificed for accurate treatment of the correlations.

Fig. 2 compares the previous DFT curves with the TF and HNC-Poisson (HNCP) screening functions. Lacking ion correlations, TF theory cannot reproduce the spatial oscillations at r=1.0au for $n_e=10^{25}cm^{-3}$. As mentioned earlier, however, this will be irrelevant in our examples. Closer to the nucleus the DFT curve falls between TF and HNCP; all curves are closely comparable.

For the examples considered here, all curves are of similar shape and magnitude, except for IS at small separations.

III. ATOMIC PHYSICS METHODS

Given one of the potentials above, we require a method of finding free-free Gaunt factors for a given electron energy. In particular, we wish to calculate unpolarized angle-integrated Gaunt factors, $G(\varepsilon, h\nu)$, where ε is the energy of the incoming electron and $h\nu$ is the photon energy. Note that $G(\varepsilon, h\nu)$ is not a plasma-averaged quantity. Multiplying $G(\varepsilon, h\nu)$ by the Kramers bremsstrahlung cross section gives the calculated value of the predicted cross section. The Kramers cross section is the well-known hydrogenic semi-classical result.[6] A review of this subject has been given by Pratt.[7]

The simplest standard method of obtaining bremsstrahlung cross sections is the Born approximation,[8] which utilizes first order perturbative corrections in the potential to plane wave incoming and outgoing electron wavefunctions. The method is restricted to high incident and final electron energies, but multiplication by the corrective Elwert factor[8] extends the validity of the approximation over the entire spectrum for large ε and small Z. The Born-Elwert (BE) Gaunt factor takes the form

$$G(\varepsilon, h\nu) = \frac{\sqrt{3}}{\pi} \sqrt{\frac{\varepsilon+h\nu}{\varepsilon}} \int_{q_{min}}^{q_{max}} dq\, q \left[\int_0^\infty dr\, r\, U(r) \sin(qr) \right]^2 , \quad (3)$$

where

$$q_{\substack{max\\min}} = \sqrt{2(\epsilon \pm h\nu)} + \sqrt{2\epsilon} \quad \text{(a.u.)}$$

are the maximum and minimum momentum transfers. The quantity q^{-1} provides an estimate of the depth of the potential probed by the electron.

The BE method is not valid when ϵ is small. An approach that makes an attempt to find cross sections valid in this regime was introduced by Lamoureux and Pratt.[9] This simplified classical (SCL) method makes use of the Larmor formula for energy loss by charged particle deceleration integrated over judiciously chosen large impact parameters. The method assumes that the most important contribution to the free-free cross section comes from large angle scattering, which allows simplification in the integration. (The usual Coulomb logarithm, not valid in strong screening, is not obtained.) The result is a simple method, both in concept and in practice, of obtaining free-free Gaunt factors.

Both of these methods make extreme, if somewhat justifiable, approximations. A method that does not take these steps, but instead attempts as complete a solution as possible is the partial wave expansion (PWE) method.[10] The continuum electron radial wavefunctions are found from the Dirac equation in the potential $U(r)$. In addition, a full multipole evaluation of the cross section is made. For a more detailed description of the method, the reader is referred to Ref. 10. The PWE method has been found to be accurate in comparison with experiments[11] on neutral atoms; it is used here as the standard against which the other two methods are assessed.

IV. RESULTS

Results for the three numerical potentials employed within the three atomic physics methods are presented in Fig. 3 for a plasma with a temperature of 0.5 keV and a density of $10^{24} cm^{-3}$ (no DFT curve is available in the PWE method). An incoming electron energy of 1 keV is assumed. Even for this high density, it is clear that, in this example, the selection of different potentials leads to small errors within a given method. This is due to the choice of a high electron energy (1 keV); all potentials are similar in the "probing region" of r=0.08 to 0.24au. The ion-sphere results however, lie outside the groups. This behavior does not change for ε=0.5 keV, but for much lower electron energies, the more fully screened regions of the potentials are important. Then the IS Gaunt factors are expected to be closer to the DFT results than the DH values.

Gaunt factors for an aluminum plasma one order of magnitude more dense are presented in Fig. 4. In Fig. 5, the DFT Gaunt factors from the three methods are compared. An immediate conclusion, comparing with Fig. 4, is that the choice of method accounts for a more significant difference than the choice of potential within a given method. Also, the SCL results are much closer to the PWE results than BE, which are in error by as much as 30% (50% in the case of Fig. 3).

For a 1 keV plasma with a density of $10^{23} cm^{-3}$, the difference between Gaunt factors of distinct potentials is very small; however, differences between the methods becomes evident. This implies again that the choice of potential is not as significant as the choice of method. Fig. 6 shows the Gaunt factors produced using the Debye potential. The SCL results are clearly superior to BE, but are still in error by 50-60% at the low photon energy end. Work is presently underway to improve the SCL theory. In addition, a more detailed presentation of these and other results is in preparation.[12]

We can conclude that for the strongly coupled aluminum plasmas considered here, the choice of potential generally makes $\lesssim$ 10% error in the Gaunt factor as long as the IS potential is not used for high electron energies. Thus, using the simple DH potential is adequate to this accuracy. The choice of method is more important, with SCL being a better choice than BE.

As displayed in Table I, it is evident that the time savings (at least for free-free emission) can be considerable if one is willing to accept errors of 10% to 50%.

TABLE I

Potentials	Computation Time
DH	10^{-5} s
IS	10^{-5} s
TF	1-10 s
HNCP	$10-10^2$ s
DFT	10^4 s

Atomic Methods	
BE	1 s
CL	1 s
PWE	10^5 s

TABLE I: Order of magnitude estimates of VAX CPU times for the calculation of one potential curve and the calculation of one Gaunt factor spectrum.

The table contains estimates of computation times for the production of a potential curve and given a potential curve, the generation of an entire spectrum. If the stated errors in the bremsstrahlung spectra are acceptable, the classical method employing a simple analytic potential can be efficiently used inline in hydrodynamic codes which model the evolution of dense plasmas over a wide range of conditions.

[*]Permanent address: Lawrence Livermore National Laboratory, Livermore, CA

REFERENCES

1. U. Gupta and A.K. Rajagopal, Phys. Rep. **87** 259 (1982); M.W.C. Dharma-wardana and F. Perrot, Phys. Rev. A **26** 2096 (1982) and A **29** 1378 (1984).

2. R.M. More, UCRL 84991, Parts I and II (March, 1981), unpublished.

3. R. Cauble, M. Blaha, and J. Davis, Phys. Rev. A **29**, 3280 (1984).

4. C. Deutsch, Phys. Lett. **60A**, 317 (1977).

5. J.P. Hansen and I.R. McDonald, Phys. Rev. A **23**, 2041 (1981).

6. C.W. Allen, _Astrophysical Quantities_ (The Athlone Press, London, 1955), p. 99.

7. R.H. Pratt, in _Fundamental Processes in Energetic Atomic Collisions_, ed., H.O. Lutz, J.S. Briggs, and H. Kleinpoppen (Plenum Press, New York, 1983).

8. H.A. Bethe and E.E. Salpeter, _Quantum Mechanics of One- and Two-Electron Atoms_, (Springer-Verlag, Berlin, 1959), p. 323.

9. M. Lamoureux and R.H. Pratt, University of Pittsburgh Report PITT-312 (1984); Proceedings of the 2nd International Conference on the Radiative Properties of Hot Dense Matter (World Scientific Publishing Co., Singapore, 1985).

10. H.K. Tseng and R.H. Pratt, Phys. Rev. A **3**, 100 (1971).

11. M. Semaan and C. Quarles, Phys. Rev. A **26**, 3152 (1982).

12. M. Lamoureux, R. Cauble, L. Kim, F. Perrot and R.H. Pratt, submitted to Phys. Rev. A (1985).

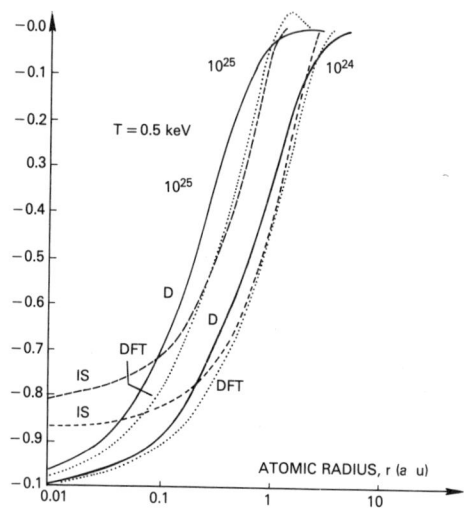

Figure 1. Screening functions for Debye-Hückel (D), ion-sphere (IS) and density functional theory (DFT) approximations in 0.5 keV Aℓ plasma at $n_e = 10^{24}$ and 10^{25} cm^{-3}.

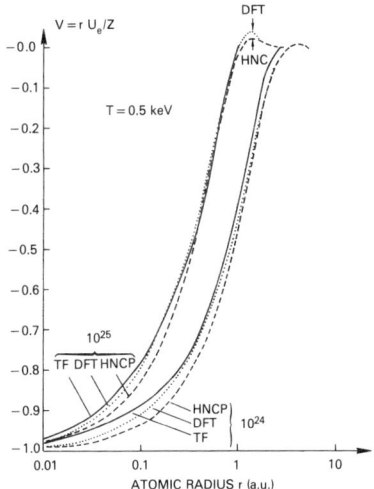

Figure 2. Screening functions for hypernetted chain (HNC), Thomas-Fermi (TF) and DFT models in 0.5 keV Aℓ plasma at $n_e = 10^{24}$ and 10^{25} cm^{-3}.

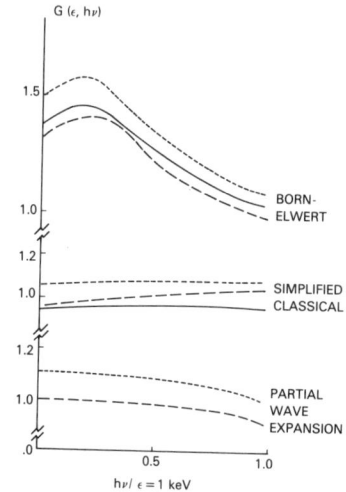

Figure 3. Gaunt factors as a function of electron impact energy ε and emitted photon energy $h\nu$ in a 0.5 keV Aℓ plasma at $n_e = 10^{24} \text{cm}^{-3}$ using DFT, HNC and TF potentials with Born-Elwert (BE), simplified classical (SCL), and partial wave expansion (PWE) methods. $\varepsilon = 1$ keV.

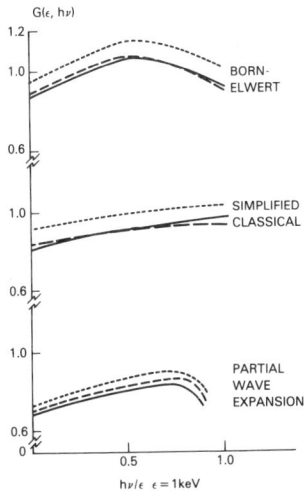

Figure 4. Same as Fig. 3 except the density is $10^{25} cm^{-3}$.

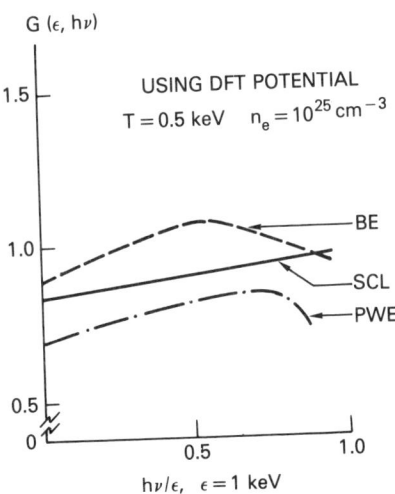

Figure 5. DFT results extracted from Figure 4.

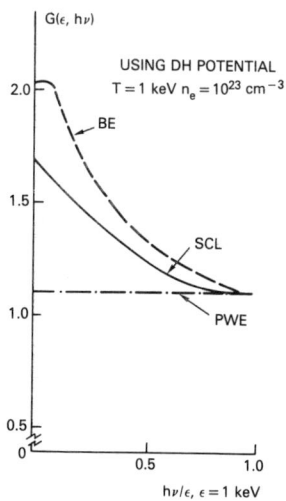

Figure 6. Gaunt factors for $n_e = 10^{23}$ cm^{-3}, 1 keV Aℓ plasma using BE, SCL, and PWE methods with Debye potential. ε = 1 keV.

CONTINUUM EMISSION (BOTH ISOTROPIC AND POLARIZED) IN A LASER-PLASMA DESCRIBED BY A FOKKER PLANCK SIMULATION

M. Lamoureux, J.P. Matte[+], C. Möller and R.Y. Yin[++]

Laboratoire de Spectroscopie Atomique et Ionique,
Université Paris-Sud, 91405 ORSAY, France

and

Gréco Interaction Laser-Matière, Ecole Polytechnique,
91128 PALAISEAU, France

ABSTRACT

We study the background continuum radiative properties of the underdense region of a laser-heated plasma and emphasize their non-Maxwellian features. In that region of electron density N_e smaller than the critical density N_c, the laser beam penetrates and the Inverse Bremsstrahlung heating mechanism leads to strongly non-Maxwellian electron distributions which in turn affect the evaluated continuum emission. We show how in a uniform plasma the isotropic emission due to either Direct Radiative Recombination or Bremsstrahlung is characterized by the only parameter $\alpha = Z\, v_{osc}^2/v_{th}^2$ where v_{osc} and v_{th} are the quiver and thermal velocities respectively. We also comment on the diagnostic possibilities offered by the continuum spectrum. We then examine to which extent the involvment of transport may alter these general properties by using detailed Fokker Planck simulations. Finally, we evaluate in a few examples the anisotropy of the continuum spectrum.

I - **INTRODUCTION**

In laser heated plasmas, the laser beam can penetrate only in the underdense region (electron density $N_e <$ critical density N_c). There, Inverse Bremsstrahlung absorption, also called classical absorption is the dominant heating mechanism and severely disturbs the electron distributions from the standard Maxwellian ones over the whole energy range.[1,2,3] On the contrary, the laser beam does not penetrate into the overdense region where the energy is therefore carried by the transport processes only; the electron distribution function is non-Maxwellian only at large energies, which contribute little to the total emission. For all these reasons, we limit ourselves to the underdense region when studying the non-Maxwellian features exhibited by the background continuum emission due to Bremsstrahlung (Br) and Direct Radiative Recombination (DRR). This problem has been dealt with only in the extreme non-Maxwellian situation [4,5] where e-e collisions were neglected altogether. Here, we extend the study to the general case where these collisions are partly restoring the Maxwellian distribution. By making use of a simple analytical form [6] for the distribution functions, the dependence of the isotropic emission, (when taking photon energy over kT_e as the energy variable), can be characterized by the only parameter $\alpha = Z\, v_{osc}^2 / v_{th}^2$ where v_{osc} and v_{th} are the quiver and thermal velocities respectively.

This characterization is rigorous in uniform plasmas, and remains valid in realistic plasmas except close to the critical surface. General expressions are given in a completely ionized plasma for the emissivity coefficients relative to Br or DRR into a given shell. We also indicate how the continuum spectrum could be used to diagnose the plasma, as regards not only the electron temperature but also the electron distribution. We evaluate what precision is lost by using the analytical isotropic distribution functions instead of the exact numerical ones for the case study of a Be plasma irradiated by a Nd laser and treated by an elaborate Fokker Planck code.[3] Finally, we evaluate in a few examples the polarization degree of the continuum emission caused by the higher orders of the electron distribution obtained in a still more elaborate version of this code.

II - ISOTROPIC CONTINUUM EMISSION IN A UNIFORM PLASMA

A - General emissivity formulae

The isotropic emission depends on the isotropic distribution functions and on the atomic Gaunt factors, besides the evident dependence on the electron and ion densities N_e and N_i. The expressions given below in relations (2) and (3) for the emissivity coefficients are general and do not depend on the choice of the system of units. These coefficients[7] have the dimension Energy/volume and are given here <u>per steradian and per mode</u> (the aspects per unit of time and per unit of angular frequency ω cancel out as regards dimensions). They are valid for a fully ionized plasma consisting of bare ions of atomic number Z. As usual, c is the velocity of light, α_{FS} the fine structure constant (not to be confused with the plasma parameter α) and M_e the electron mass. We define the constant K by

$$K = (2 \alpha_{FS}^3 \hbar^3) / (3 \sqrt{3} M_e^2) , \qquad (1)$$

it amounts to 1.497×10^{-7} in a.u. The distribution functions are normalized so that $\int_0^\infty f_o(v) v^2 \, dv = 1$. We remind that for each atomic process considered, the Gaunt factors are the ratios of the actual cross sections (integrated here over angle and polarization) over the semi-classical Kramers ones which contain most of the dependence on the angular frequency ω and on the incident electron velocity v. With all these notations, the emissivity coefficients for DRR into the n shell and Br (per steradian and per mode) write

$$j_R^n(\omega) = N_e N_i K Z^4 \frac{c^2 \alpha_{FS}^2}{n^3} G_R^n(\omega) f_o(v) \qquad (2)$$

with $\hbar\omega = I_n + (1/2) M_e v^2$ and

$$j_{Br}(\omega) = N_e N_i K Z^2 \int_{v_{min}}^\infty G_{Br}(\omega, v) f_o(v) \, v \, dv \qquad (3)$$

with $(1/2) M_e v_{min}^2 = \hbar\omega$.

Further literal expressions in this paper will also be given independently of the unit system. When coming to numerical evaluations, it has been convenient to turn to atomic units, and when giving numerical results, we will do it in SI as recommended (though it leads to very small orders of magnitudes !).

B - **Analytical distribution functions**

The shape of the distribution function $f_o(v)$ is determined[1] by the competition between the laser classical absorption which preferentially heats the slower electrons and the electron-electron collisions which tend to restore the Maxwellian distribution. Qualitatively, the extent of the non-Maxwellian character varies with the ratio of the heating rate (depending on the e-ion collisions and the laser field) over the thermalization rate depending on the e-e collisions.[8] This competition between the two processes is described by

$$\alpha = \frac{\nu_{ei} \; M_e v_{osc}^2 /2}{\nu_{ee} \; M_e v_{th}^2 /2}$$

where v_{osc} and v_{th} are the quiver and the thermal velocities, and ν_{ei} and ν_{ee} the e-ion and e-e collision frequencies respectively. <u>This key parameter α depends on the laser intensity and wavelength, and on the kinetic temperature T_e according to</u>

$$\alpha = 3.7 \; Z \; \frac{I_o}{(10^{16} \; W \; cm^{-2})} \; \frac{\lambda^2}{(\mu m)^2} \; \frac{(keV)}{kT_e} \; . \tag{4}$$

In uniform plasmas where the temperature and the density are constant and where there is thus no heat transport at all, the electron distribution depends rigorously on the value of the Langdon parameter α only.[1] Recently [9,10,11], it has been shown that a simple analytical form[12] could be fitted to the numerical distribution functions obtained from a one dimensional Fokker-Planck code. In the uniform plasma, $f_o(v)$ can be written :

$$f_o(v) = \left(\frac{M_e}{2kT_e}\right)^{3/2} \frac{m\, a_m^{3/2}}{\Gamma(3/m)} \exp -\left(\frac{M_e v^2}{2kT_e} a_m\right)^{m/2} \qquad (5)$$

with $\quad a_m = \dfrac{2}{3} \dfrac{\Gamma(5/m)}{\Gamma(3/m)}$.

The generally non-integer value of m is related to the Langdon parameter by[6]

$$m(\alpha) = 2 + \frac{3}{1 + 1.66/\alpha^{0.724}} \qquad (6)$$

as plotted in fig.1. As expected, $\alpha = 0$ corresponds to the Maxwellian distribution (m = 2) whereas $\alpha \rightarrow \infty$ corresponds to the distribution with m = 5, which has been found[1] when e-e collisions are neglected.

C - **Parametrized description of the DRR and Br emission**

Taking advantage of the analytical distributions (5), we can give general expressions for the emissivity

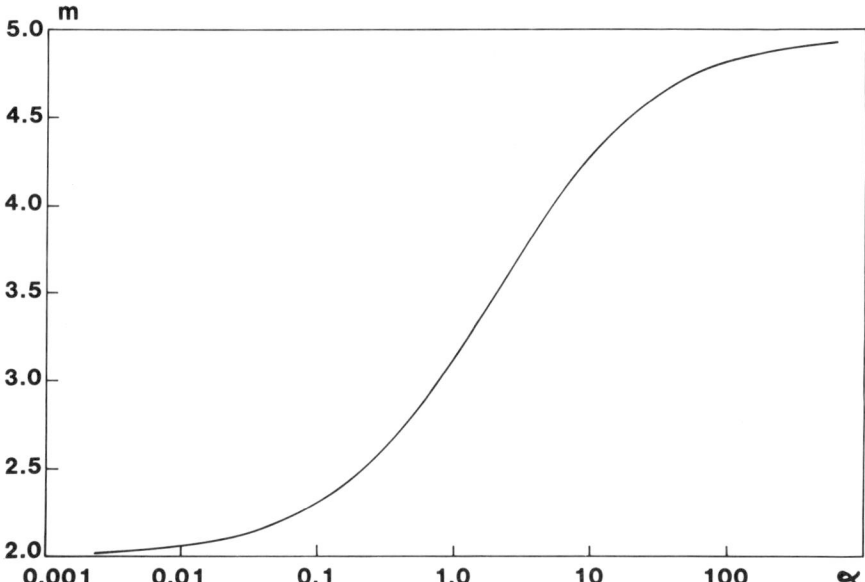

Fig. 1 : Parameter m characterizing the analytical distribution functions vs. Langdon parameter α.

coefficients evaluated by (2) and (3). From now on, we assume Gaunt factors to be equal to 1 (if they are not, we have to introduce them as multiplicative factor in expressions (8) and (11), which destroys their general character). We use the same reduced coordinates as we did before[4] for the extreme case m = 5. For DRR, we thus define the reduced emissivity coefficients

$$y_R = j_R(\omega)(kT_e)^{3/2} n^3/(N_e N_i Z^4) \tag{7}$$

and have

$$y_R\left(\frac{\hbar\omega - I_n}{kT_e}\right) = R_m(M_e c^2 \alpha_{FS}^2) \exp-\left(\frac{\hbar\omega - I_n}{kT_e} a_m\right)^{m/2}. \tag{8}$$

The constant R_m amounts to

$$R_m = K M_e^{1/2} \left[\frac{m}{\Gamma(3/m)} \left(\frac{a_m}{2}\right)^{3/2}\right] \tag{9}$$

with K given in Eq.(1). Similarly, we define the reduced emissivity coefficient for Br by

$$y_{Br} = j_{Br}(\omega)(kT_e)^{1/2}/(N_e N_i Z^2), \tag{10}$$

we then find

$$y_{Br}(\hbar\omega/kT_e) = R_m \int_{\hbar\omega/kT_e}^{\infty} \exp-(xa_m)^{m/2} dx. \tag{11}$$

The integration in (11) may be performed analytically for $\omega = 0$ only and in that case

$$y_{Br}(\omega = 0) = K\sqrt{M_e}\ \Gamma(2/m) \sqrt{\frac{\Gamma(5/m)}{3\ \Gamma^3(3/m)}}.$$

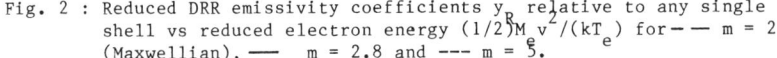

Fig. 2 : Reduced DRR emissivity coefficients y_R relative to any single shell vs reduced electron energy $(1/2)M_e v^2/(kT_e)$ for $--$ m = 2 (Maxwellian), $—$ m = 2.8 and $---$ m = 5.

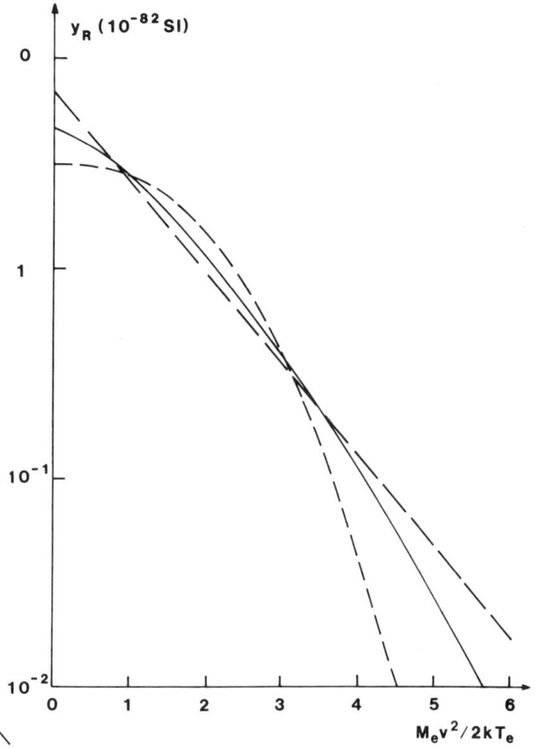

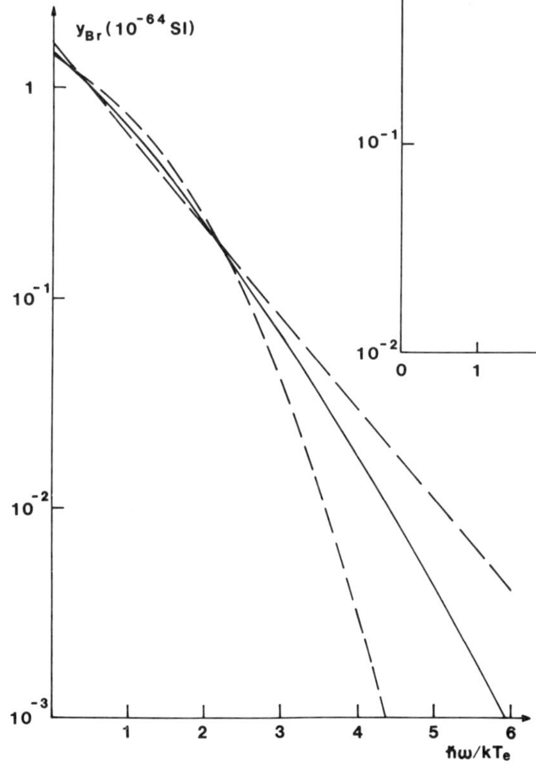

Fig. 3 : Reduced bremsstrahlung emissivity coefficient y_{Br} vs reduced photon energy $\hbar\omega/(kT_e)$ for $--$ m = 2 (Maxwellian), $—$ m = 2.8 and $---$ m = 5.

Let us also mention that y_R and y_{Br} do not have the same dimension because y_R is obtained by multiplying j_R by $(kT_e)^{3/2}$ whereas y_{Br} is obtained by multiplying j_{Br} by $(kT_e)^{1/2}$. This difference is evident also from expressions (8) and (11) since $M_e c^2 \alpha_{FS}^2$ has indeed the dimension of energy and is namely equal to one Hartree.

It is clear from (8) and (11) that the reduced emissivity coefficients for DRR and Br depend only on m and on the reduced variables ($\hbar\omega - I_n$)/kT_e and $\hbar\omega/kT_e$ respectively. In figs. 2 and 3, we show ln y for m = 2 and m = 5, as well as for m = 2.8, which corresponds to a realistic Be plasma[3] which has been studied in details and to which we will turn in the next paragraph. For DRR, the evolution of the emissivity immediately follows the evolution of the distribution because of the direct proportionality of y_R to f_o in (2). We find emissivity coefficients smaller than in the Maxwellian case both at low and at very large energies. The typical plateau behaviour (fig. 2) right at the threshold ($\hbar\omega - I_n = 0$) has been observed for an Al^{13+} plasma in time and space resolved spectra.[13] For Br, the non-Maxwellian character is less pronounced because of the integration in (3 and 11), except at large photon energies where only the depleted distribution tail is involved.

D - Diagnosing the plasma

The quantities t_e defined by $kt_e(\omega)=-d(\hbar\omega)/d\ln j(\omega)$ is equivalent to a temperature but is equal to the kinetic temperature T_e itself only for Maxwellian plasmas. They remain useful quantities to refer to when trying to diagnose the non-Maxwellian laser-plasma for which both T_e and m (or in other terms α) are now unknown, provided we assume the exponential dependence of the distribution functions of (5). It is possible to easily interpret the continuum spectrum in situations where one can isolate the emission due to a single atomic process, or where one process is greatly predominant. This is the case for DRR into the n = 1 shell of high Z ions, and for Br with low Z ions at low photon energies.

For DRR the following relation could be used to deduce the value of m from two points along the spectrum :

$$\frac{t_e(\omega_1)}{t_e(\omega_2)} = \left(\frac{\hbar\omega_2 - I_n}{\hbar\omega_1 - I_n}\right)^{m/2 - 1} . \quad (12)$$

The kinetic temperature T_e would then be obtained from the relation

$$\frac{m}{2} a_m^{m/2} \left(\frac{\hbar\omega - I_n}{kT_e}\right)^{m/2 - 1} = \frac{T_e}{t_e(\omega)} . \quad (13)$$

For Br, the diagnostic is more elaborate because of the integration over the incident electron energies.

Three detailed data along the spectrum are now needed to get the estimates of m and T_e from the relation

$$\ln\left[\left(\frac{d\omega}{dj(\omega)}\right)_{\omega_2} \left(\frac{dj(\omega)}{d\omega}\right)_{\omega_1}\right] = \left(\frac{a_m}{kT_e}\right)^{m/2} \left[(\hbar\omega_2)^{m/2} - (\hbar\omega_1)^{m/2}\right] . \quad (14)$$

However, it may be questionable whether the j's can be measured with enough precision to allow for the present type of diagnostics. In that respect, one should at least check that the α calculated by (6) from the "experimental" value of m is consistent with the α evaluated from the "experimental" temperature through (4) (given Z and the laser intensity).

III - EMISSION IN A NON UNIFORM PLASMA (INFLUENCE OF HEAT TRANSPORT)

Laser-heated plasmas actually present temperature and density gradients and are non-uniform so that heat transport comes into play. Electron heat flow has been dealt with recently in Fokker-Planck simulations by Albritton[2] and then by Matte et al[3] who also included the effects of ablation. The main role of transport in the underdense region is to cool the plasma by transmitting part of the laser beam energy to the overdense region so that the temperature does not get as high as it would in a supposedly uniform plasma with the same laser exposure.

This important aspect being taken care of, the general characteristics of the isotropic emission presented in II remain valid. We will show this in III A by comparing emissivity coefficients obtained from numerical distributions determined by a detailed simulation[3] (including f_o and f_1) to the coefficients obtained from the analytical distributions of Eq.(5) relative to the uniform plasma having the same local temperature and density. Another consequence of transport on the emission is to somewhat polarize it as we will see in III-B.

A - **Accuracy of the isotropic emissivity coefficients evaluated from the analytical distribution functions**

We will show that the emissivity coefficients evaluated with the analytical distribution functions (5) are still basically correct. However, the α parameter has now to be evaluated from the total laser intensity (both incident I_+ and reflected I_-) and from the so-called swelling factor. The suitable α parameter is determined by multiplying the value of the "uniform α" given by Eq.(4) by the supplementary factor

$$\left[(I_+ + I_-)/I_o\right] / (1 - N_e/N_c)^{1/2}.$$

We have taken the example of a completely ionized beryllium plasma[3] produced by a 1.06 μm Nd laser of intensity 3.10^{14} W/cm^2. The Fokker-Planck simulation includes inverse

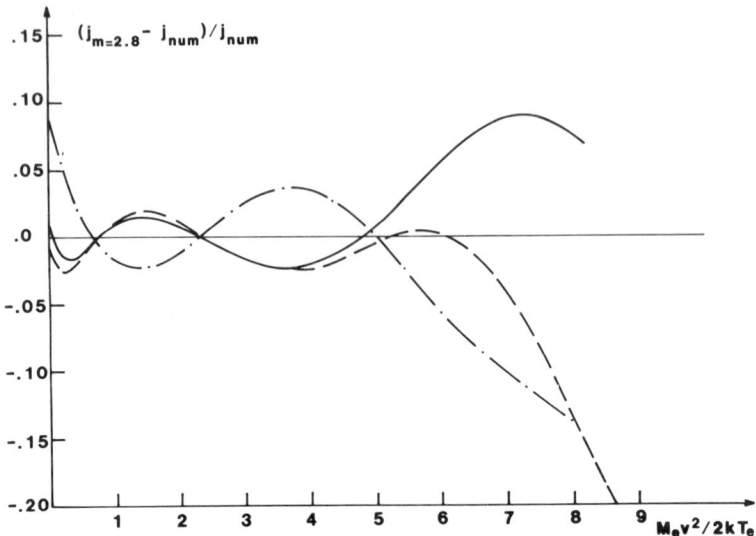

Fig. 4 : Relative error introduced in DRR emissivity coefficients when taking analytical distribution functions with m = 2.8 instead of numerical ones at three distances from the high density boundary in the underdense region : --- 678 μm(N_e/N_c = 0.19, T_e = 2.13 keV), —— 517 μm (N_e/N_c = 0.40, T_e = 2.27 keV) and —·— 402 μm(N_e/N_c = 0.80, T_e = 2.31 keV).

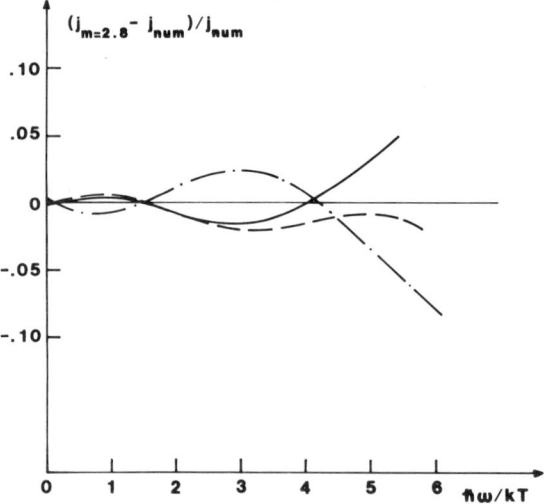

Fig. 5 : As in Fig. 4 but for bremsstrahlung.

bremsstrahlung absorption, electron-electron and electron-ion collisions, ablation and the electric field as presented in detail elsewhere.[6] After 600 ps the temperature in the underdense region is nearly uniform ($\sim$ 2.3 keV).

Away from the critical surface analytical distribution functions with a single value of m will prove to describe the underdense region correctly. We consider a point at a distance of 517 μm from the high density boundary (See Fig. 1 of Ref. 3). For that position, the density is 0.4 N_c, temperature 2.27 keV and laser intensity $I_+ + I_- = 1.6 \, I_o$. This leads to $\alpha = 0.47$ and therefore m = 2.8. Figures 4 and 5 show the relative error introduced in the DRR (note that it is the same error as for the distribution functions themselves) and Br emissivity coefficients when taking f(m = 2.8) instead of exact numerical distribution functions. These figures also show that the agreement remains satisfactory at other positions in the underdense plasma, with electron densities ranging from 0.2 to 0.8 N_c, while still keeping the value m = 2.8 for these other positions. This is due the quasi-uniformity of α in that region. As shown in fig. 4 for DDR the error introduced stays within 4 % for photon energies up to 5.5 kT_e (with reservations for the very slow electrons in the highest density case) ; for higher energies the use of the analytical distribution becomes less reliable. For bremsstrahlung (fig. 5) the relative error is on the whole less than for DRR, staying below 2.5 % for photon energies

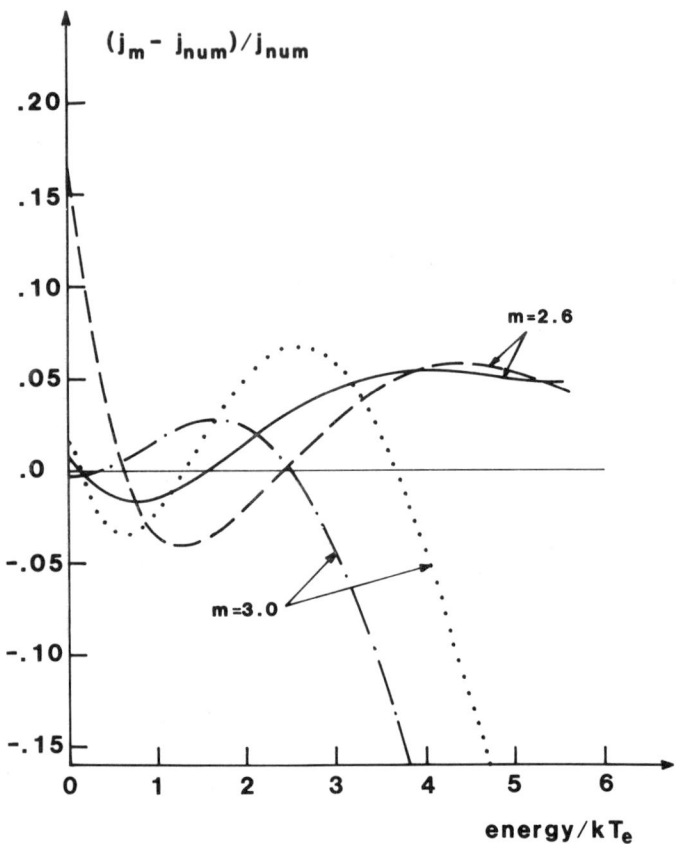

Fig. 6 : Relative error introduced when taking analytical distribution functions instead of the numerical one, both for DRR, (--- and ...) and Br (—— and —·—) very near the critical surface (N_e/N_c = 0.97, T_e = 2.32 keV at 380 μm from the high density boundary) vs reduced energy.

up to 4.8 kT_e and less than 8 % for the larger energies considered here. We also checked that for each photon energy we get the same imprecision using Kramers approximation as in section II, or more sophisticated Gaunt factors as appropriate for this Be plasma[6].

Closer to the critical surface, m = 2.8 becomes less adequate. This is the case for example at the distance of 380 μm from the high density boundary, position for which the density is N_e/N_c = 0.97. The m value derived from α, i.e. m = 3, leads to correct evaluations of the emissivity coefficients at low energies both for Br and DRR as can be seen in fig. 6, but when considering the whole energy range, m = 2.6 appears more suitable. This can be explained by the fact than the slow electrons are determined by the local value of α, while the fast electrons are already sensitive to the overdense region owing to their large mean free-path. (Let us remember that, for our example, bremsstrahlung emissivity coefficients at low energies are an order of magnitude larger than DRR emissivity coefficients so that the 15% or so error on j_R, introduced by taking m = 2.6 reduces to a mere additionnal 1.5 % when considering both processes simultaneously.)

B - **Polarization of the emission**

Besides restricting the increase of the temperature in the corona and thereby affecting the isotropic

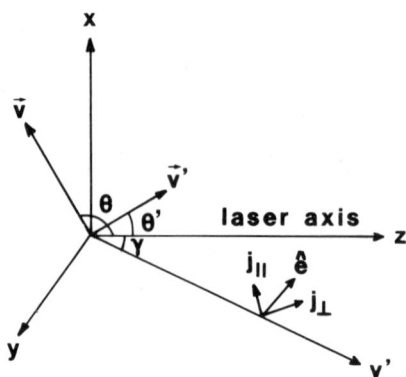

Fig. 7 : Geometry of observation of emitted radiation. Line of sight Oy' lies in the yz plane with the polarization vector ê being perpendicular to that direction. $\vec{v}$ and $\vec{v}'$ are the velocity vectors of the incident and outgoing electrons (last one only for Br).

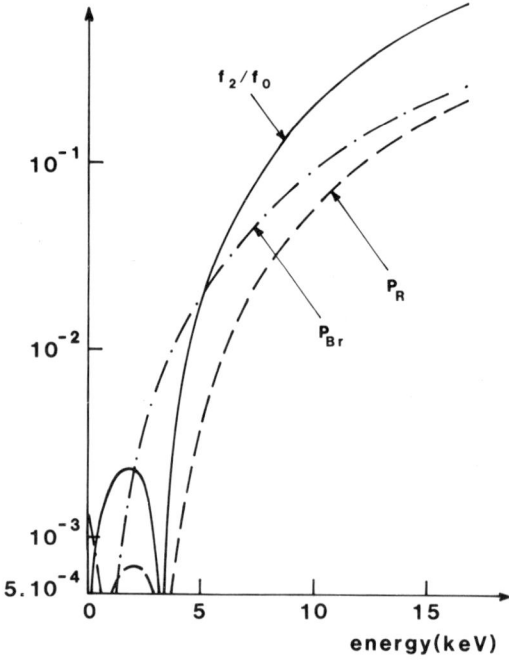

Fig. 8 : Anisotropy of the electron distribution (for $N_e = 0.42\ N_c$, $T_e = 2.39$ keV) f_2/f_0 vs $1/2\ M_e v^2$ and degree of polarization P of the DRR into the $n^o = 1$ shell and Br emission vs photon energy $\hbar\omega$.

part $f_o(v)$ of the distribution $f(\vec{v})$, the involvment of transport adds anisotropic terms to it. Calling θ the pitch angle, i.e. the angle of the incident electron velocity $\vec{v}$ with the laser axis Oz, the distribution function is written

$$f(\vec{v}) = f_o(v) + \sum_{i \geq 1} f_i(v) \, P_i(\cos\theta) \qquad (15)$$

where the $P_i(\cos\theta)$ are the Legendre polynomials. In the process of integrating over the azimuthal angle θ, only even Legendre polynomials will lead to non-zero contributions owing to the axial symmetry Oz defining the polar axis in spherical coordinates. For that reason, the Fokker Planck simulation[3] corresponding to our Be plasma, which included only the f_o and f_1 terms, has been redone taking f_i terms beyond the first order, actually till the fifth order. Notice that including the higher order terms hardly affects the hydrodynamic evolution of the plasma and the isotropic distributions $f_o(v)$. The degree of polarization of the radiative emission is equal to $(j_{\parallel} - j_{\perp})/(j_{\parallel} + j_{\perp})$. As visualized in fig. 7, these two orthogonal directions are defined in the plane perpendicular to the observation direction Oy' (at γ angle with Oz), the polarization vector ê lying in that plane.

For DRR the collision is characterized by the velocity vector $\vec{v}$ relative to the incident electron, and the atomic cross section for the elementary process is proportional to $(1 - \beta/2) + (3\beta/2)\left[\frac{(\vec{v}.\hat{e})}{v}\right]^2$ where β is the asymmetry parameter relative to the shell considered. This quantity has to be multiplied by $f(\vec{v})$ v and integrated over the azimuthal and polar angles of the incident electron. Calculations are carried out at the second order[15] and give

$$P_R = \frac{3 \beta \sin^2\gamma \, f_2}{20 f_o + \beta f_2 (3 \sin^2\gamma - 2)} .$$

For Br, the process involves also the velocity vector of the outgoing electron $\vec{v}'$. We used the Born cross section for the elementary collision[16]:

$$\sigma_{Br} \propto \frac{v'}{v} \frac{[(\vec{v} - \vec{v}').\hat{e}]^2}{(v^2 - v'^2) |\vec{v} - \vec{v}'|^4} .$$

This has to be multiplied by $f(\vec{v})v$, and then integrated over the angular coordinates of both the incident and outgoing electrons, that integration being feasible analytically. The final summation over the incident electron velocities v is then performed numerically.

Results are given in figure 8 for the observation angle of 90°. They have been evaluated from f_o and f_2 data relative to the distance of 506 μm from the high density boundary, i. e. for N_e/N_c = 0.42. We mention that the ratio f_2/f_o is significantly different whether expressed in the laboratory or ion frames, and that we used last ones as physically appropriate. Figure 8 shows the ratio f_2/f_o versus electron energy, and the degree of polarization both of the DRR into the n = 1 shell, and of the Br emission versus photon energy.

IV - **CONCLUSION**

We have shown that the evaluation of the isotropic continuum emission is made easier with the use of analytical distribution functions of the form $\exp - (v/v_m)^m$. This shape has been found[6] for uniform plasmas submitted to the competitive actions of classical heating by the laser beam and electron collisions. The m parameter amounts to $m = 2 + 3/(1 + 1.66 \alpha^{-0.724})$ where $\alpha = Z v_{osc}^2 / v_{th}^2$, v_{osc} and v_{th} being the quiver and thermal velocities. When Kramers data are appropriate, general formulae depending only on m (or in other terms α) have been established for the emissivity coefficients of Bremsstrahlung (Br) and of Direct Radiative Recombination (DRR) into each shell. The current temperature diagnostic based on the continuum spectrum has been reformulated and could serve to trace back both the kinetic temperature and the value of m.

A comparison has then been made with emission results calculated from elaborate accurate distribution functions determined by a Fokker Planck code[3] including transport and ion motion. It concerns a Be plasma irradiated by a Nd laser of intensity $3.10^{14} W/cm^2$. Though the role of transport is essential in modulating the temperature profile of the plasma, it hardly affects the above parametrization, provided v_{osc} is estimated taking into account the reflected beam intensity and the swelling factor. At electron densities smaller than around 0.8 times the critical density, the agreement between the emissivity coefficients evaluated from the analytical and numerical distribution functions is excellent, and remains within a few percent both for DRR and Br even when using the same value of m all throughout. The precision is not as good closer to the critical surface. Nevertherless, it is still more adequate to use the analytical distribution functions (with the option of taking different values of m for the slow and fast electrons) than to assume the Maxwellian distribution. So, in the underdense region of the plasma, the analytical isotropic distribution functions (available for any velocity at little expense) with 2 < m < 5 are very appropriate and of great usefulness. They could be taken advantage of to calculate quickly and efficiently other atomic properties involving them, and have been used indeed to evaluate ionization and excitation rates.[14] Finally, the polarization of the emission due to the anisotropy of the electron distribution was shown to be slight. This corresponds to the fact that, when submitted to the perturbing laser beam, the plasma environment restores the isotropic character much more efficiently than the Maxwellian character.

Acknowledgments :

We thank A. Bekkali for his help in the angular integrations and Drs. P. Jaeglé, J. Delettrez, J. Virmont, T.W. Johnston, A. Langdon, J. Albritton and finally R. More for stimulating conversations.

+ Institut National de la Recherche Scientifique -Energie-
Université du Québec, CP 1020, Varennes, Québec, JOL 2 PO
CANADA.

++ On leave from the University of Pittsburgh, Department of
Physics and Astronomy, Pittsburgh, PA 15260, U.S.A.

1 - A.B. Langdon, Phys. Rev. Lett. **44**, 575 (1980).

2 - J.R. Albritton, Phys. Rev. Lett. **50**, 2078 (1983).

3 - J.P. Matte, T.W. Johnston, J. Delettrez and R.L. McCrory
Phys. Rev. Lett. **53**, 1461 (1984).

4 - M. Lamoureux, C. Möller and P. Jaeglé,
Phys. Rev. A**30**, 429 (1984).

5 - M. Lamoureux, C. Möller and P. Jaeglé,
J. Quant. Spectrosc. Radiat. Transf. **33**, 127 (1985).

6 - J.P. Matte, M. Lamoureux, C. Möller, R.Y. Yin,
J. Delettrez, J. Virmont and T.W. Johnston
(To be published).

7 - G. Bekefi, _Radiation Processes in plasmas_ (Wiley,
New-York, 1966) Chap. 3.

8 - W.L. Kruer, in _Laser Plasma Interaction 2_ ed. by
R.A. Cairns (pub. by the Scottish Universities Summer
School in Physics, 1983), p. 196.

9 - M. Lamoureux, J.P. Matte, C. Möller and R.Y. Yin
Internal report of Laboratoire de Spectoscopie Atomique
et Ionique, November 1984.

10 - J.P. Matte, T.W. Johnston, J. Delettrez, R.L. McCrory,
M. Lamoureux, R.Y. Yin and C. Möller, 15th Anomalous
Absorption Conference, June 23-28, 1985, Banff,
Alberta, Canada.

11 - M. Lamoureux, C. Möller, R.Y. Yin, J.P. Matte and
J. Delettrez, 14th international Conference on the
Physics of Electronic and Atomic collisions, Stanford
University, Palo Alto, California, USA, July 1985.

12 - P. Mora and H. Yahi, Phys. Rev. A$\underline{26}$, 2259 (1982).

13 - D.L. Matthews, R.L. Kauffman, J.D. Kilkenny and
R.W. Lee, Appl. Phys. Lett. $\underline{44}$, 586 (1984).

14 - P. Alaterre, J.P. Matte and M. Lamoureux
(To be published).

15 - H.M. Milchberg and J.C. Weisheit, Phys. Rev. A$\underline{26}$,
1023 (1982).

16 - H.A. Bethe and E.E. Salpeter,
Quantum Mechanics of One -and Two- Electron Systems
(Springer Verlag 1957) Sect. 77.

COMPARISON OF MODELS FOR THE FREE-FREE GAUNT FACTOR
AT LOW TEMPERATURES AND FREQUENCIES

L. A. Collins and A. L. Merts
Group T-4 Los Alamos National Laboratory
Los Alamos, NM 87545

ABSTRACT

We perform calculations for the free-free Gaunt factor at electron and photon energies below 1 Ry in the dipole approximation to the radiation field for a variety of representations of the scattering potential. We consider the static-exchange, static-exchange + model polarization, model exchange, and static models. Within each model, the resulting Schrödinger equation is solved exactly using a linear algebraic prescription. We investigate the rare gas and alkali systems. We find great sensitivity to the models for energies below four electron volts (4 eV). Above this energy, the Gaunt factors for the various models come into better agreement.

I. INTRODUCTION

The free-free absorption or inverse Bremsstrahlung process is important to the proper description of a variety of both laboratory and astrophysical plasmas. The mechanism, which involves the absorption of a photon with the resulting increase in energy of the scattered electron, is mediated by a heavy particle such as an atom or molecule and is described by the following reaction:

$$h\nu + e^-(E) + A \rightarrow e^-(E') + A \quad ,$$

where $E(E')$ is the initial (final) energy of the colliding electron, A represents a target atom, and $h\nu (= \hbar\omega)$ is the photon energy. The process has been studied using a variety of techniques and approximations[1-18] from simple classical models to elaborate quantum mechanical constructions. In an earlier paper,[13] we investigated the free-free absorption process for various models for photon and electron energies above about ten electron volts (10 eV). We extend this study in this article to a much lower energy regime and to a wider variety of models. In Section II, we give a brief outline of the general theory and of the models employed. We follow this exposition by a description of the results in Section III and reserve Section IV for a few concluding remarks. A more comprehensive description of these calculations is given in Reference 19.

II. THEORETICAL FORMULATIONS

In this section, we derive the basic equations used to calculate the free-free absorption parameters. All our models are based on a full quantum mechanical treatment of the interaction of the photon and electron with the atomic target. We first introduce the basic quantities, such as the free-free Gaunt factor, in terms of the dipole matrix elements and continuum wavefunctions for the colliding electron. We then present a brief description of the methods used to solve for these continuum functions. In all calculations, we employ the <u>dipole approximation</u> to the radiation field.

II.A. GAUNT FACTOR

The free-free absorption coefficient, $a(E,\omega)$, in units of cm^5 is related to a dimensionless quantity, $g(E,\omega)$, called the Gaunt factor, through the expression

$$a(E,\omega) = a_K g(E,\omega) \quad . \tag{1}$$

The scaling term a_K is the Kramer's form of the semiclassical free-free absorption coefficient for an electron interacting with a point charge Ze and is given by[15]

$$a_K = \frac{4\pi Z_c^2 e^6}{3\sqrt{3}\, m_c^2 h\nu\nu^3} \quad [cm^5] \quad , \tag{2}$$

where m(v) is the mass (velocity) of the incident electron, e is the unit electron charge, c is the speed of light, h is Planck's constant, and ν is the frequency of the absorbed radiation. For the semiclassical formulation, the quantity Z_c represents the point charge with which the electron interacts. However, for a neutral system, this identification is not appropriate. Since a_K is proportional to Z_c^2 and $g(E,\omega)$ to Z_c^{-2}, the dependence on Z_c cancels out of the determination of $a(E,\omega)$. Thus, we may make any choice for Z_c provided we are careful in choosing the same convention when comparing to other calculations.

For most applications, we are interested in the Gaunt factor averaged over a Maxwell-Boltzmann distribution as

$$\bar{g}(T,\omega) = \frac{\int_0^\infty g(E,\omega)\, f(E)\, dE}{\int_0^\infty f(E)\, dE} \quad , \tag{3}$$

where

$$f(E)dE = \frac{2}{\pi^{1/2}} (k_B T)^{-3/2} \exp[-E/k_B T] E^{1/2} dE \quad , \tag{4}$$

k_B is the Boltzmann constant, T is the electron temperature, and E is the energy of the electron (= $\frac{1}{2} mv^2$). Upon performing the indicated integrals, we can simplify Eq. (3) to the form

$$\bar{g}(T,\omega) = (k_B T)^{-1} \int_0^\infty g(E,\omega) \exp[-E/k_B T] dE \quad . \tag{5}$$

The integration over energy is usually performed with an n_e-point Gauss-Laguerre quadrature scheme. The averaged Gaunt factor is also related to the absorption coefficient per unit pressure per atom, κ, which is more commonly used in astrophysical models, by:

$$\kappa = C_5 \, z_c^2 \, (\Delta k^2)^{-3} \, \Theta^{3/2} \, \bar{g}(T,\omega) \quad , \quad [cm^4/dyne] \tag{6}$$

where Δk^2 is the photon energy in Rydbergs, Θ is given by 5040/T, and C_5 is a constant equal to 2.0991×10^{-28}.

The Gaunt factor in atomic units (e = $\hbar$ = m = 1) is expressed in terms of a quantum mechanical matrix element $M^{(i)}$ by

$$g(E,\omega) = \frac{\sqrt{3}(\Delta k^2)^4}{8 z_c^2 \pi k k'} M^{(i)} \quad , \tag{7}$$

where $k(k')$ is the wave number ($k = mv/\hbar = 2mE/k^2)^{1/2}$ of the incoming (outgoing) electron. The quantity $M^{(i)}$ is, in turn, related to the dipole matrix element $d_{\ell\ell'}^{(i)}(k|k')$ through a complicated expression involving angular momentum coupling coefficients.[15] The dipole matrix element is given by

$$d_{\ell\ell'}^{(i)}(k|k') = \int_0^\infty f_{k\ell}(r) \, O_i \, f_{k'\ell'}(r) dr \quad , \tag{8}$$

where

$$O_i = \begin{cases} r & i = 1 \\ d/dr & i = 2 \\ dV/dr & i = 3 \end{cases},$$

the continuum wavefunction for an electron scattering from an atomic target (see II.B) is given by $f_{k\ell}(r)$, and $\ell(\ell')$ represents the initial (final) orbital angular momentum of the scattered electron. The first (second) expression for the dipole term is designated the length (velocity) form while the final relationship in terms of the interaction potential is termed the acceleration form. For a local scattering potential, all three expressions should yield the same value for the dipole matrix element. We generally employ the dipole length ($i = 1$) form; a detailed description of the method of calculation is given in Reference 19.

In this report, we shall restrict our attention to elastic scattering from 1) a local potential, 2) a closed-shell target atom or ion, or 3) an atom with a single s-electron outside a closed shell (alkali). Within this restriction, the angular algebra can be greatly simplified, and the quantity $M^{(i)}$ can be expressed as

$$M^{(i)} = \sum_{\ell\ell'}^{\ell_m} \ell_{max} d_{\ell\ell'}^{(i)}(k|k')^2 \quad , \tag{9}$$

where ℓ_{max} is the maximum value of ℓ and ℓ'. Technically, the sum in Eq. (9) should extend over an infinite range of the orbital angular momentum variables. However, in practical applications, the dipole matrix element decreases rapidly with increasing ℓ, and the sum may be truncated at a finite value ℓ_m.

II.B CONTINUUM SOLUTIONS

In the previous section, we presented the principal expressions used in calculating most free-free absorption parameters of interest. All of these expressions were related to a dipole matrix element $d_{\ell\ell'}^{(i)}$ between the continuum solutions for the incoming and outgoing electrons and a particular operator

$O_{(i)}$. In this section, we give the prescriptions for calculating the scattering wavefunction as well as the various approximations invoked.

For elastic scattering from a target atom, the scattering wavefunction $f_{k\ell}$ is a solution of the following radial Schrödinger equation:

$$Lf_{k\ell}(R) = \int_0^\infty W(R|R')\, f_{k\ell}(R')\, dR' \quad, \tag{10}$$

where

$$L = d^2/dR^2 + \ell(\ell+1)R^{-2} + k^2 \quad, \tag{11a}$$

and

$$W(R|R') \equiv V(R)\,\delta(R-R') + K(R|R') \quad. \tag{11b}$$

We have divided the "potential" term W into local and nonlocal parts. The local part usually represents the static or direct interaction while the nonlocal term corresponds to exchange and polarization-correlation effects. The latter interaction arises from virtual excitations to the excited states of the atomic system. In the next subsection (IIC), we shall consider various approximate representations of W.

In all of the derivations of the Gaunt factor in Section II.A, we have assumed the following asymptotic form for the continuum solution:

$$f_{k\ell}(R) \underset{R\to\infty}{\sim} \sin(kR + \ell\pi/2 + \eta_{k\ell}) \quad, \tag{12}$$

where $\eta_{k\ell}$ is the phase shift. The programs that calculate the solutions employ a slightly different asymptotic behavior

$$f^K_{k\ell}(R) \underset{R\to\infty}{\sim} [\hat{j}_\ell(kR) + \hat{n}_\ell(kR)K]k^{-1/2} \quad, \tag{13}$$

where $\hat{j}_\ell(\hat{n}_\ell)$ is the Ricatti-Bessel (-Neumann) function of order ℓ, and K is the reactance matrix ($K = \tan(\eta_\ell)$). The reactance matrix form can be converted to the behavior of Eq. (12) by dividing f^K by $\lfloor(1 + K^2)/k\rfloor^{1/2}$.

We have solved Eq. (10) by a linear algebraic[20] (LA) scheme. In the LA approach, we convert the differential form of Eq. (10) to an integral equation. This integral formulation is in turn transformed to a set of algebraic equations by introducing a discrete quadrature of n_p points for the integrals and wavefunctions. The resulting set of LA equations can be solved using standard linear systems techniques. The approach is non-iterative and can take full advantage of the vector architecture of the new super computers. In addition, nonlocal terms introduce no additional difficulties.

II.C MODEL POTENTIALS

In this section, we review the various approximations we make to the form of the interaction potential given in Eq. (11b). In general, the "potential" is rather complicated, involving contributions from not only the elastic channels but from all virtual excited states of the compound system. In order to properly represent all these effects, we must employ a multichannel or optical potential formulation[20] of the scattering problem. We have not performed such calculations in this report although we have compared our various models with the results from more elaborate close-coupling calculations.

1) Static-Exchange (SE)

In the SE approximation, we neglect all virtual excitations but consider the full effect of the Pauli principle on the composite system of ground-state atom and continuum electron. The resulting interaction has the form of Eq. (11b) with both a direct (static) and a nonlocal contribution.[15,20,21] In this case, the construction of the exchange kernel $K(R|R')$ is confined to orbitals representing the ground state of the target atom or ion.

2) Static-Exchange + Polarization (SEP)

We attempt to enhance the SE model by considering several approximations to the polarization-correlation term, which arises in elastic collisions from the virtual transitions to the excited states of the system. The first two approximations are based on a simple truncation of the long-range form of the polarization potential and have the form

$$V_c(R) = V_p(R) [1 - \exp(-(R|R_o)^6)] \qquad (14a)$$

and

$$V_c(R) = V_p(R) [1 - \exp(-(R|R_o))]^6 \qquad (14b)$$

where

$$V_p(R) \equiv -\frac{\alpha}{2} R^{-4} \qquad (14c)$$

with α the dipole polarizability of the atom. We term the first form SEP1a and the second SEP1b. We adjust the parameter R_o to make the phase shifts agree as close as possible with those of more elaborate calculations.

A second set of models [SEP2] is derived from a free-electron-gas [FEG] representation of the short-range component of the polarization-correlation contribution.[22,23] The approximate potential is given by

$$V_c(R) = \begin{cases} V_{corr}(R) & R < R_o \\ V_p(R) & R \geqslant R_o \end{cases} \qquad (15)$$

where V_{corr} is taken from Reference 23. The R_o is the point at which the short- and long-range contributions become equal and is <u>not</u> an adjustable parameter.

The full SEP potential is formed by adding V_c to V_s, the static interaction (see II.C.3), in Eq. (11b) and retaining the exchange contribution.

3) Static

The static or direct interaction potential is given by

$$V_s(\vec{R}) = \int^{\infty} \rho(\vec{r}) |\vec{R} - \vec{r}|^{-1} d\vec{r} - Z_N R^{-1} \qquad , \qquad (16)$$

where Z_N is the nuclear charge and the atomic charge density $\rho(\vec{r})$ has the form

$$\rho(\vec{r}) = \sum_{i=1}^{n_o} n_i |\phi_i(\vec{r})|^2 \quad , \tag{17}$$

with n_o being the number of occupied orbitals, and $n_i(\phi_i)$ being the occupation number (wavefunction) for the i-th bound orbital. For a closed-shell target or for one with a single s-type electron outside a closed shell, Eq. (16) is an exact representation of the static potential. For other atomic targets, this expression represents an "average" of the direct interaction. For the static (S) approximation, we replace V in Eq. (11b) with V_s and neglect K.

4) FEG Exchange

In the two FEG forms, we represent V by V_s and make a local approximation to the exchange term K based on free-electron-gas models.[24] Both forms are proportional to the one-third power of the charge density of Eq. (17) and are distinguished by their asymptotic behavior. The Hara model (HFEGE) contains the ionization energy in its definition of the local momentum while the asymptotically-adjusted expression (AAFEGE) does not.

5) Screened Coulomb and Point Charge

The screened Coulomb or Yukawa potential (Y) has a particularly simple form

$$V_Y(R) = -\frac{Z}{R} \exp[-\lambda R] \tag{18}$$

as does the point charge (PC)

$$V_{PC}(R) = -\frac{Z}{R} \quad . \tag{19}$$

The Y potential is used to study plasma screening effects in a very crude fashion while the PC form is employed to test various schemes for scattering from ions. In both cases V_Y or V_{PC} replace V and exchange is ignored.

We can rank the various models according to accuracy by the following scheme:

1) SEP

2) SE

3) AAFEGE/HFEGE

4) S

5) Y, PC .

The top ranking given the SEP model is only valid in those cases in which we have elaborate close-coupling calculations to test the choice of R_o.

III. RESULTS AND DISCUSSION

In this section, we compare the various model potentials discussed above for the determination of the free-free Gaunt factor. Before presenting the detailed comparison, we first investigate the validity of the various models by comparing with other calculations.

III.A COMPARISON WITH OTHER METHODS

As a first test, we treated atomic hydrogen at infrared frequencies and at temperatures of a few thousand degrees at the SE level. In Table 1, we compare our absorption coefficient, corrected for stimulated emission,[25] with those of John[8] and of Doughty and Frazer.[10] The agreement is quite good over the entire range of temperatures and frequencies. In addition, our results compare well with those of Bell et al,[26] who have probably performed the best calculations to date. Even though these authors include correlation-polarization effects through the solution of multi-channel equations, their results are within 20% of the SE for the range under consideration. In addition, we have obtained similar agreement for He and Ne with other more elaborate calculations[27,28,29] by using the SEP2 model. Finally, in order to check the higher energy regimes, we have found excellent agreement with Green[16] for the screened Coulomb potential. These comparisons indicate that the basic formalism and numerical procedures are being evaluated correctly.

III.B COMPARISON OF MODELS

In this section, we compare the averaged free-free Gaunt factors for the various models for a range of photon energies and electron temperatures extending down from about 10 eV. We make the comparison for several atomic species including the rare gases (He, Ne, Ar) and the alkalis (Li, Na).

In Fig. 1, we make a comparison of the SEP2, SE, AAFEGE, and S models for helium (He). We used the near-Hartree-Fock wavefunction of Clementi[30] to represent the target atom. We used 90 points for the Gauss-Legendre mesh in the LA calculation of the continuum wavefunction and distributed there points in three subregions as follows: 30/0.0 - 1.0/, 30/1.0 - 3.0/, and 30/3.0 - 10.a_o/. To calculate the integral over energy in Eq. (4), we employed a five-point (n_e = 5) Gauss-Laguerre quadrature; the sum over partial waves in Eq. (9) was carried to an ℓ_m equal to four (4). In applying the SEP2 polarization-correlation potential, we let α be 1.38 a_o^3 and determined R_o to be 1.773 a_o.

We observe from Fig. 1 that all models perform reasonably well for He. The S case is approximately a factor of eight too high at the lower energies but comes into good agreement at about 0.5 Ry. The SE and AAFEGE models remain close throughout the entire energy range (0.03 - 1.0 Ry), and both are within fifteen percent or better of the SEP2 result, which is used as a standard.

In Fig. 2, we make a similar comparison for argon (Ar). The parameters employed are as follows: n_p = 120, 40/0.0 - 1.0/, 40/1.0 - 3.0/, 40/3.0 - 20./; n_e = 5; ℓ_m = 4; α = 11.0 a_o^3 and R_o = 2.9177 a_o (SEP2); and Z_c = 1. We note much more striking differences among the models. At the lowest energy (0.03 Ry), the SE and SEP2 models are over a factor of five apart. As the energy rises, the differences between these two cases narrow but still remain near fifteen percent (15%) at 0.3 Ry. The AAFEGE again follows the SE rather closely, the largest difference being about 20%. On the other hand, the S model is in error by over an order of magnitude at the lowest energy and does not reach reasonable agreement with the SEP2 until an energy of about 0.3 Ry or 4 eV. In order to further caution against the cavelier use of models below 2 eV, we performed additional calculations using the SEP1a model. The cut-off parameter R_o was adjusted to reproduce the low-energy s- and p-wave scattering results of Thompson,[31] who employed a full treatment of exchange and a model polarization potential. We found an R_o of 3.0 a_o gave the best agreement with Thompson. For energies above 0.1 Ry (1.5 eV), the two SEP models agree to better than 15 %; however, at the lowest energy (0.03 Ry ~ 0.4 eV), they differ by almost 40 %. Thus, even two models selected to reproduce the best elastic scattering data for Ar and to properly simulate the electronic interaction can be in substantial disagreement below about one electron volt (1 eV). We observe similar results for neon (Ne).

In Fig. 3, we display the results for Li and observe even larger differences among the various models. The parameters employed are as follows: $n_p = 90$, $30/0.0 - 1.0/$, $30/1.0 - 3.0/$, and $30/3.0 - 10.0/$; $n_e = 5$; $\ell_m = 4$; $\alpha = 164\ a_o^3$ and $R_o = 4.1\ a_o$ (SEP1a); and $Z_c = 1$. As with the previous two systems, we have used the bound orbitals of Clementi.[30] The SEP2 model produces much too strong a polarization potential for e-Li collisions. Instead, we have employed the SEP1a model and adjusted R_o to give the best fit to the parameters of the lowest $^3p^o$ resonance determined from a two-state close-coupling calculation. We again observe differences of almost 50% between the SE and SEP1a models at the lowest energy; above 0.1 Ry, they agree to within 20 %. The AAFEGE case exhibits much greater departs from the SE result for Li than for the rare gases. The results for Na are similar to those for Li and re-enforce the need to employ highly accurate representations of the collision mechanism at energies below a few volts.

IV. CONCLUDING REMARKS

We have calculated averaged free-free Gaunt factors within the dipole approximation to the radiation field for a variety of models to the interaction potential experienced by the continuum electron for photon and electron energies below one Rydberg (1 Ry). We have investigated several atomic species including the rare gases (He, Ne, Ar) and the alkalis (Li, Na). We find great sensitivity of the Gaunt factor to the model employed for energies below four electron volts (4 eV). As the energies rise, the agreement among the various models improves. We judge that for accurate results below 4 eV, elaborate close-coupling or optical potentials schemes must be used. We find that for ions, the sensitivity to the model potential is much less than for the neutrals.

ACKNOWLEDGMENTS

The authors wish to acknowledge useful conversations with W. Huebner, N. Magee, N. T. Padial, J. Mann, R. Clark, and N. Delamater. Work performed under the auspices of the U. S. Department of Energy through the Theoretical Division of the Los Alamos National Laboratory.

REFERENCES

[1] J. A. Gaunt, "V. Continuous Absorption," Philos. Trans. R. Soc. London, Ser. A **229**, 16 3 (1930),

[2] J. A. Wheeler and R. Wildt, "The Absorption Coefficient of the Free-Free Transitions of the Negative Hydrogen Ion," Astrophys. J. **95**, 281 (1942).

[3] S. Chandrasekhar and F. H. Breen, "On the Continuous Absorption Coefficient of the Negative Hydrogen Ion III," Astrophys. J. **104**, 430 (1946).

[4] W. J. Karzas and R. Latter, "Electron Radiative Transitions in a Coulomb Field," Astrophys. J. Supp. **6**, 167 (1961).

[5] J. Green, "Fermi-Dirac Averages of the Free-Free Hydrogenic Gaunt Factor," Research Memorandum RM-2580-AEC, The Rand Corp. (1960).

[6] S. Geltman, "Continuum States of H^- and the Free-Free Absorption Coefficient," Astrophys. J. **141**, 376 (1965).

[7] T. Ohmura and H. Ohmura, "The Free-Free Transitions of the Negative Hydrogen Ion in the Exchange Approximation," Astrophys. J. **131**, 8 (1960).

[8] T. L. John, "The Free-Free Transitions of the Negative Hydrogen Ion in the Exchange Approximation," Mon. Not. R. Astron. Soc. **128**, 93 (1964).

[9] T. L. John, D. J. Morgan, and A. R. Williams, "The Free-Free Transitions of Li^- by a Multichannel Asymptotic Method," J. Phys. B **7**, 1990 (1974).

[10] N. A. Doughty and P. A. Fraser, "The Free-Free Absorption Coefficient of the Negative Hydrogen Ion," Mon. Not. R. Astron. Soc. **132**, 267 (1966).

[11] K. L. Bell, P. G. Burke, and A. E. Kingston, "Free-Free Transitions of an Electron in the Presence of an Atomic System," J. Phys. B **10**, 3117 (1977).

[12] M. S. Pindzola and H. P. Kelly, "Free-Free Radiative Absorption Coefficient for the Negative Argon Ion," Phys. Rev. A **14**, 204 (1976).

[13] L. A. Collins and A. L. Merts, "Comparison of Models of the Gaunt Factor for Free-Free Absorption," J. Quant. Spectrosc. Radiat. Transfer **26**, 443 (1981).

[14] P. I. Richards, "Summary Report on Investigations of Radiative and Chemical Calculations," Technical Operations Research report TO-1362-24 (1966).

[15] I. I. Sobelman, *Atomic Spectra and Radiative Transitions* (Springer-Verlag, Berlin, 1977), Chap. 9.

[16] J. Green, "Boltzmann Averages of the Free-Free Gaunt Factor in the Screened Coulomb Potential," R and D Associates reports, RDA-TR-4900-007, (1974).

[17] H. R. Griem, *Plamsa Spectroscopy* (McGraw-Hill, New York, 1964), p. 488.

[18] W. A. Lokke and W. H. Grasbergen, "XSNQ-U-a Non-LTE Emission and Absorption Coefficient Subroutine," Lawrence Livermore Laboratory report UCRL-52276 (1977).

[19] L. A. Collins and A. L. Merts, "Free-Free Gaunt Factors: Comparison of Various Models," Los Alamos Technical Report LA-10553-MS (1986).

[20] L. A. Collins and B. I. Schneider, "Linear Algebriac Approach to Electron-Molecule Collisions: General Formulation," Phys. Rev. A $\underline{24}$, 2387 (1981); ibid "Linear Algebraic Approach to Electronic Excitation of Atoms and Molecules by Electron Impact," Phys. Rev. A $\underline{27}$, 101 (1983).

[21] I. Percival and M. J. Seaton, "The Partial Wave Theory of Electron-Hydrogen Atom Collisions," Proc. Cambridge. Philos. Soc. $\underline{53}$, 654 (1957).

[22] J. R. O´Connell and N. F. Lane, "Nonadjustable Exchange-Correlation Model for Electron Scattering from Closed-Shell Atoms and Molecules," Phys. Rev. A $\underline{27}$, 1893 (1983).

[23] N. T. Padial and D. W. Norcross, "Parameter-Free Model of the Correlation-Polarization Potential for Electron-Molecule Collisions," Phys. Rev. A $\underline{29}$, 1742 (1984).

[24] M. E. Riley and D. G. Truhlar, "Approximations for the Exchange Potential in Electron Scattering," J. Chem. Phys. $\underline{63}$, 2182 (1975).

[25] The absorption coefficient per unit pressure per atom, κ, is corrected for stimulated emission by multiplying by $[1 - \exp(-C \ominus \Delta k^2)]$ with C_1 = 31.32625.

[26] K. L. Bell, A. E. Kingston, and W. A. McIlveen, "The Total Absorption Coefficient of the Negative Hydrogen Ion," J. Phys. B $\underline{8}$, 358 (1975).

[27] T. L. John, "The Free-Free Transitions of He$^-$," Mon. Not. R. Astron. Soc. $\underline{138}$, 137 (1968).

[28] K. L. Bell, K. A. Berrington, and J. P. Croskery, "The Free-Free Absorption Coefficient of the Negative Helium Ion," J. Phys. B $\underline{15}$, 977 (1982).

[29] T. L. John and A. R. Williams, "The Continuous Absorption Coefficient of Ne$^-$," Phys. Lett. A $\underline{43}$, 227 (1973).

[30] E. Clementi, "Tables of Atomic Wavefunctions," IBM J. Res. Dev. $\underline{9}$, 2 (1965).

[31] D. G. Thompson, "The Elastic Scattering of Electrons from Nitrogen, Neon, Phosphorous, and Argon Atoms," J. Phys. B $\underline{4}$, 468 (1971).

Table 1. Comparison of Absorption Coefficients for H⁻ at the SE Level[a,b]

Δk^2	Θ	$\bar{g}_s$	$\bar{g}_t$	$\bar{g}$	κ	κ_E	κ_J	κ_{BKM}
0.05	0.5	4.828(-2)	2.773(-2)	3.287(-2)	1.952	1.060	1.04	
	1.0	3.180(-2)	1.099(-2)	1.620(-2)	2.720	2.152	2.17	2.08
	2.0	2.090(-2)	4.766(-3)	8.799(-3)	4.179	3.997	4.11	3.54
0.10	0.5	7.115(-2)	3.839(-2)	4.658(-2)	0.346	0.274	0.271	
	1.0	5.501(-2)	1.842(-2)	2.757(-2)	0.579	0.554	0.567	0.545
	2.0	4.239(-2)	9.770(-3)	1.792(-2)	1.064	1.062	1.11	0.977
0.20	0.5	1.222(-1)	6.408(-2)	7.860(-2)	0.073	0.070	0.073	
	1.0	1.087(-1)	3.754(-2)	5.532(-2)	0.145	0.145	0.158	
	2.0	9.217(-2)	2.348(-2)	4.065(-2)	0.302	0.302	0.339	

---

[a]Nomenclature: Δk^2 is the photon energy in rydbergs; $\Theta = 5040/T$ (kelvins); $\bar{g}_s$ ($\bar{g}_t$) is the averaged Gaunt factor for singlet (triplet) scattering; $\bar{g} = (3\bar{g}_t + \bar{g}_s)/4$; κ is the pressure absorption coefficient; κ_E is this coefficient corrected for stimulated emission, κ_J from Ref. 8 (κ's in units of 10^{-26} cm⁴/dyne) and κ_{BKM} from Ref. 26; and $Z_c = 1$.

[b]Parameters: Calculations performed in the LA approximation with $n_p = 90$ and a mesh /0.0 - 1.0/1.0 - 3.0/3.0 - 10.0/ with 30 points per region; $\ell_m = 4$; and $n_e = 5$.

FIGURE CAPTIONS

Figure 1: Comparison of Averaged Gaunt Factor $\bar{g}(T,\omega)$ for various models for He.
Nomenclature: Line – SEP2, chain line – S, circle – SE, and cross – AAFEGE

Figure 2: Comparison of $\bar{g}(T,\omega)$ for Various Models for Ar.
Nomenclature: Line – SEP2, chain line – SE, circle – AAFEGE, and cross – S.

Figure 3: Same as Fig. 2 except for Li; the line represents SEP1b.

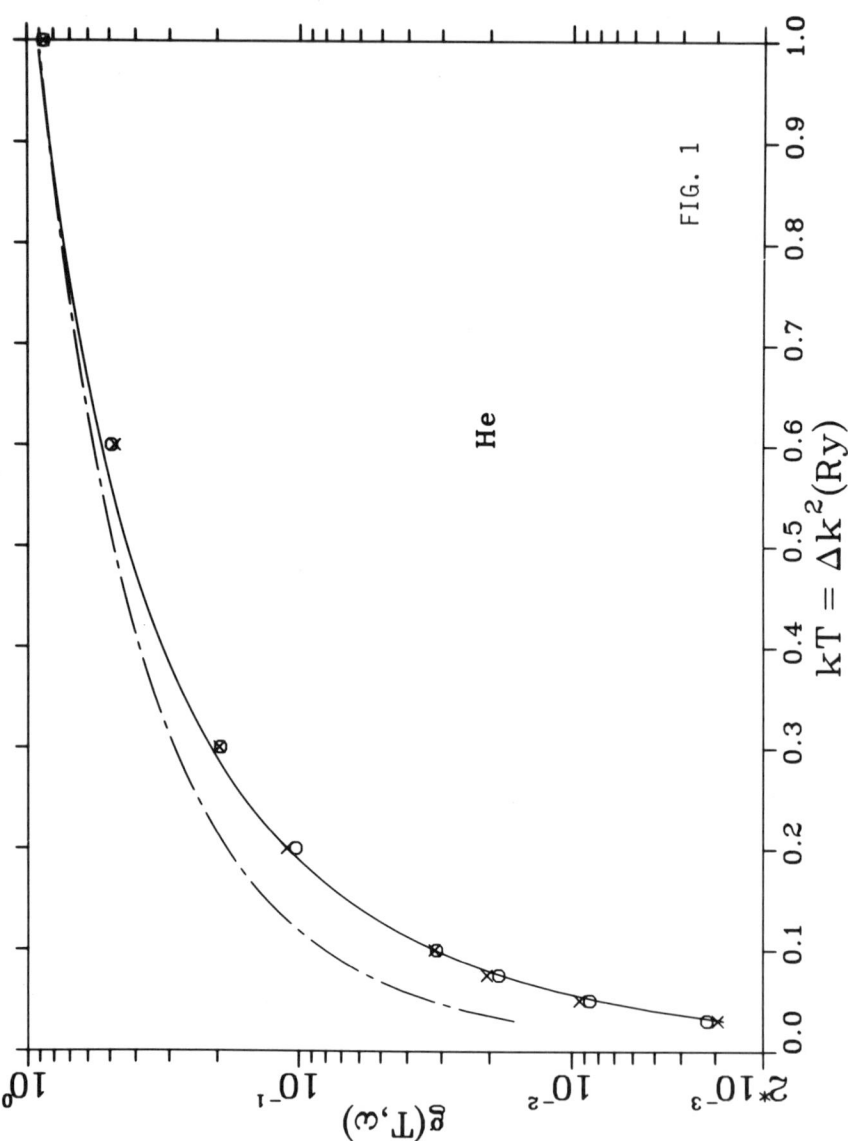

FIG. 1

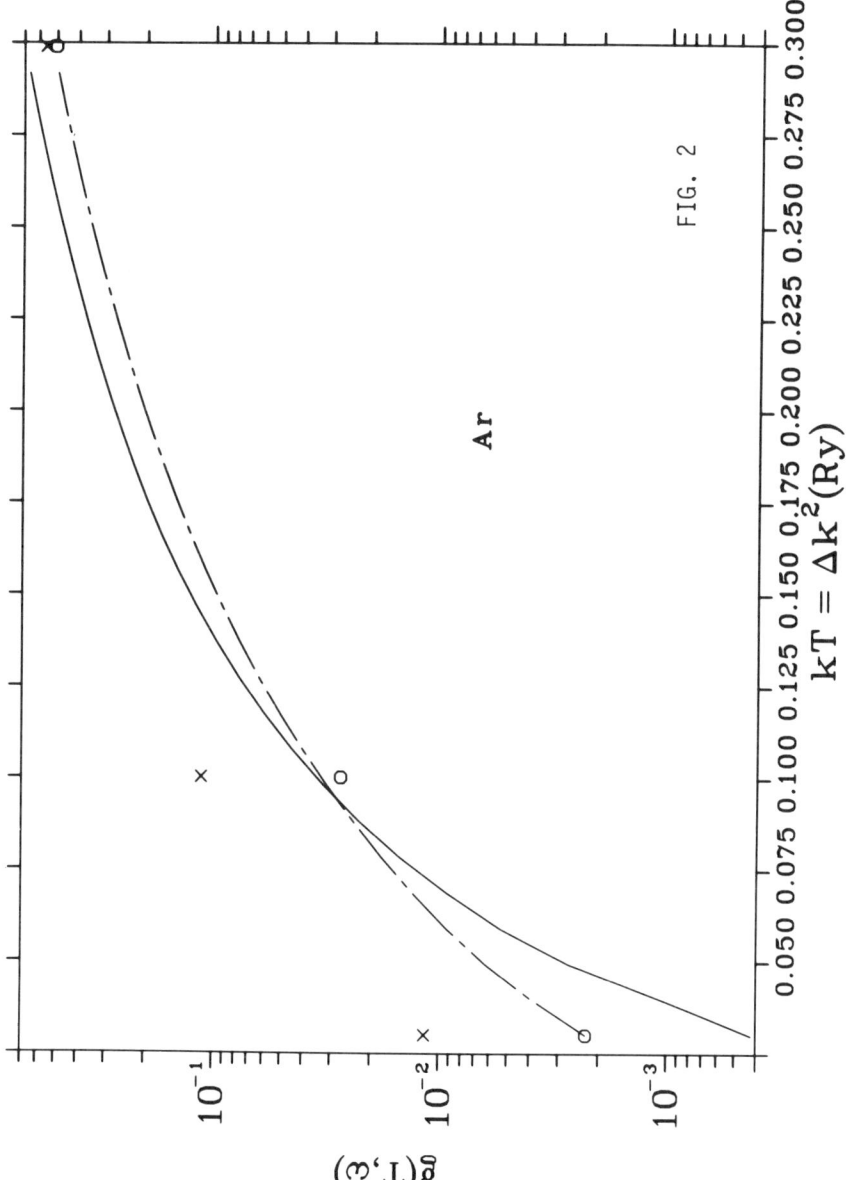

FIG. 2

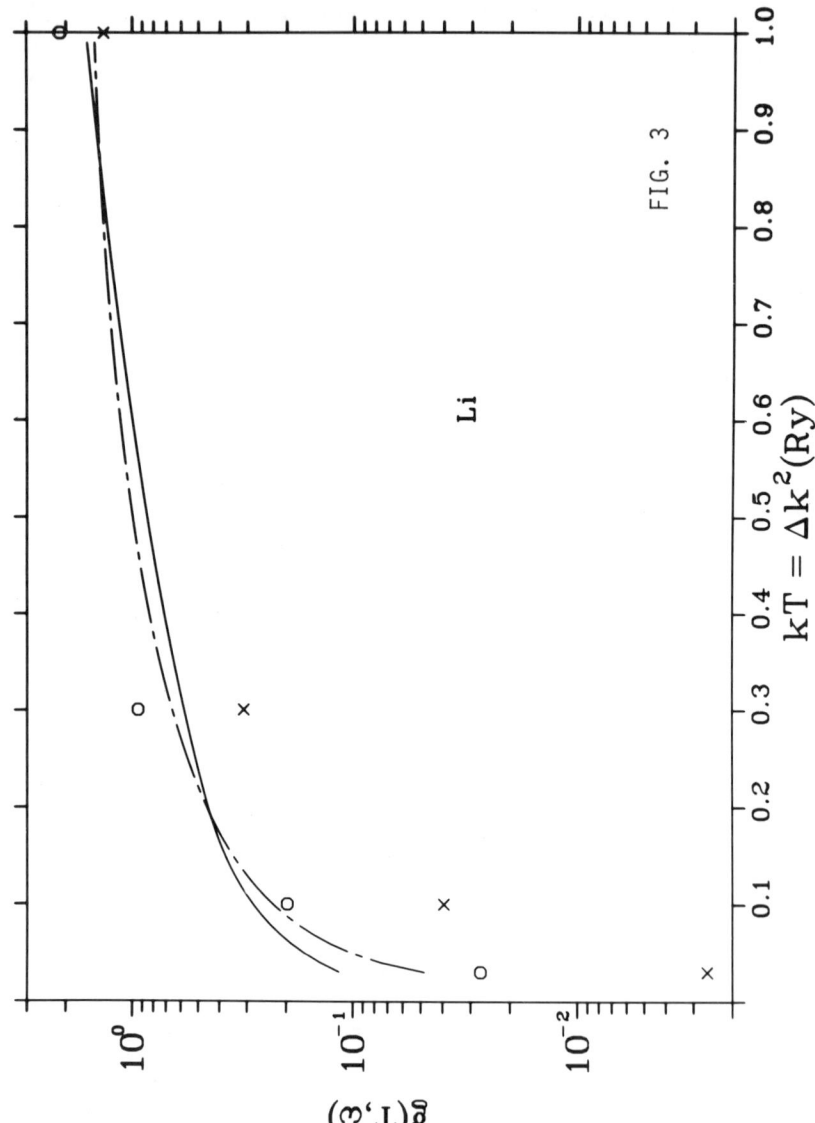

FIG. 3

COMPUTATION OF ANGLE AVERAGED CROSS SECTIONS IN A DEGENERATE COMPTON SCATTERING MEDIUM

J. Michael Pochkowski
Air Force Institute of Technology, Dayton, Ohio 45433

George H. Nickel
Los Alamos National Laboratory, Los Alamos, New Mexico 87545

William F. Bailey
Air Force Institute of Technology, Dayton, Ohio 45433

An accurate yet simple analytic method is developed for calculating Compton cross sections for monoenergetic photons interacting with an isotropic and degenerate distribution of relativistic electrons.

I. INTRODUCTION

The Compton scattering kernel[1]

$$\sigma_s(E_\nu, E'_\nu; f_e(E_e))dE'_\nu$$

gives the probability that a photon of energy E_ν, upon scattering from a collection of electrons characterized by the distribution function $f_e(E_e)$, scatters into the energy range $(E_\nu, E_\nu' + dE_\nu')$. A previously reported analytical approximation to σ_s[2] has been extended to include degenerate electron distribution functions. In such a distribution, the electrons are at a low enough temperature and/or sufficiently high density so that the number of final energy states available to the scattered electrons is limited. As before, the method is easy to calculate and gives good accuracy for multikilovolt photons scattered by electrons having similar energies.

The quantities which define the geometry of a Compton scattering (along with their dimensionless notation as used in the method) are:

$E_\nu = \alpha(m_e c^2)$ Initial photon energy (laboratory frame LF)

$E'_\nu = \alpha_s(m_e c^2)$ Final photon energy (LF)

ϕ Angle between directions of initial photon and electron (LF)

θ, τ Polar and azimuthal angle changes, respectively, of the photon in the electron rest frame.

$v_e = \beta c$ velocity of electron before scattering (LF)

We prefer to use the symbol α for dimensionless photon energies, in contrast with much of the literature on this subject, thus "freeing" the symbol γ for its conventional role as the relativistic parameter $\gamma = (1-\beta^2)^{-\frac{1}{2}}$.

A brief review of the analytical model is given here, including correction of some algebraic errors; readers are referred to the proceedings of the previous conference[2] for more details.

The kernel, σ_s, is an average over the three angles (ϕ, θ, τ) and the electron velocity β, each variable being weighted by its own probability distribution function. This procedure averages over the angles only, postponing the average over β for a subsequent calculation. Given a set of parameters (α, β, α_s), energy and momentum conservation define a surface in (ϕ, θ, τ)-space. This is the surface of constant final photon energy for a given electron velocity and initial photon energy. If the integral of scattering probability over the entire (ϕ, θ, τ)-space is normalized, the integral of that portion of (ϕ, θ, τ)-space "under" the surface represents the cumulative probability F of scattering to final energy α_s (or less). Differentiation of F gives σ, the probability distribution function, but since photon distributions are usually defined on a discrete grid of photon energy, differences of F at upper and lower edges of a photon "bin" give the scattering to that "bin" directly. Values of α_s giving a surface just "tangent" to the (ϕ, θ, τ)-space give the maximum and minimum scattered photon energies α_{smin} and α_{smax}. A useful transformation of the parameters $(\alpha, \beta, \alpha_s)$ to (z, ξ, a)

$$z = \frac{1}{\gamma^2 \beta} \left(\frac{\gamma_s}{\alpha} - 1\right) \tag{1}$$

$$\xi = \frac{\alpha_s}{\gamma \beta} - \beta \tag{2}$$

$$a = \frac{\alpha_s}{\gamma} \tag{3}$$

was guessed from analysis of detailed numerical calculations and later refined by algebraic considerations. F has a dependence on (z, ξ, a) which is dominated by z, having "first order" variations proportional to ξ and only a weak dependence on a.

The result of the simplest approximation for F is

$$F(z,\xi,a) \simeq \tfrac{1}{2}\binom{+}{-} h_0(r_{min}) + z g_0(r_{min}) + \xi g_{\cdot}(r_{min}) \begin{pmatrix} z<0 \\ z \geq 0 \end{pmatrix} \tag{4}$$

where

$$h_0(r) = \frac{3}{8}r - \frac{3}{16}r^2 + \frac{1}{16}r^3 \tag{5}$$

$$g_n(r) = \frac{3}{16\sqrt{2}} \int_r^2 (2-2r+r^2) r^{n-\frac{1}{2}} dr \tag{6}$$

$$r = 1 - \cos\phi \tag{7}$$

$$r_{min} = \frac{+X + \sqrt{X^2 + z^2 Y}}{Y} \qquad \text{if } Y \neq 0 \qquad (8)$$

$$X = 1 - \beta^2 - \xi z \qquad (9)$$

$$Y = \beta^2 - \xi^2 - a^2 \qquad (10)$$

$$r_{min} = \frac{z^2}{2X} \qquad \text{if } Y = 0 \qquad (11)$$

Equation 4 provides reasonable results for electron temperatures and photon energies into the tens of kilovolts. Attempts at extending this scheme to even higher energies is planned for discussion in subsequent publications.

II. RESULTS AND DISCUSSION

Differential Compton cross sections were calculated for electron temperatures of 1 and 20 keV; number densities of 10^{21}, 10^{25}, 10^{27}, and 10^{29} electrons per cubic centimeter; and incident photons of energies 5, 10, 20, 40, and 60 keV. The cross sections were computed by evaluating

$$\sigma_{Compton} = \left\{ \frac{F(\alpha, \alpha_{i+1}; \beta) - F(\alpha, \alpha_i; \beta)}{\alpha_{i+1} - \alpha_i} \right\} \sigma_{KN}^{T} \qquad (12)$$

where σ_{KN}^{T} is the temperature averaged Klein-Nishina cross section and has been averaged over one hundred equally probable electron energies.

Figures 1 and 2 show the differential cross section in millibarns keV. The solid lines are the cross sections in the nondegenerate limit. These profiles agree with other published results[1]. At low incident photon energies, the profile of the Compton cross sections have some characteristic width controlled by Doppler broadening. When the energy of the photon is increased, the profile of the cross section is driven by the Compton scattering formula which represents a rectangular step function in the electron rest frame. This rectangular step function broadens at the higher incident photon energies and begins to dominate the Doppler effect in influencing the shape of the cross section profiles.

When the electron gas is degenerate, the cumulative scattering probability is multiplied by a factor of $1-n(e)$. $n(e)$ is the probability of final state occupation for the electron, and through the use of Fermi statistics is determined to be:

$$n(e) = \frac{1}{e^{(\varepsilon-\mu)/kt}+1} \quad . \tag{13}$$

The chemical potential, μ, is then computed to be

$$\mu = E_F \left\{ 1 - \frac{\pi^2}{12} \left(\frac{T}{T_F}\right)^2 + \frac{\pi^4}{720} \left(\frac{T}{T_f}\right)^4 \right\} \tag{14}$$

when $\mu \leq 0$. T_F and E_F are the Fermi temperature and energy respectively. In the other limit, $\mu < 0$, the chemical potential is

$$\mu = -kT \log \left\{ \frac{gV}{N} \left. \frac{m_0 kT}{2\pi\hbar^2} \right)^{3/2} \right\} \tag{15}$$

The dashed lines in Figures 1 and 2 show the effects of degeneracy on the cross section profiles. With the number of final states limited, the instances where the electron delivers most of its energy are increased. If a Compton scattering event does occur, the photon will most likely downscatter. Thus, a slight shift to lower energies is seen in the profile. In addition, the peaks of the profiles decrease due to the overall decrease in the scattering probability.

A deeper appreciation can be given to this development by noting the reduction in the computer run times required to determine the Compton cross sections. Exact evaluation of the kernel took approximately one hour of Cray time for each given photon energy, electron temperature, and number density. Other methods use as much, if not more computer time. The results cited in this report required a little over two and one half seconds on a CDC 7600 using the same parameters.

REFERENCES

1. G. C. Pomraning, The Equations of Radiation Hydrodynamics, Pergammon Press, New York (1973).

2. Davis, Hooper, Lee, Merts, and Rozsnyai Proceedings of the Second Conference on the Radiative Properties of Hot Dense Matter, World Scientific Publishing Company, (1985) pp. 210-220.

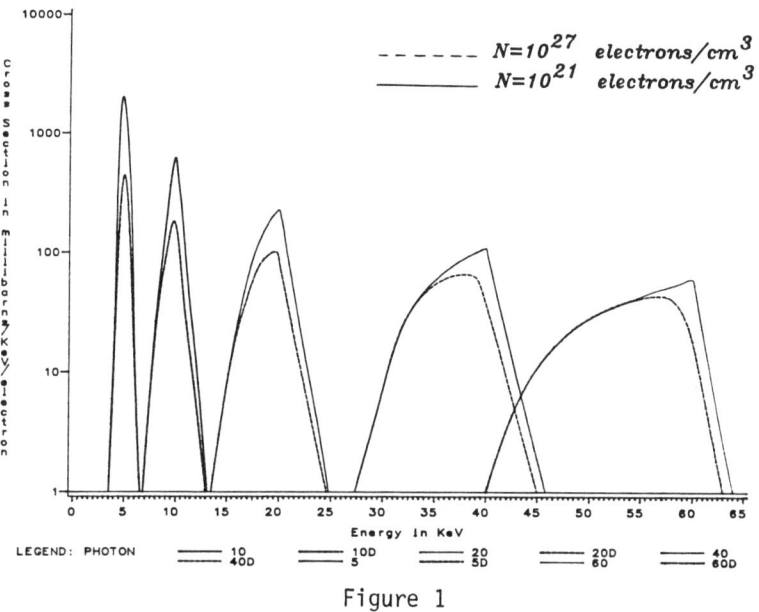

Figure 1

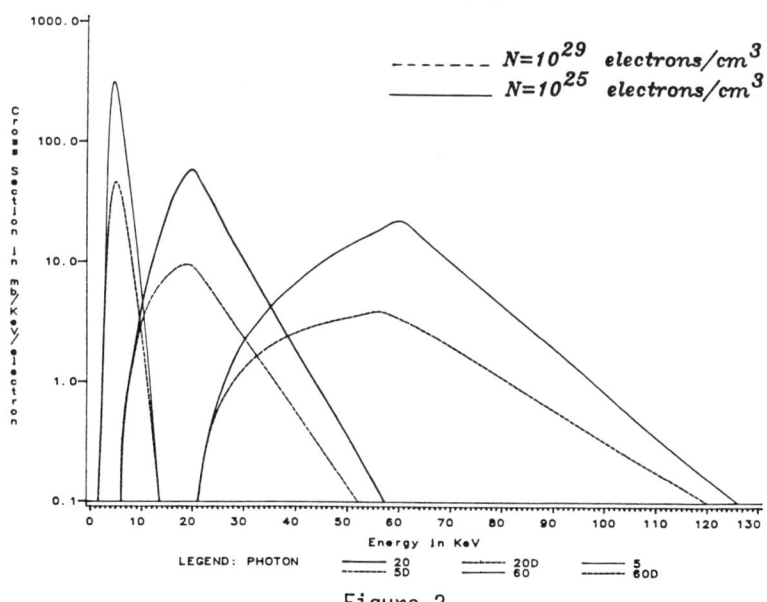

Figure 2

Optical Potential Approach for Scattering of Electrons by Ions

Dipak H. Oza
Atomic and Plasma Radiation Division
National Bureau of Standards
Gaithersburg, MD 20899

Joseph Callaway
Department of Physics
Louisiana State University
Baton Rouge, LA 70803-4001

Abstract

Currently, limited basis set close-coupling calculations represent the "state of the art" for the scattering of electrons by atoms and ions in the intermediate energy range. An addition of a matrix optical potential to a close-coupling scheme is described here and it is suggested that it should lead to improved results, particularly for scattering from excited states. The optical potential is evaluated with the use of a pseudostate basis of expansion. The resulting set of coupled integrodifferential equations are solved by an integral equation linear-algebraic method. Results for the excitation of singly-charged helium ion from ground state to the 2s and 2p states by electron impact are presented.

I. INTRODUCTION

Our main interest here is the calculation of scattering of electrons by excited ions in the low and intermediate energy range. At these energies, the perturbative methods such as Born approximation and their variants have limited and often questionable validity. For ions in the excited states, particularly for hydrogenic ions, the coupling between various states is very strong. A direct calculation using the close-coupling method with a large basis of target states, as has been successfully done in the case of the scattering by ions in the ground state, may be inadequate due to the need to include a large number of channels. The primary motivation for the evaluation of the scattering data from excited states of ions stems from its need in the line broadening calculations in laboratory and astrophysical plasmas. Due to its simplicity, the scattering of electrons from hydrogenic ions is considered here.

For the scattering from a discrete state, at electron energies of our interest, many scattering channels are open. It is practically not possible to include all open channels explicitly in a close-coupling calculation. Therefore, it becomes necessary to include the effects of a large many of these channels in an approximate way. The physical effects of channels not explicitly included are incorporated via a complex, energy dependent, matrix optical potential. This optical potential embodies the energy dependent polarizability of the target states and accounts for the loss of flux to open channels not included explicitly.

The basic concept of this approach can be described in the following manner. The close-coupling method gives rise to a coupled set of integro-differential equations. Of these, we specify the few channels which are of our interest. We designate these channels to be within group P and let all other channels be part of group Q. We intend to consider the channels in group P explicitly and the effects of channels in group Q approximately. The solutions for the channels in group Q can be formally related to the ones for the channels in group P by use of a Green's function technique. Next, we drop the coupling of the channels in group Q with each other. This enables us to replace the formal solution of channels in group Q in the equations for channels in group P leading to potentials which are non-local in character, energy-dependent and complex. Finally, we solve these equations using a linear algebraic integral equation method. The theory and method of calculations are described in the next section. For further details the reader is referred to the paper by Callaway and Oza[1] and references therein.

Besides the inherent theoretical interest in this work, the scattering information produced by these calculations would find applications in theoretical line-broadening calculations in plasmas. A recent review of the relevance of accurate collision descriptions to the shifts of ion lines in plasmas is given by Kelleher and Cooper.[2] They discuss the so-called "plasma polarization" shift and reach the conclusion that data from an accurate quantum mechanical scattering calculation should be able to shed new light on a number of unresolved issues. It is suggested that with emitter-perturber correlation being adequately accounted for, it would be inappropriate to add the so-called "plasma polarization" terms.

As an example for the method described, we consider the excitation of hydrogen-like helium ion from the ground state to the 2s and 2p states for incident electron energies up to 7.2 Rydbergs (1 Ry $\equiv$ 13.6 eV). The target state description is chosen to be an eighteen-state expansion with all the exact states up to n=3, supplemented by exact 4f state and eleven pseudostates. The channels corresponding to the exact states through n=3 are placed in the P space; the remaining channels define the Q space. Hence we perform a six-state close-coupling (with exchange) calculation supplemented by an optical potential approximately representing higher bound and continuum states.

II. THEORY

We closely follow the notations of our previous paper.[1] We consider a hydrogen-like target ion of nuclear charge Z. Let $\vec{r}$ denote the coordinates of the scattering electron and $\vec{x}$ be the coordinates of the atomic electron. The total wave-function of this two electron system can be expanded in terms of the complete set of target atom (ion) states.

$$\Psi_a(\vec{r}, \vec{x}) = [1+(-1)^S P_{12}] \sum_j \phi_{ja}(r) u_j(x) Y^{M\pi}_{L\ell_1'\ell_2'}(\hat{r}_1, \hat{r}_2) \qquad (1)$$

In this equation, $u_j(x)$ is a member of the complete set of target states. $Y^{M\pi}_{L\ell_1'\ell_2'}$ is a two-particle spherical harmonic for total angular momentum L, z-component M, parity π and individual angular momenta ℓ_1' and ℓ_2'. The expansion coefficient functions $\phi_{ja}(r)$ describe the radial motion of the scattering electron. The subscript "a" on the left-hand-side implies that there is an incident particle in channel "a" only. The expression in the square parenthesis represents the electron-exchange operator and allows for the exchange of identical particles for a total spin of S.

Expansion (1) is the mathematical statement for the "close-coupling" scheme. In principle, the sum should run over an infinite set of bound and continuum target states. That would require the determination of an infinite number of scattering functions : a practical impossibility. Therefore, in practice, one replaces the sum by a few exact target states and the omitted bound and continuum states are approximated by a few judiciously chosen pseudostates. The functions $u_j(x)$ are required to diagonalize the target Hamiltonian H_T.

$$\int d^3x \, u_k^*(x) \, Y_{\ell_k m_k}^*(x) \, H_T(x) \, u_j(x) \, Y_{\ell_j m_j}(x) = E_j \delta_{jk} \tag{2}$$

where

$$H_T(x) = -\nabla_x^2 + \frac{Ze^2}{x} \tag{3}$$

The total two-particle Hamiltonian is

$$H(\vec{r},\vec{x}) = H_T(\vec{r}) + H_T(\vec{x}) + \frac{e^2}{|\vec{r}-\vec{x}|} \tag{4}$$

We require that the Schroedinger equation be exactly satisfied in that part of the function space spanned by the target functions. Mathematically expressed, it is

$$\int u_i^*(x) \, Y_{L\ell_1\ell_2}^{M\pi}(\hat{r},\hat{x}) \, [H(\vec{r},\vec{x}) - E] \, \Psi_a(\vec{r},\vec{x}) \, d\hat{r} d^3\hat{x} = 0 \tag{5}$$

where E is the total energy of the two electrons.

The use of Eqs. (1)-(4) in (5) reduce to a set of coupled integro-differential equations:

$$[H_{\ell_1}(r) - k_i^2] \phi_{ia}(r) + \sum_j \int dx\, x^2\, K_{ij}(r,x) \phi_{ja}(x) = 0 \qquad (6)$$

where the radial Hamiltonian is

$$H_{\ell_1}(r) = -\frac{1}{2}\frac{1}{r^2}\frac{d}{dr}\left(r^2 \frac{d}{dr}\right) + \frac{\ell_1(\ell_1+1)}{r^2} - \frac{(Z-1)e^2}{r} \qquad (7)$$

and the kernel, K_{ij}, consists of local (direct) and non-local (exchange) components.

$$K_{ij}(r,x) = \left[V_{ij}(r) - \frac{e^2}{r}\delta_{ij}\right]\frac{\delta(r-x)}{r^2}$$

$$+ (-1)^S u_i^*(x)\left[W_{ij}(x,r) - (k_i^2 - E_j)\delta_{\ell_1 \ell_2'}\delta_{\ell_2 \ell_1'}\right] u_j(r) \qquad (8)$$

The direct potential matrix elements are given by

$$V_{ij}(r) = \int d^3x\, d\hat{r}\, u_i(x)\left[Y^{M\pi}_{L\ell_1\ell_2}(\hat{r},\hat{x})\right]^* \frac{e^2}{|\vec{r}-\vec{x}|} u_j(x)\, Y^{M\pi}_{L\ell_1\ell_2}(\hat{x},\hat{r}) \qquad (9)$$

and the exchange potential matrix elements are given by

$$W_{ij}(x,r) = \int dr dx [Y^{M\pi}_{L\ell_1\ell_2}(\hat{r},\hat{x})]^* \frac{e^2}{|\vec{r}-\vec{x}|} Y^{M\pi}_{L\ell_1\ell_2}(\hat{r},\hat{x}) \tag{10}$$

The set of Eq. (6) needs to be solved in the close-coupling approximation. However, depending on the number of target states included in expansion (1), one may have a very large set of equations. It is at this point that we construct a matrix optical potential for the channels we do not wish to include explicitly in the calculation. We recall that the channels we wish to calculate explicitly form a space P and the rest form the space Q.

We first solve formally for a channel i in Q space. Using a Green's function to solve for the inhomogeneous equation (6), we have

$$\phi_{ia}(r) = -\sum_{j\epsilon P} \int dy\, y^2 \int dx\, x^2\, \gamma_i(r,y)\, K_{ij}(y,x)\phi_{ja}(x) \quad ; \quad i\epsilon Q \tag{11}$$

where the Green's function satisfies

$$(H_{\ell_1}(r) - k_i^2)\, \gamma_i(r,y) = \frac{1}{y^2} \delta(r-y) \tag{12}$$

In the formal solution (11), we have made the approximation that the sum is carried out only for channels in P space, i.e. we have neglected the channel couplings between channels in space Q. Having this formal solution for the channels in P space, we go on to solve eq. (6) for channels in P space.

Let i in eq. (6) now belong to set P. The sum over j extends to both P and Q spaces. Making the substitution (11) for j belonging to Q space, after rearrangement, we have

$$\left(H_{\ell_1}(r) - k_i^2\right) \phi_{ia}(r)$$

$$+ \sum_{j \in P} \int dx \, x^2 \left[K_{ij}(r,x) + U_{ij}(r,x)\right] \phi_{ja}(x) = 0 \qquad (13)$$

where the optical potential matrix elements are

$$U_{ij}(r,x) = - \sum_{m \in Q} \int y^2 dy \int z^2 dz \, K_{im}(r,y) \, \gamma_m(y,z) \, K_{mj}(z,x) \qquad (14)$$

Now instead of a large set of coupled integrodifferential equations (6), we need to solve for a smaller set of eq. (13).

The construction of optical potential requires the Green's functions, γ_m, which can be specified analytically for the radial operator $H_\ell(r)$ which contains a pure Coulomb potential. The details of the Green's function are given in Ref. 1.

For channels belonging to Q space which are energetically open, the corresponding Green's functions, γ_m, are complex, thereby making the optical potential complex. This feature physically accounts for the loss of flux. Another important physical feature of the optical potential is that in the low energy limit for the elastic scattering, it reduces to the ordinary polarization potential with non-adiabatic and energy-dependent corrections.

III. METHOD OF SOLUTION

For the numerical solution of the set of coupled integro-differential equations, Eq. (13), we have chosen the integral equation linear algebraic method.[3,4] We observe that Eq. (13) is an inhomogenous equation. We convert the set of partial differential equations to a set of integral equations by the use of appropriate Green's function. The Green's function is expected to satisfy Eq. (12) and we employ K-matrix boundary conditions. The Green's function for an open channel is

$$\Omega_i(r,x) = \frac{1}{k_i} \frac{F_{\ell_1}(k_i r) G_{\ell_1}(k_i r)}{rx} \tag{15}$$

where F_ℓ and G_ℓ are regular and irregular radial Coulomb functions.

The solution can be expressed as

$$\phi_{ia}(r) = \frac{F_{\ell_1}(k_i r) \delta_{ia}}{r} - \sum_{j \in P} \int dx\, x^2\, \Omega_j(r,x) \int dy\, y^2 \left[K_{ij}(x,y) + U_{ij}(x,y) \right] \phi_{ja}(y) \tag{16}$$

One can introduce a coordinate grid and cast this set of equations in a matrix form. After some rearrangement, we have

$$\sum_{j,n} \left(\delta_{ij} \delta_{mn} + B_{im,jn} \right) \phi_{ja}(r_n) = \frac{1}{r_m} F_{\ell_1}(k_i r_m) \delta_{ia} \tag{17}$$

where the elements of B embody the details of the potentials, Green's functions and the quadrature scheme.

Next, we turn our attention to the algorithms used for numerical integration. An integral can be approximated using a numerical quadrature scheme as

$$\int_a^b y(x)dx = \sum_{i=1}^{N} \omega_i \, y(x_i) + \text{Error term} \quad (18)$$

The order of the error term depends on the quadrature scheme used. In order for the error term to be well-defined, the integrand y(x) must be continuous over the closed interval [a,b] and differentiable m times over the open interval where m depends on the specific order of approximation for the quadrature.

In equation (16), the integrand contains the Green's function which has a discontinuous first derivative at r=x. Also, the non-local kernels exhibit similar derivative discontinuities. An analysis presented in Ref. 4 shows that the Gaussian quadratures are not optimal in this problem and favors the use of Newton-Cotes type formulas with evenly spaced mesh with occasional jumps in the grid spacing. We have used a five-point Newton-Cotes quadrature scheme for which the error is of the order of h^7. Details of the procedure are described in Ref. 4 with simple illustrative examples.

After a numerical solution is obtained from the matrix equation (17), the reactance matrix (K-matrix) is determined by fitting the solution at large distances to the asymptotic form

$$\phi_{ia}(r) \sim \frac{1}{k_i^{1/2} r} \left[F_{\ell_1}(k_i r) \delta_{ia} + K_{ia} G_{\ell_1}(k_i r) \right] \tag{19}$$

The exchange potential Eq. (10) gives rise to short range terms. The direct potential, Eq. (9), on the other hand, has long-range character with off-diagonal terms proportional to $\frac{1}{r^2}$, $\frac{1}{r^3}$ etc. on account of the degeneracy of the levels with respect to ℓ for a given n. Since the optical potential is constructed from direct potentials, it exhibits long-range polarization potential proportional to $\frac{1}{r^4}$. These long-range potentials necessitate that we carry the calculations out to large r values. For this purpose, we have followed the procedure of Henry et al.[5] for propagating K-matrices as implemented in the program ASYM3.

The K-matrix is in general complex because of the use of complex optical potential and from it the scattering matrix and other collisional information can be computed in the usual way.

$$S = \frac{1+iK}{1-iK} \tag{20}$$

IV. TARGET STATES

The target states of the ion are described as a linear combination of Slater-type-orbitals.

$$u_j^{(\ell)}(r) = \sum_k C_{jk}^{(\ell)} r^{n_k^{(\ell)}} \varepsilon^{-\zeta_k^{(\ell)} r} \tag{21}$$

The superscript, ℓ, is added to denote the angular momentum. Specifying the set of parameters $n_k^{(\ell)}$ and $\zeta_k^{(\ell)}$, uniquely determines the expansion coefficients $c_{jk}^{(\ell)}$ by the diagonalization requirement stated in Eq. (2).

The set of parameters used in the present calculation are given in Table I. Inclusion of the parameters of Slater-type-orbitals specifying the exact eigenstates reproduces corresponding exact states in expansion (21). The states in expansion (21) which do not correspond to exact radial functions of the target ion are termed as "pseudostates".

The basis set consists of 7 s states, 5 p states, 3 d states, 2 f states and 1 g state. It has six exact states up to n=3, exact 4f state and eleven pseudostates. This basis set was used for electron scattering by neutral hydrogen atom in Ref. 1. The set contains good approximations to the 4s, 4p and 4d states. This is determined by calculating the overlap of the lowest pseudostates with respect to the exact states. This fact is also reflected by the energy-values of the spectrum in Table I.

It is well-known that the polarizability of the excited state, n, is mostly contributed by the n+1 states. For an accurate description of the process of low energy elastic scattering, it is essential to embody the polarization potential in the calculation. Keeping these facts in mind and recalling that the basis set contains a good description of all the exact states up to n=4, it is possible to obtain good results for the scattering from the states in the n=3 manifold. The scattering data thus obtained here should be useful in line-broadening calculations of Lyman-alpha and Balmer-alpha lines. As pointed out in Ref. 2, there are a number of unanswered issues concerning the Balmer-alpha line of the He II ion. For other transitions, involving states higher than n=3, we would need to include additional states in our target state description. With the implementation of the concept of optical potentials we would be able to keep the calculations computationally tractable. Such calculations are planned for the future.

In the present calculations, we have used channels corresponding to the states through n=3 in the P space, i.e. we have included them explicitly in the calculation. The remaining channels define the Q space and are used to construct the optical potential. The net result is that of performing a six-state close-coupling calculation with a matrix optical potential representing the effects of the rest of the bound and continuum states. The assumption is made here that the 12 states contributing to Q space are complete enough to permit sufficiently accurate evaluation of the optical potential.

Including all the channels (corresponding to all the 18 states) would have made the number of equations in (6) to be 18 for the partial wave with total angular momentum L=0 and 39 for the partial waves with L > 4. By the introduction of the optical potential with the above choice for P space, we have 6 coupled integro-differential equations for L=0 and no more than 10 equations for any partial wave in (13).

V. RESULTS

We present the results for the excitation of the 2s and 2p states from the ground state for the He^+ ion here. The results are based on the 18-state basis described in the previous section. As mentioned before, with this basis set, it would be possible to achieve good results for the scattering of electrons from the states in the n=3 manifold.

Here we discuss the results of our analysis for the excitation from the ground state. Similar calculations for the scattering of electrons from the ground state of neutral hydrogen[1] have yielded quite satisfactory results for the processes of elastic scattering, excitation, total reaction cross-section[6] and spin-asymmetry in elastic scattering.[7] We are in the process of analyzing our data for the scattering from excited states of neutral hydrogen and singly-charged helium ions.

The cross-sections for excitations to the 2s and 2p states are displayed in Figures 1 and 2, respectively, for electron energies from just above excitation threshold to 7.2 Ry. The optical potential calculations described in this paper are performed for the partial waves up to the total angular momentum of $L=6$. In the results shown in Figures 1 and 2, higher partial waves contributions are included approximately using the unitarized Born approximation with exchange (UBX).

For the purpose of comparison, the excitation cross-sections calculated by Henry and Matese[8] using a five-state pseudostate close-coupling approximation are shown in the figures. Their target is described by 1s, 2s and 2p eigenstates and one s and one p pseudostates. Although the agreement is not perfect, the two sets of results obtained using different sets of target state descriptions are quite close. The agreement may be contrasted with the cross-sections from a 3-state close-coupling calculation[9] (1s, 2s and 2p eigenstates). Just above ionization threshold, where an infinite number of channels are open and channel-couplings are strong, the 3-state close-coupling results are about 50% larger than the present results. This demonstrates the need to allow for the loss of flux either via explicit pseudostates or via an absorptive potential such as the complex optical potential described in this work. The cross-sections for the excitation of the 2s state of another six-state pseudostate close-coupling calculation[10] (1s, 2s, 2p eigenstates and 1s and 2p pseudostates) agree well with the present results (not shown in the figure) above the ionization threshold. The three-state close-coupling excitation cross-sections for the 2s state remain about 50% larger at higher energies up to twice the ionization threshold. The excitation cross-sections for the allowed transition get within about 20% agreement at higher energies in the figure. The improvement at higher energies for the allowed transition may be partly accounted for by the fact that at higher energies, higher partial waves (larger total angular momentum) begin to contribute. The high partial waves are less sensitive to the target state description.

VI. CONCLUSIONS

We have developed and tested a procedure for the construction of a matrix optical potential from a pseudostate basis set describing the target ion. We have successfully implemented this optical potential approach in conjunction with close-coupling scheme for the calculations of electron scattering by atoms and ions. The complex optical potentials account for such important physical effects as the polarizability of the target states and absorption of scattering flux to the omitted bound and continuum states. The introduction of the optical potential reduces the size of the set of coupled integrodifferential equations, making it computationally tractable.

In this paper, we have discussed our results for the excitation of singly-charged helium ions to the 2s and 2p states in the intermediate energy range. Employing a six-state close-coupling approximation with exchange and matrix optical potential constructed from twelve pseudostates, we find strong absorptive effects above the ionization threshold which are not accounted for by a pure three-state close-coupling approximation.

We have successfully utilized our own version of integral equation linear algebraic method in order to obtain the numerical solutions of the set of coupled integro-differential equations.

We are in the process of analyzing our data for scattering processes from excited states and applying it to line-broadening calculations in plasma environment.

ACKNOWLEDGMENT

This work was supported in part by the AFOSR.

REFERENCES

1. J. Callaway and D. H. Oza, Phys. Rev. A $\underline{32}$, 2628 (1985).

2. D. E. Kelleher and J. Cooper, in <u>Spectral Line Shapes</u> Vol. 3 (Ed. F. Rostas) Walter de Gruyter and Co., p. 85 (1985).

3. L. A. Collins and B. I. Schneider, Phys. Rev. A $\underline{24}$, 2387 (1981), and references therein.

4. D. H. Oza and J. Callaway, J. Comp. Phys. (1986), in press.

5. R. J. W. Henry, S. P. Rountree, and E. R. Smith, Comput. Phys. Commun. $\underline{23}$, 233 (1981).

6. J. Callaway and D. H. Oza, Phys. Rev. A (submitted).

7. D. H. Oza and J. Callaway, Phys. Rev. A $\underline{32}$, 2534 (1985).

8. R. J. W. Henry and J. J. Matese, Phys. Rev. A $\underline{14}$, 1368 (1976).

9. P. G. Burke, D. D. McVicar, and K. Smith, Proc. Phys. Soc. (Lon.) $\underline{83}$, 397 (1964).

10. S. A. Wakid and J. Callaway, J. Phys. B: Atom. Molec. Physics $\underline{13}$, L605 (1980).

Table I

n_k	ζ_k/Z	E_j/Z^2
	$\ell=0$	
0	1.0	-1.0
0	0.5	-0.25
1	0.5	-0.111111
0	1/3	-0.06238
1	1/3	-0.01723
2	1/3	0.19588
0	0.2	2.03964
	$\ell=1$	
1	1.0	-0.25
1	0.5	-0.111111
1	1/3	-0.06219
2	1/3	+0.02744
1	0.2	0.90912
	$\ell=2$	
2	0.5	-0.11111
2	1/3	-0.06245
2	0.2	+0.04187
	$\ell=3$	
3	0.5	-0.0625
3	0.25	+0.03313
	$\ell=4$	
4	0.4	0.0

Table 1. Parameters and energies (in Ry) of the pseudostate basis. Note that the energies refer to the combination, Eq. (21), rather than to the individual components. Here, for He^+, Z=2.

FIGURE CAPTIONS

Figure 1 Excitation cross-section to the 2s state. Solid curve - present results; chain curve - 5-state pseudostate close-coupling results (Ref. 8); dashed curve - 3-state close-coupling results (Ref. 9).

Figure 2 Excitation cross-section to the 2p state. Curves have the same meaning as in Fig. 1.

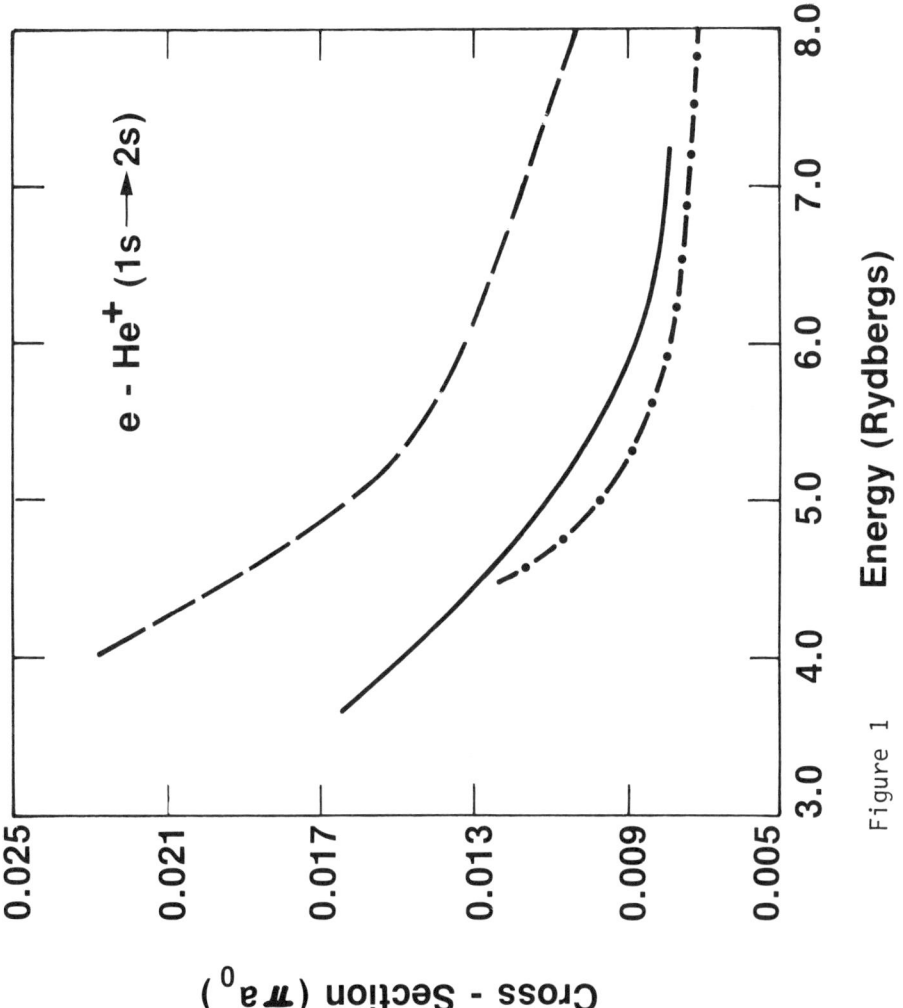

Figure 1

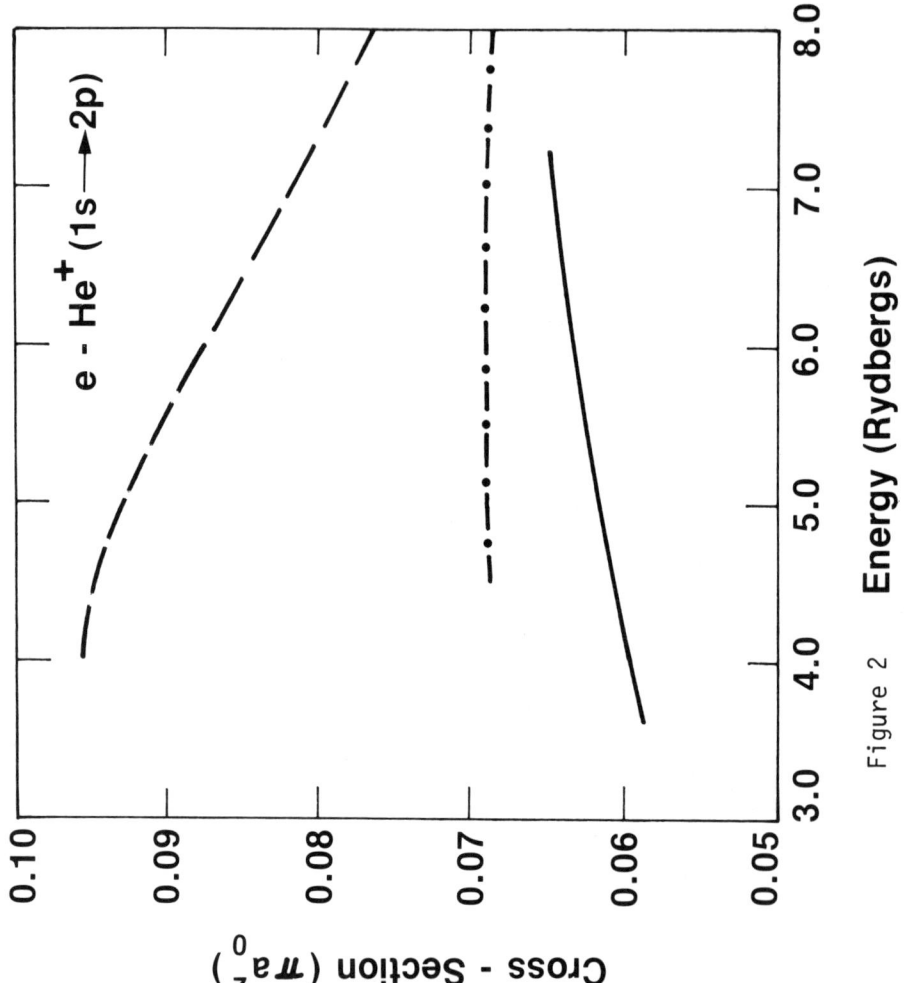

Figure 2

II. X-RAY SPECTROSCOPY, THEORY AND EXPERIMENT

Non-Thermal Effects in a Hot Dense Plasma

by

L. A. Jones
Los Alamos National Laboratory

Collaborators: D. R. Kania, B. A. Hammel, E. Källne, M. Maestas, J. McGurn, R. Shepherd

INTRODUCTION:

Observations made on collapsing gas shell Z-pinches have produced some results which are difficult to explain by using the usual thermal model.[1,2] These pinches produce a fast burst of x-rays at the time of pinch usually by exciting the closed shell resonance lines, i.e. He-, Ne-, or Ar- like emission. This emission comes from so called "hot spots" which form at irregular intervals along the pinched column axis. These "hot spots" are peculiar in that their diameter is always smaller than the smallest diameter observed for the pinching plasma column.

In this paper I will briefly describe a hollow gas shell Z-pinch device and show how some initial observation lead to the conclusion that there is an energetic electron beam produced along the axis of the collapsing gas shell. I will then summarize an experiment that directly measured some of the characteristics of this runaway electron beam. Finally, the results of an experiment which observed a new effect will be presented along with a model that uses a runaway electron beam to explain this new effect.

EXPERIMENTAL APPARATUS:

The Los Alamos High Density Z-pinch is shown schematically in Figure 1. It consists of a 72 kJ - 600 kV Marx bank made up of twelve parallel, low inductance capacitor modules. This Marx bank pulse charges a 1 Ω water transmission line to a voltage of roughly 1.5 times that produced by the Marx bank. Once peak voltage is reached, an in-line switch between the water line and load chamber over-volts, delivering the current to the gas load.

In all the experiments that follow the Z-pinch was operated at a derated level in which the Marx bank produced only 400 kV. This developed 600 kV across the gas load which produced a peak current of 600 kA with a rise time of 200 ns. Figure 2 shows a schematic of the gas load, indicating the position of the in-line switch, the fast gas valve, and the supersonic nozzle which produces the hollow gas jet that is driven to the axis by the fast rising current. The hollow gas shell is made up of argon and has an inside diameter of 2 cm, an outside diameter of 2.5 cm, and is 0.9 cm long. The line density of argon gas is chosen so that the shell reaches the axis at peak current. This requires 9×10^{16} argon atom/cm.

A series of visible framing pictures is shown in figure 3. These pictures have an exposure time of 10 ns and there is 50 ns between frames. These pictures were taken looking radially at the injected gas column, a schematic of the electrodes is shown at the bottom of the figure to orient the reader. Sometime between frames 1 and 2 the current began to flow. Frame 3 and 4 show the hollow gas shell moving in toward the axis with the outer boundary straightening up with time. Between frames 4 and 5 the gas collided with itself on the axis producing a dense hot plasma. Frames 5 and 6 are bloomed from the intense visible emission at the time of pinch and after. From this type of data the radius as a function of time (figure 4) and hence the collapse velocity can be obtained.

EVIDENCE FOR A NON-THERMAL ELECTRON DISTRIBUTION:[3]

A bent crystal x-ray spectrum of the emission of the collapsed argon shell was taken and is shown in figure 5. As can be seen the spectrum consists of not only the $1s^2$ - 1s 2p resonance and intercombination line of He-like argon but also displays seven satellite lines. Another part of the spectrum (not shown) also contained some weak H-like lines from argon. The seven satellite lines were identified as the 1s-2p transition of the ion species indicated in figure 5. Presumably these low ionization states lose a 1s electron and then these weak lines are formed when the atom relaxes by undergoing the 1s-2p transition. Also shown in the figure is a calculated 1s-2p spectrum by Cowan[4] showing where one would expect these satellites to occur. The agreement is excellent.

The occurrence of 1s vacancies in an argon plasma means that there are energetic electrons capable of providing the 3-4 keV of energy required to ionize an argon atom from the 1s level. These energetic electrons cannot, however, form a thermal distribution because the lower ionization states, Ar X - XV, could not exist at such a high temperature.

The temperature can be estimated from the data in many ways. One way is to assume that we have a thermal plasma and use line ratios to deduce the temperature. Figure 6 gives a summary of the two methods commonly used to estimate the temperature. By assuming the Ar XV structure is a dielectronic satellite a temperature of 1.1 keV can be obtained. Also by taking the line ratio of H-like to He-like resonance lines and assuming coronal equilibrium one can obtain a temperature of ~1 keV. Both these methods give approximately the same temperature but neither explain how energetic excitations can be made in regions in the plasma where low ionization states exist. Figure 7 shows that the temperature range required for the existance of argon X through argon XVII ground state ions is 150 - 300 eV. This is considerably less than the temperature required to do the resonance line excitation, ~1 keV.

One possible way to consistantly explain the observations is to postulate that the collapsed plasma column consists of a dense low temperature (150-300 eV) plasma and that _all_ the excitation observed in the x-ray spectrum are caused by an energetic beam of electrons. If this assumption is made some of the relevant beam parameters can be estimated. Since the transitions observed all require excitations in the 3-4 keV range it can be said that the beam must have a significant number of electrons with energy above about 4 keV. The beam current can be estimated as shown in figure 8 from the ratio of H-like to He-like emission and from the absolute intensity of the Li-like satellite line. Using typical plasma parameters, i.e. $N_e \sim 1 \times 10^{20}/cm^3$ and $T \simeq 350$ eV, 18.4 and 5.4 kA are obtained by the two methods respectively. Considering that the bank initially provides 600 kV across the electrodes and that a peak current of ~ 600 kA flows through the plasma, neither of these estimates is outside the realm of possibility.

DIRECT BEAM MEASUREMENT:[5]

Once having postulated a beam it can be asked how to go about directly detecting and measuring the beam. Figure 9 shows the configuration of our Z-pinch load chamber which allowed the installation of a Faraday cup detector and a filter wheel. A vacuum line-of-sight was added so that energetic electrons produced at the time of pinch would drift down a tube, go through a filter, and be detected in the Faraday cup. This configuration yielded an electron beam energy measurement in three ways: 1) time-of-flight, 2) attenuation of signal caused by placing an absorber in the path, and 3) change in time-of-flight due to energy loss of electrons which pass through the absorber and make it to the detector.

This device was tested on the collapsing gas shell Z-pinch. We found a slow low amplitude positive signal with a fast negative going signal superimposed on it sometime around pinch. When a permanent magnet was placed across the drift tube the fast negative signal went away. We concluded that the slow positive signal was due to VUV and X-ray photons going down the drift tube and knocking electrons off of the Faraday cup. The fast negative signal was due to the runaway electrons produced at the time of pinch.

Upon examining the electron signal, we found that it consisted of two parts, a slow large component and a small fast component. We found that the slow component had an average energy below 4 keV, as measured from the time-of-flight, and was completely attenuated by an 8 μm thick aluminum filter. See figure 10. The fast component, although much smaller than the slower signal when no filter was present, became the only signal when the 8 μm thick aluminum filter was added. We believe that the late arriving large signal is generated when the high energy electron beam expands and strikes the walls of the drift tube thereby driving off low energy secondary electrons.

Figure 11 shows how the energy determination from the three measurements agree once the corrections have been made for filter attenuation and change in time-of-flight. The conclusion is that the energy of the runaway electrons produced in the Z-pinch at the time of pinch is 20 keV ± 10 keV. The amplitudes of the three measurements agreed within error bars once the attenuation of the filters was taken into account. The error bars represent the statistical fluxuations of the measured signal shot-to-shot on the Z-pinch. The conversion from measured signal to absolute beam current, as was done for figure 11, is uncertain since it relies on a model for the beam attenuation due to losses in the drift tube. We used a free expansion model which led to an attenuation of over 10^6. Even so the measured beam current agrees very well with the estimated current required to cause the x-ray emission as described earlier.

X-RAY STREAK CAMERA MEASUREMENTS:

The next step was to see how the 4 keV x-ray radiation from the collapsed gas shell Z-pinch was distributed along the z-axis as a function of time. To accomplish this we used the experimental set up shown in figure 12. The idea was to calibrate the x-ray streak camera[6] by simultaneously taking time integrated x-ray pinhole pictures and time resolved x-ray emission from a specific part of the column. By fitting the emissions to each other and knowing the magnification of the x-ray pinhole camera and the temporal scale on the x-ray InP:Fe detector, we were able to calibrate the x-ray streak camera. See figure 13.

An interesting new phenomenon was observed[7] in the x-ray streak pictures. Figure 14 shows ~ 6 mm of the pinched column length in the keV x-ray region. The emission is streaked from left to right and the calibrated timescale is included on the bottom of the figure. If the x-ray emission is scanned at a z-location as a function of time, as is represented by the dashed line, it is found that the emission "winks", i.e. turns on for a few nanoseconds, turns off for a few nanoseconds, and then turns back on for a few nanoseconds. By making some reasonable assumptions, we found that we could explain this temporal behavior by a simple model which included a runaway electron beam at the time of pinch. This model made specific requirements on the runaway electron beam in order to quantitatively describe the observed time behavior.

The model assumes that the kilovolt x-ray emission from the pinch is produced entirely by runaway electrons exciting the ions. It begins to describe what happens after the hollow plasma shell has collapsed into a homogeneous plasma cylinder. It assumes that the current density is uniform, that the electric field and background plasma temperatures are constant, and that the collapse and expansion velocity are equal and constant. The number of runaway electrons, N_r, is controlled by a rate equation,

$$\frac{dN_r}{dt} = N_e \lambda \nu - N_r \frac{v_r}{\ell} ,$$

where ν is the collision frequency, λ is the runaway electron production fraction per collision time[8], N_e the electron density, v_r the runaway electron velocity, and ℓ the length of the plasma column. This equation says that the rate of change of the number of runaway electrons is equal to the production rate of runaway electrons out of the thermal distribution minus the loss rate of runaway electrons out the end of the pinched column. The production fraction, λ, was obtained from ref. 8 and is a strong function of the applied field, the electron density, and the electron temperature. This equation was solved under the constraint that the pinched column has the characteristic measured in previous experiments. Previously it was found that the minimum column diameter was measured to be 1 mm with 1 ns temporal resolution, thus the trajectory of the outer boundary of the plasma was forced to pass through two points 1 ns apart that represented a column diameter of 1 mm. The ultimate munimum radius, r_o, and hence the final collapse velocity were left as adjustable parameters. Another consideration is that runaway electrons can only be produced within a gyroradius of the axis, singular orbits[9], and if they interact with the ions to produce the hard x-rays, then the radius of the column that radiates, r_{rad}, will, in general, be smaller than the pinch radius at that time.

One can see that a double pulsed x-ray emission is possible using this model. As the plasma column compresses, the runaway electron number density and ion number density increase causing an increase in the x-ray emission. As peak compression is approached the electron number density gets large and hence the electron mean free path gets so short that runaway electrons can no longer be produced. This leads to a decrease in the x-ray emission. Finally, as the plasma column expands, the production of runaway electrons resumes along with the hard x-ray emission. This model also has the advantage that it explains the observation that the x-ray radiating column size is much smaller than the actual plasma column size.

The rate equation was solved and used along with the electron gyroradius to calculate the time integrated radiative radius as a function of the applied electric field. The results are plotted in figure 15. The shaded area represents the radiative radius as observed from the x-ray pinhole pictures. The observed radiative radius yields an applied field of ~ 65 kV/cm. Figure 16 contains a plot of the calculated temporal separation of the x-ray emission peaks versus the final minimum radius, r_o. This relationship varies as a function of the applied field. The observed emission peak separation is represented by the shaded area along with the average. As can be seen at ~ 65 kV/cm, the predicted final radius is .035 cm. Using these parameters, the temporal development of the x-ray emission was calculated. A comparison of the calculated emission (solid line) and the observed emission (dashed line) is presented in figure 17. The agreement between the two curves is quite good except in the region around $t = o$. This, of course, is the region where the simplifying assumptions we have made breakdown.

The electron-ion mean free path in the pinched column is ~ 0.35 cm. Since the model predicts an applied field of ~ 65 kV/cm, this means that on the average a runaway electron can acquire ~ 23 kV of energy. From the previously discussed Faraday cup experiment, we concluded that the electron energy was ~ 20 kV.

CONCLUSIONS:

We have found indirect and direct evidence of an energetic electron beam which permeates the plasma column at the time of pinch. We have observed a new effect we call "winking" in the temporal emission of the pinching plasma column. We have found that a model of the pinched plasma column can be made which is consistent with our observations. Since the presence of runaway electrons can explain all the kilovolt emission from the pinched plasma, temperature measurements based on spectroscopic evidence and a thermal electron distribution are suspect.

REFERENCES:
1. B. A. Hammel, Ph.D. thesis, University of Colorado, 1984 (unpublished).
2. J. Shiloh, A. Fisher, and N. Rostoker, Phys. Rev. Lett. <u>40</u>, 515 (1978).
3. B. A. Hammel and Larry A. Jones, Appl. Phys. Lett. <u>44</u>, 667 (1984).
4. R. D. Cowan, "The Theory of Atomic Structure and Spectra" (University of California, Berkeley, California (1981).
5. D. R. Kania and L. A. Jones, Phys. Rev. Lett. <u>53</u>, 166 (1984).
6. L. A. Jones, E. Källne, D. R. Kania, M. Maestas, J. S. McGurn, and R. Shepherd, J. Appl. Phys. <u>58</u>, 1711 (1985).
7. L. A. Jones, and D. R. Kania, Phys. Rev. Lett. <u>55</u>, 1993 (1985).
8. R. M. Kulsrud, Y. Sun, N. K. Winsor, and H. A. Fallon, Phys. Rev. Lett. <u>31</u>, 690 (1973).
9. M. G. Haines, J. Phys. D. <u>11</u>, 1709 (1978).

FIGURE CAPTIONS:

Figure 1. A schematic representation of the Los Alamos High Density Z-pinch machine.

Figure 2. A drawing of the gas load discharge chamber. The drawing includes the in-line isolation switch, the fast valve, and supersonic nozzle. The hollow gas cylinder is formed in the small gap between the anode and cathode.

Figure 3. A series of framing pictures which shows the formation and collapse of the hollow gas shell plasma.

Figure 4. A plot of the radius of the collapsing plasma versus time.

Figure 5. A scan of a curved crystal x-ray spectrum showing the He-like resonance and intercombination lines along with seven inner shell satellite lines. Also shown is a calculated spectrum.

Figure 6. A sample calculation of the plasma temperature assuming a thermal electron distribution and that the first satellite line is caused by dielectronic recombination.

Figure 7. A reproduction of a drawing showing the fractional abundances of various argon ions as a function of temperature.

Figure 8. Estimates of the runaway electron beam required to produce the hard x-ray spectrum.

Figure 9. A schematic of the altered gas discharge chamber so that an electron drift region and Faraday cup could be included.

Figure 10. A plot of the raw results of the signals produced in the Faraday cup as a function of time-of-flight after pinch.

Figure 11. A plot of the corrected and calibrated signals of the Faraday cup plotted against the inferred electron energy.

Figure 12. A schematic representation of the diagnostics used to calibrate the x-ray streak camera and to observe the hard x-ray emission from the pinched column as a function of axial position and time.

Figure 13. A sample of the method used to calibrate the x-ray streak camera.

Figure 14. One of the streak pictures taken of the x-ray emission from the pinched column showing the "wink" effect.

Figure 15. A plot of the model predicted radiative radius versus applied electric field. Also shown is the average and the fluctuation (shaded area) in the observed radiative radius.

Figure 16. A plot of the model predicted x-ray emission maxima separation as a function of final collapse radius for various applied electric fields. Also shown is the average and the fluctuation (shaded area) in the observed temporal separation.

Figure 17. A plot of the plasma x-ray emission as measured (dashed line) and as predicted by the model (solid line). The two crosses on the model predicted curve are 1 ns apart and represents the time resolution of the diagnostics.

HDZP PULSE POWER SUPPLY

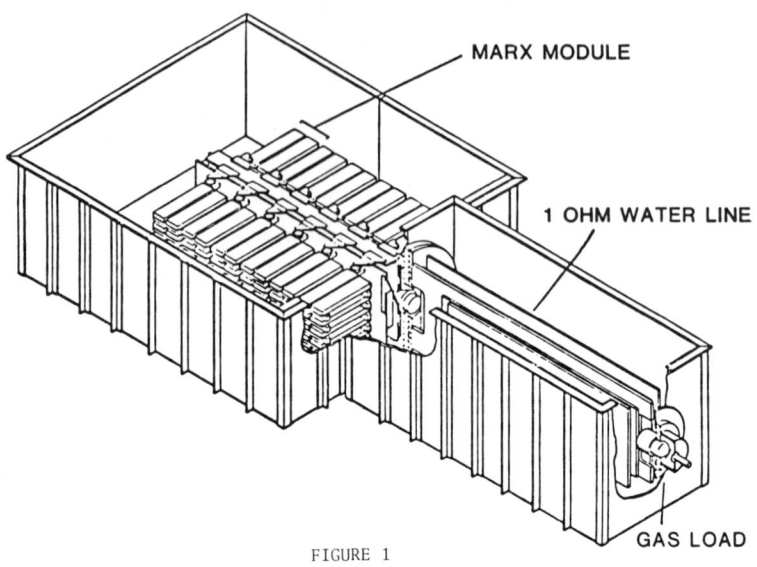

FIGURE 1

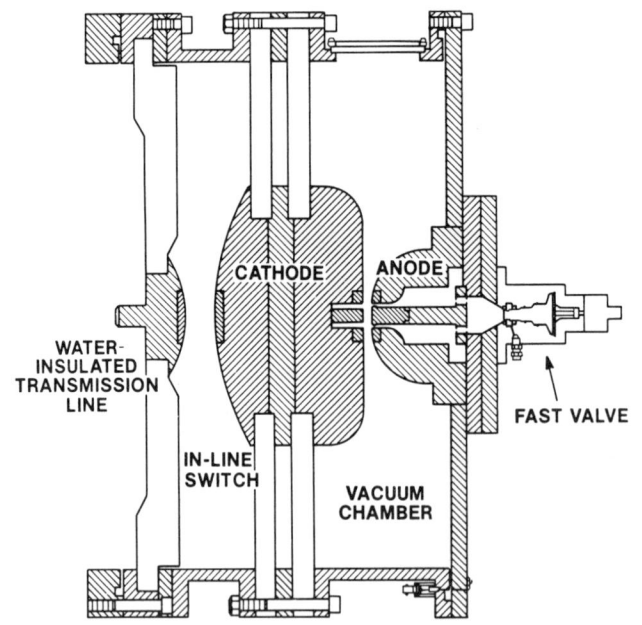

FIGURE 2

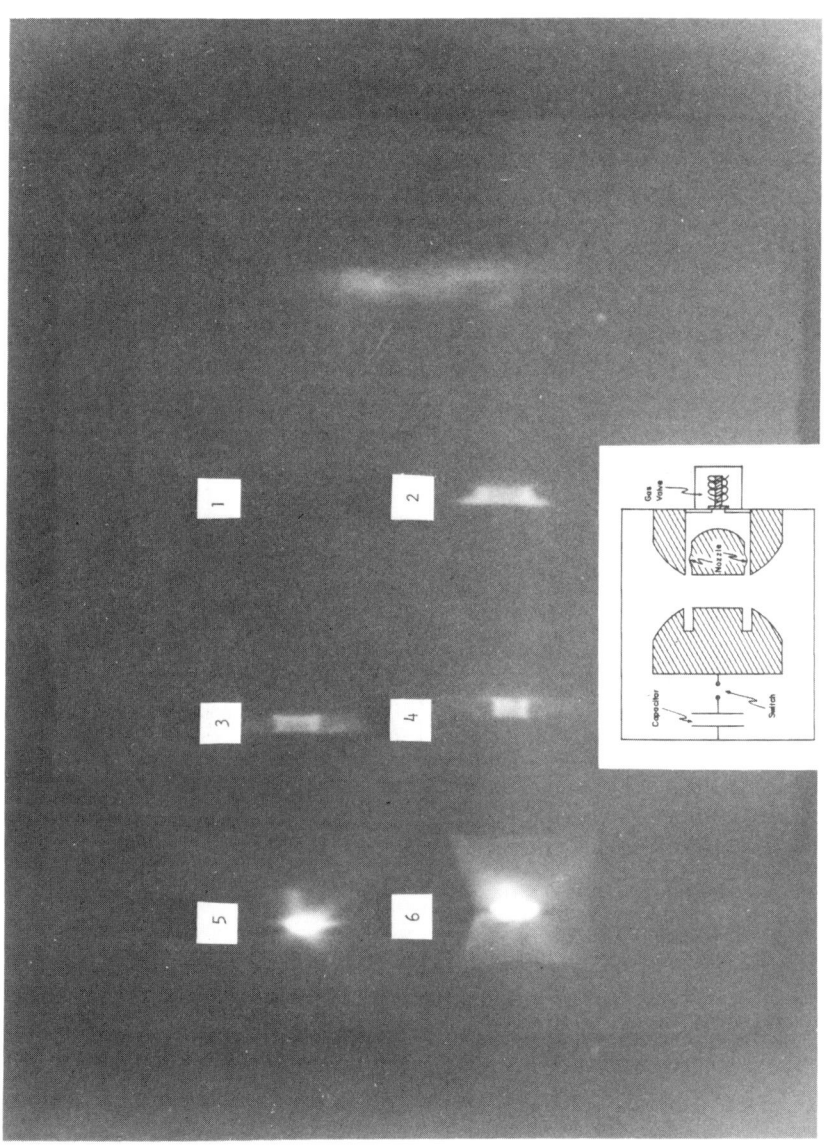

FIGURE 3

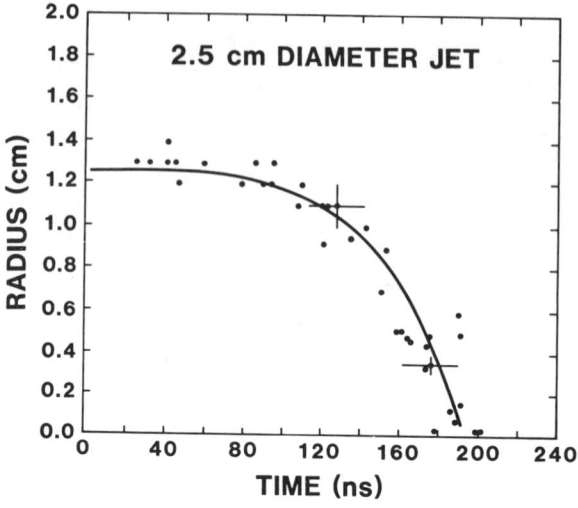

FIGURE 4

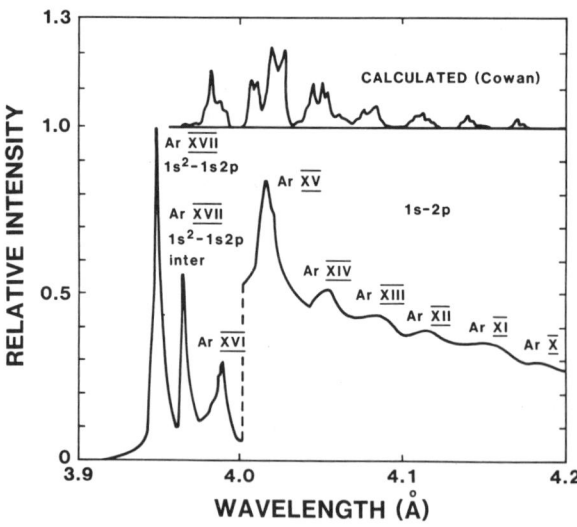

FIGURE 5

THERMAL INTERPRETATION

1. H – LIKE TO He – LIKE LINE RATIO

    CORONAL MODEL (GRIEM)

    $$\frac{I'}{I} = \frac{f'g'\lambda^3}{fg\lambda'^3} \exp\left(\frac{E'_\infty - E' - E_\infty + E}{kT}\right) \frac{S}{\alpha}$$

    $T \approx 1$ keV

2. IF Li – LIKE SATELLITE FROM DIELECTRONIC RECOMBINATION

    $\dfrac{I_{sat}}{I_R}$ (GABRIEL) $\Rightarrow T \approx 1.1$ keV

FIGURE 6

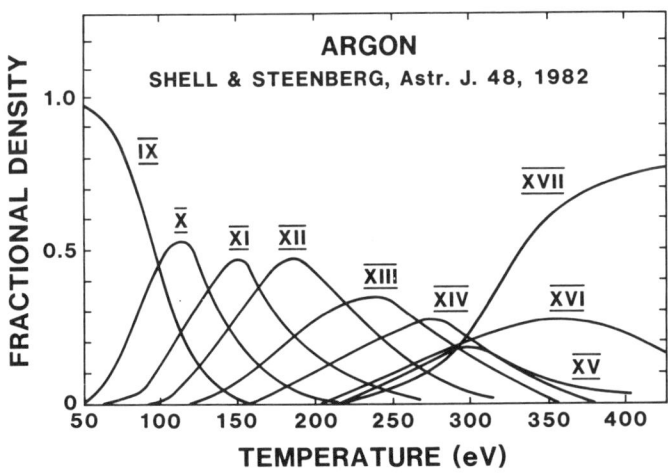

FIGURE 7

ESTIMATES OF BEAM REQUIREMENTS

1. FROM THE H – LIKE TO He – LIKE LINE RATIO

$$\frac{I_H}{I_{He}} \approx \frac{1}{100} = \frac{N^0{}_H}{N^0{}_{He}} = \frac{S_{He}}{R_H}$$

FOR A BEAM $S_{He} = \frac{I_b}{e\pi r^2} \sigma_i$

$R_H = 2.7 \times 10^{-13} \frac{Z^2}{\sqrt{T_e}} \frac{N_e}{2}$

$I_b \approx 18.4$ kA

2. FROM THE INTENSITY OF THE Li –LIKE SATELLITE LINE

I_{Li} (exp) $\approx 8.7 \times 10^{19}$ γ/cm·s

$I_{Li} = \frac{I_b}{e} \sigma_{ex} N_{Li} \eta$

$\sigma_{ex} \approx 2.4 \times 10^{-13} \frac{fg}{\epsilon \Delta E}$ cm^2

$\eta \approx$ FLUORESCENCE YIELD = 0.5

$N_{Li} \approx 3 \times 10^{18}$/cm^3

$I_b = 5.4$ kA

FIGURE 8

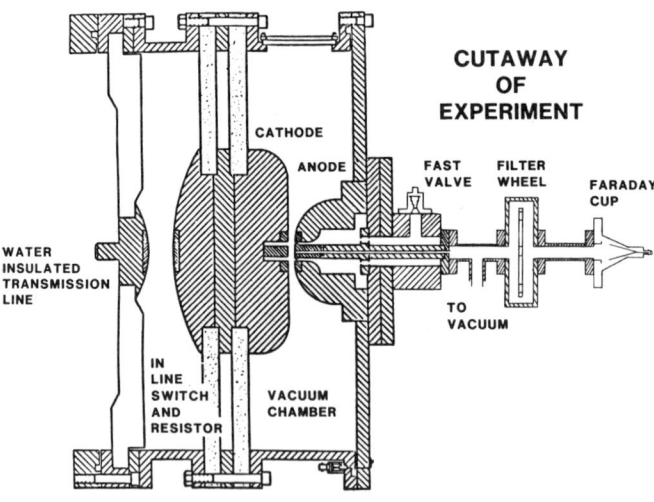

FIGURE 9

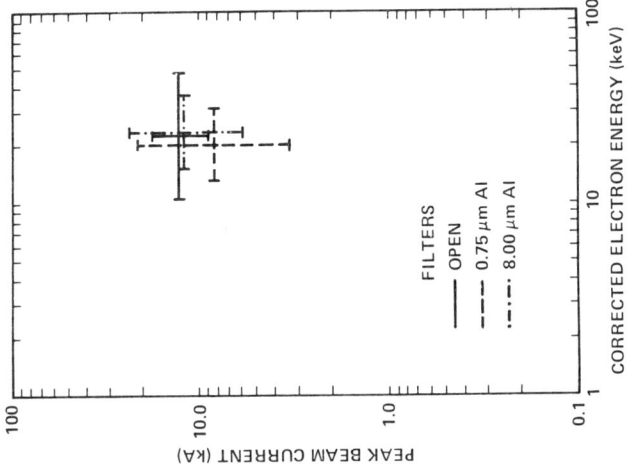

FIGURE 11

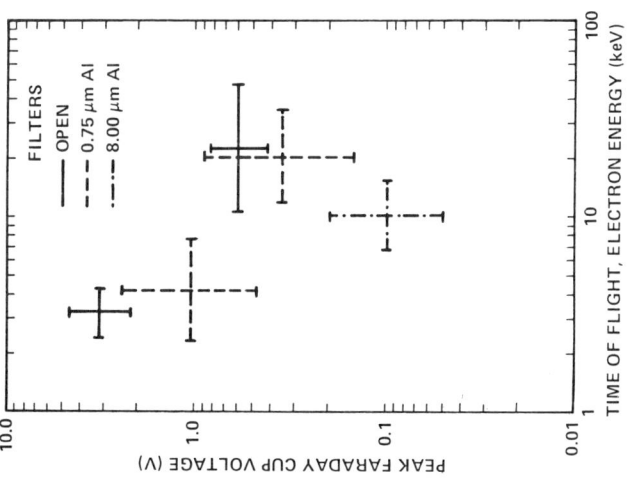

FIGURE 10

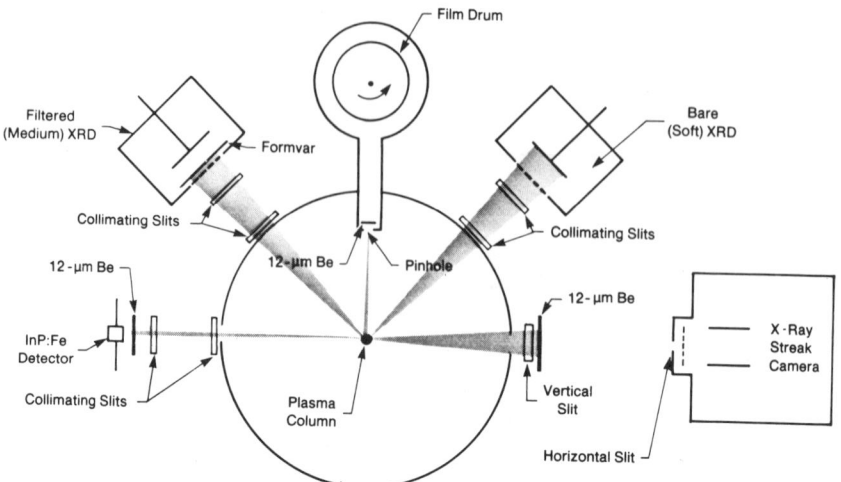

FIGURE 12

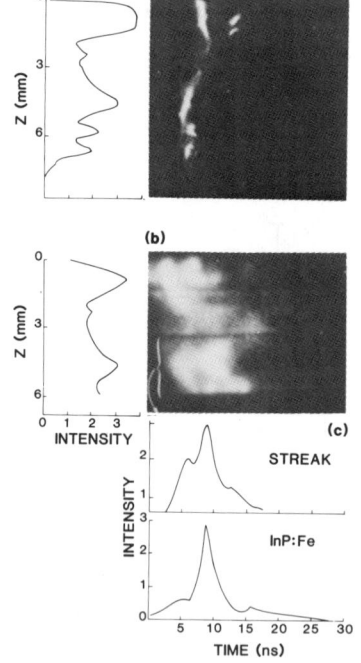

FIGURE 13

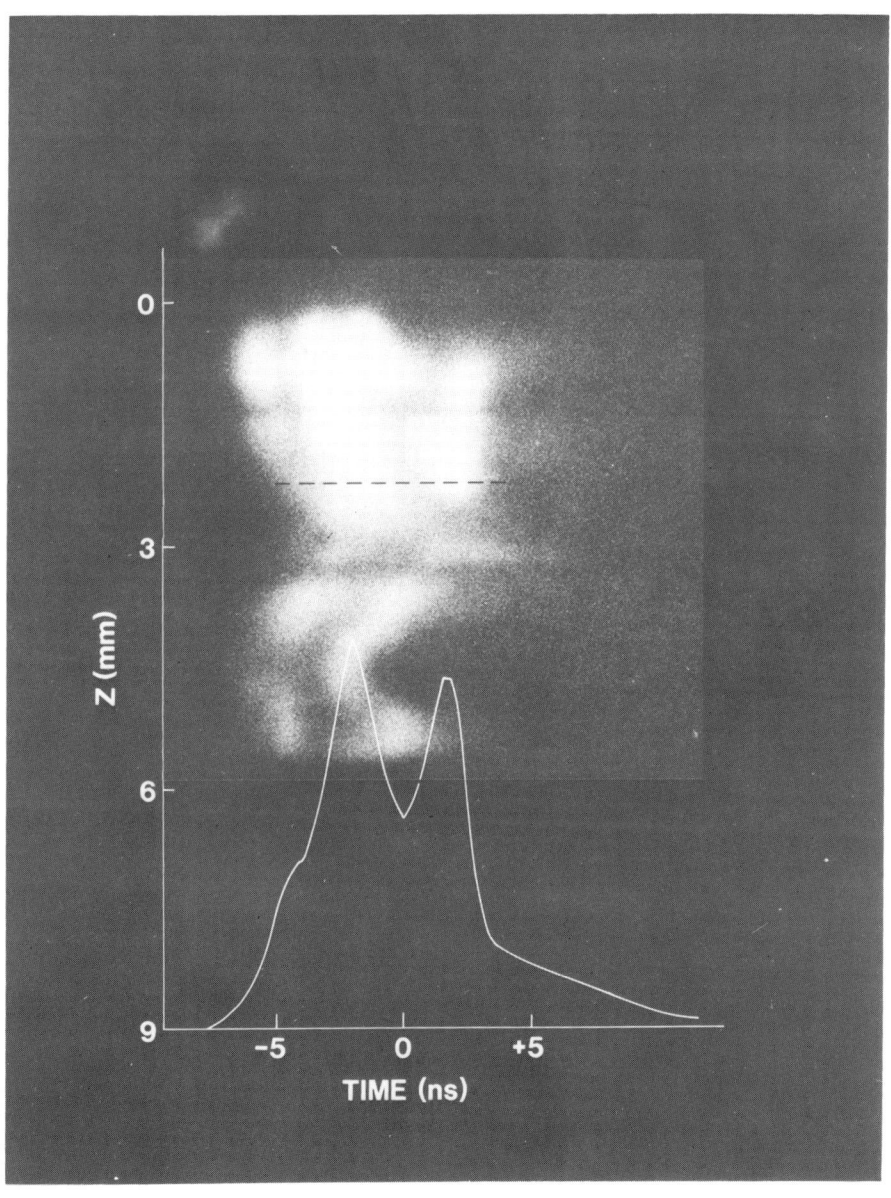

FIGURE 14

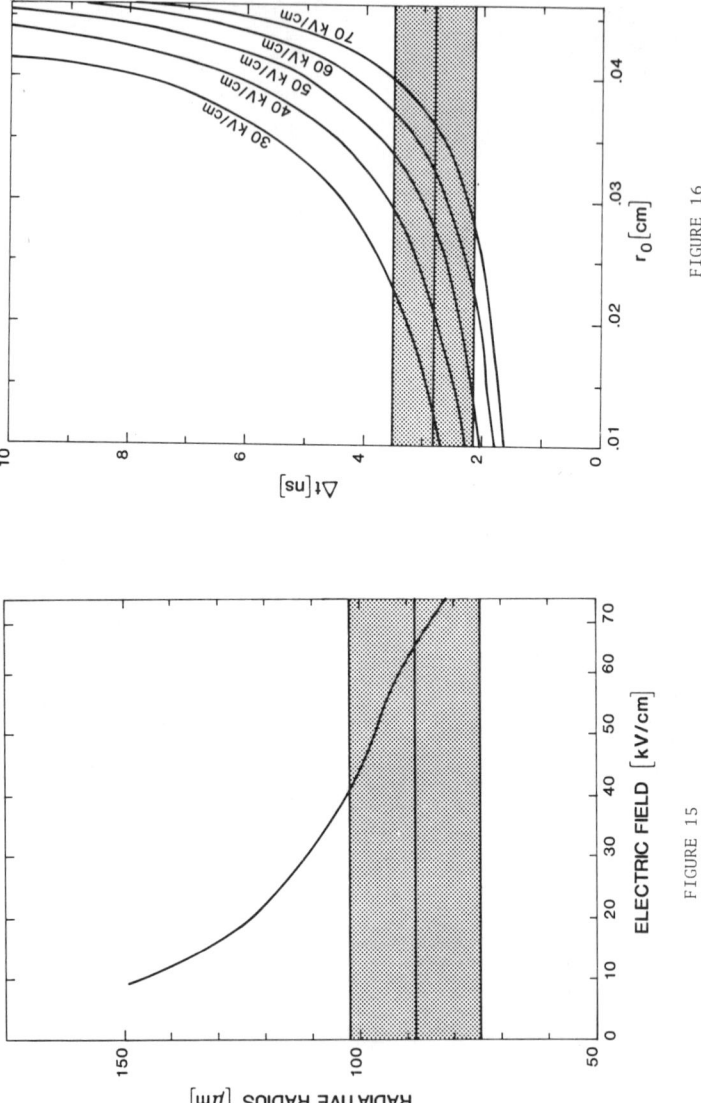

FIGURE 15

FIGURE 16

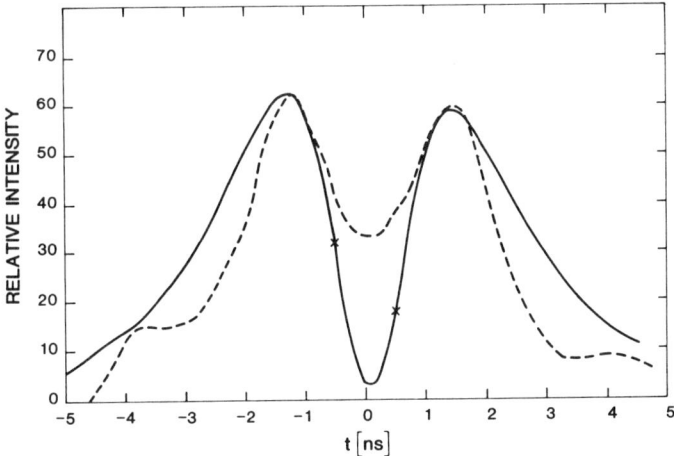

FIGURE 17

X-RAY SPECTROSCOPY AND DETERMINING PLASMA DYNAMICS

R. L. Kauffman
K. G. Estabrook
R. W. Lee

ABSTRACT

An application of plasma x-ray spectroscopy to determine plasma parameters is briefly discussed. An experimental design that takes some of the uncertainty away from the analysis is described and the results are briefly outlined.

X-ray line spectroscopy has been used to study the dynamics of exploding foils in underdense plasma experiments.[1] Time histories of the x-ray lines are calculated and correlate to experimental results within 100 ps. The computer model uses the hydrodynamics output from LASNEX[2] as input for a code which solves the time-dependent rate equations for the atomic levels. The rate equations calculate the ionization balance and atomic level population of an ion species as it moves through the time-dependent temperature and density profiles of the plasmas. The predicted line intensities of this model generally agree with the observed line intensities, although the ionization during the early ablation phase of the experiment is underestimated. This may be evidence for penetration of the high-energy tail of the electron distribution into the denser plasma.

X-ray spectroscopy on the exploding low-Z foil was done by placing higher Z tracer elements in the foil. A 4 atom % concentration of sulfur was placed in the center 300 µm diameter of the CH foil.[3] The spot size of the beam was on the order of 600 µm, overfilling the doped region. By limiting the sulfur to the center of the irradiation area, effects of the edge plasma on line intensities are minimized. The low concentration of the higher Z element should change the average charge state of the plasma by less than 20%. Simulations indicate that it produces a negligible perturbation of the hydrodynamics of the foil. The low concentration also minimizes effects of opacity on the line intensities used as diagnostics. Optical depths are estimated to be less than one, even for the resonance lines of the spectra.

The time-resolved spectra were measured using the x-ray crystal streak camera (XCSS) on Novette.[4] An example of the data is shown in Fig. 1. Data were taken with a KAP crystal at a principal Bragg angle of 10.1°. The dispersion of the data was 74 eV/mm at the slit; the resolution was 7 eV. The n=2 to n=1 (α) and n=3 to n=1 (β) features from the He-like and Lyman series are observed. The resolution is not sufficient to resolve the He-like intercombination line ($^3P_1$) from the resonance line ($^1P_1$).

The time histories for the He resonance (1s $^1S_-$ - 1s 2p ^1P) and Ly α lines are shown in Fig. 2, where the time resolution is $\sim$50 ps. Both time histories are approximately 350 ps at full width, half maximum, which is considerably shorter than the 900 ps laser pulse. These time histories show similar structure during the rising part of the pulse. On the trailing edge, however, the He-resonance intensity decreases more rapidly. This is presumably when the foil burns through, becoming underdense to the 0.53 µm laser. There is continuum emission at late time around the He-resonance line, which has been subtracted from the data. This allows only an upper limit to be set for several data points after burnthough.

The time histories calculated from the model are compared with data in Fig. 2. The data have no absolute timing marker, and the time axis has been normalized to the calculations at the half-intensity point on the leading edge of the He-resonance line. The peak of the laser pulse is at t=0. Overall agreement between the data and calculations is quite good, the widths of the time histories agree to within 100 ps. The Ly-α time-history data show a leading edge which start earlier than calculations predict and decays more rapidly than predicted.

The He-resonance-to-ly-α ratio is shown in Fig. 3. The large scatter in the data is due to the low signal levels and the inherent noise in the streak camera. The ratio is less than one early in the pulse and rises quickly to about two. When the foil burns through, it quickly increases to about four before the He-resonance signal is lost in the continuum noise. If one uses a simple collisional steady-state model assuming uniform temperature, the ratio would indicate a plasma temperature around 1.5 keV at early times and 2 keV before burnthrough. After burnthrough, the inferred temperature is 2.5-3.0 keV in approximate agreement with temperature estimates derived from Raman spectra.

The calculated ratio is also shown in Fig. 3. The calculation underestimates the ratio at early times as evidenced by the early rise in the data of the Ly-α line. After burnthrough, the line ratio plateaus as the charge states are frozen into the underdense plasma. Comparisons have been made for two flux multipliers in LASNEX. The principal difference between the two results is the higher coronal temperature in the underdense plasma, as evidenced by the higher ratio of He-resonance to Ly-α for the smaller flux multiplier. The flux multiplier, 0.03, is in better agreement with the data, although the f=0.1 curve is within the uncertainty of the data. The smaller flux multiplier also predicts a longer time for burnthrough in relation to the peak of the pulse. Since the XCSS has no absolute fiducial, the absolute burnthrough time cannot be measured. The predicted shapes of the time histories from both flux multipliers are similar and both agree with the data as seen in Fig. 2.

Although the modeling predicts the overall time history of the line intensities and ratios, the Ly-α intensity increases more rapidly than calculated. This appears to be evidence for nonlocal transport of electrons into the denser solid. The early onset of the Ly-α intensity is related to the ionization of the sulfur as it moves through the heat front. As the ion moves through the steep gradients, it requires a finite time, which is related to the collision frequency, to equilibrate with the local plasma temperature and density. This collision frequency is related to the ionization energy for He-like and H-like sulfur which are 3.2 and 3.5 keV, respectively, and electron energies two to three times these ionization energies are needed for ionization. As the ion moves through the ablation front, calculations assume a Maxwellian distribution of the thermal electrons at local temperature. Only in the underdense corona, where the electron temperature is greater than 1 keV, do the electrons have enough energy to efficiently ionize the K-shell; however, the plasma density decreases rapidly in the corona, and the ion quickly decouples from the plasma. Energetic electrons, on the other hand, with energies greater than 6 keV can penetrate through the ablation front and ionize the sulfur. This increases the time that the sulfur samples the energetic electrons allowing faster equilibration. For example, the range of a 10 keV electron in cold matter is comparable to the total thickness of the foil. Penetration of these electrons from the tail of electron distribution can significantly pre-ionize the sulfur. This indicates that better nonlocal electron transport models are needed to estimate this effect.

In summary, time histories of the x-ray lines from sulfur comfirms the hydrodynamics modeling of the thin foil experiments. Line ratios favor the modeling using a flux multiplier of 0.03, although a flux multiplier as high as 0.1 cannot be ruled out from the data. The data show evidence of nonlocal electron transport, and better electron transport modeling may be needed to predict ionization of the ions through the ablation region.

REFERENCES

1. Turner, R.E., K.G. Estabrook, R.L. Kauffman, et al., Phys.Rev.Lett. $\underline{54}$, (1985).

2. Zimmerman, G.B., and W.L. Kruer, Comments Plas.Phys. $\underline{2}$, 51 (1975).

3. Laser Program Annual Report 83, LLNL, Livermore, California, UCLR-50021-83, (1984) pp. 4-11 to 4-13.

4. IBID, pp. 5-21 to 5-22.

FIGURE CAPTIONS

1. Time-resolved x-ray spectrum from a thin, sulfur-doped CH foil.

2. Time histories of the He-resonance and Ly-α lines compared with computer simulations.

3. Time history of the Ly-α/He-resonance line ratios from the CH foil compared with calculations for two different flux multipliers.

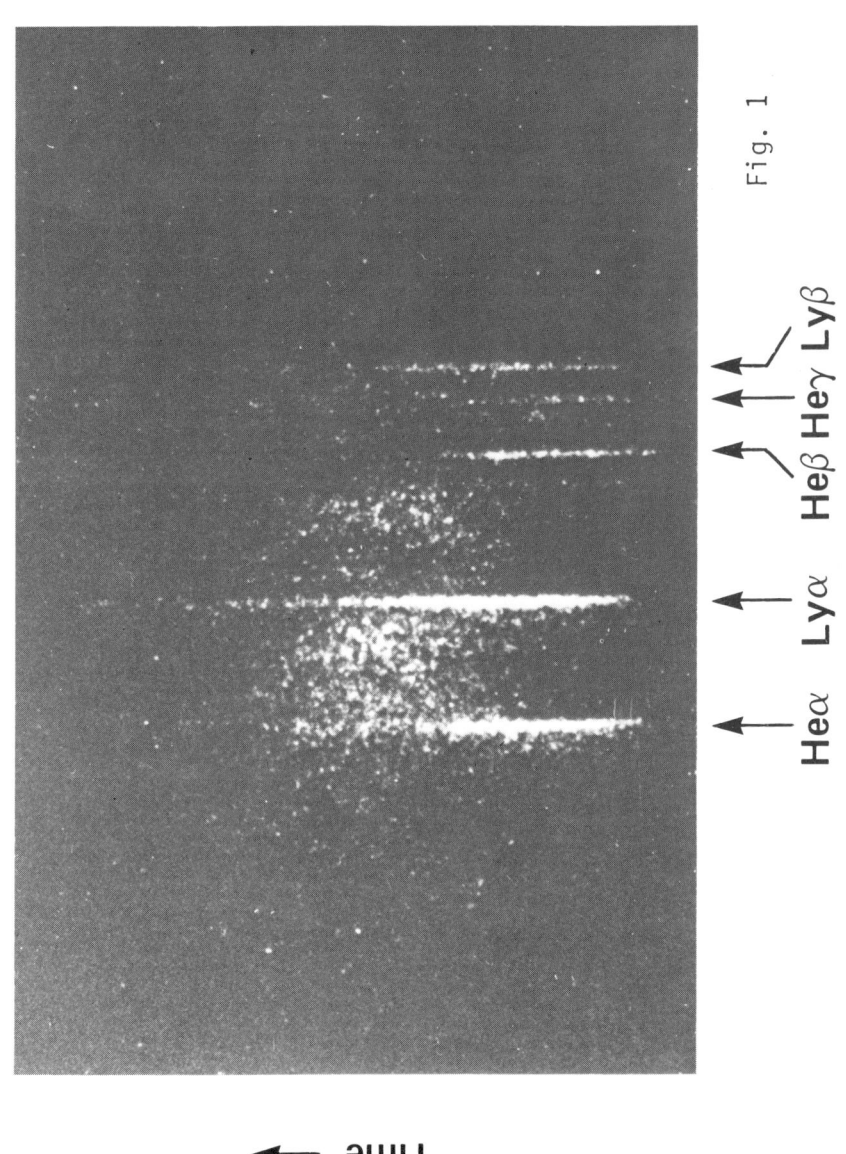

Fig. 1

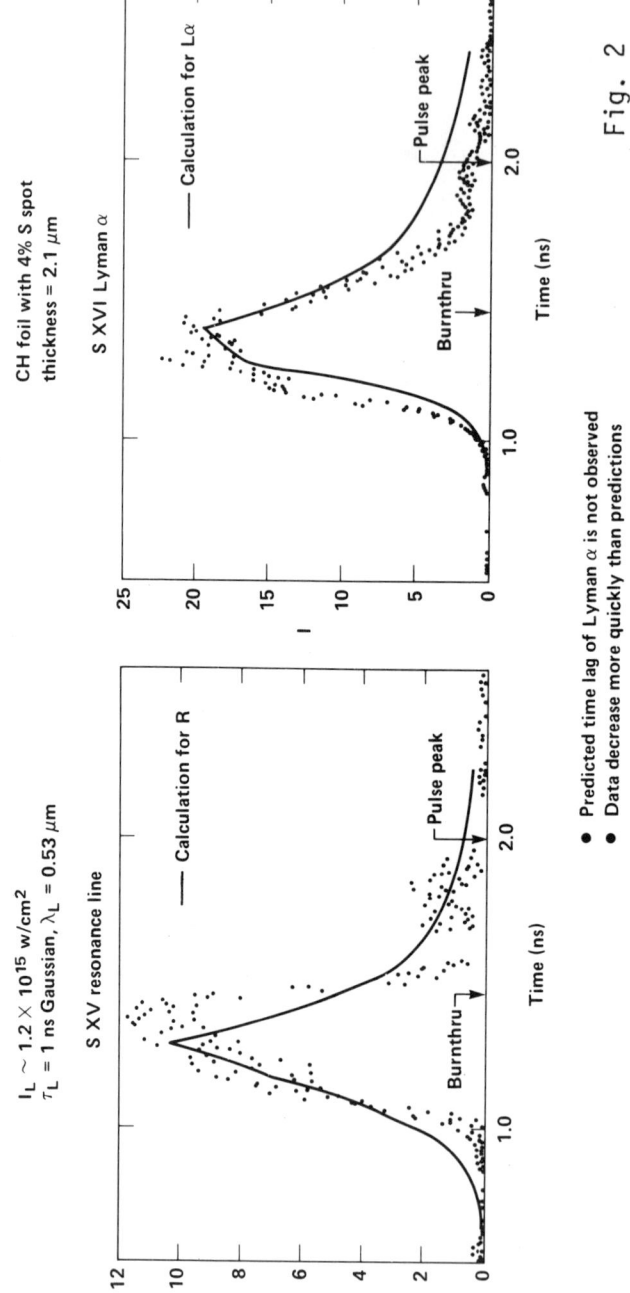

Fig. 2

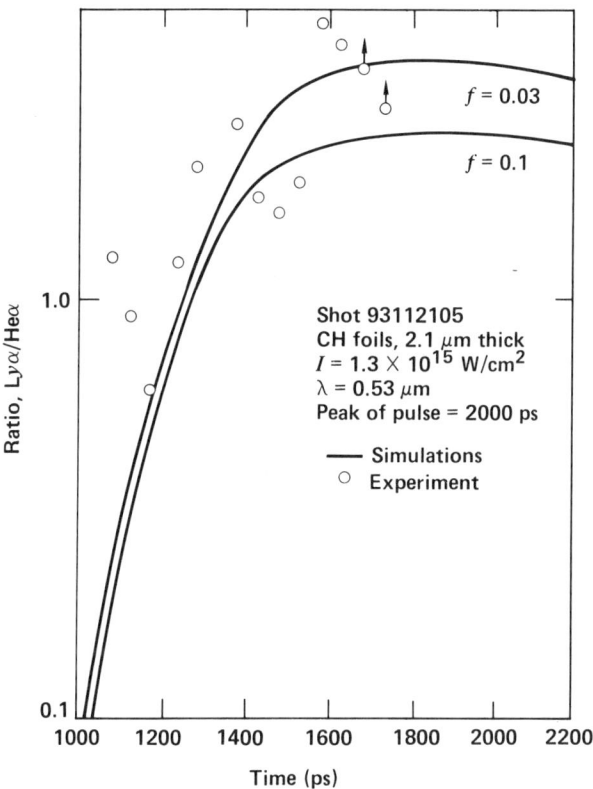

Fig. 3

Sub-keV x-ray Emission from Laser-produced Plasmas

K. Eidmann, T. Kishimoto, G.D. Tsakiris,
R.F. Schmalz, R. Sigel, and S. Witkowski

Max-Planck-Institut für Quantenoptik
D-8046 Garching, FRG

Abstract

The radiation emitted by laser irradiated planar targets was measured in the energetically important soft x-ray spectral range 50 eV $\leq h\nu \leq$ 1000 eV using a transmission grating spectrometer with x-ray film, bolometer or x-ray streak camera as detector. The target was irradiated by a frequency doubled Nd-laser (λ = 0.53 μm, 10 J/3 ns) or by an iodine laser (λ = 1.3 μm, 100 J/300 ps). Absolutely measured spectra from targets of different material with atomic numbers in the range 4 to 82 are presented. The conversion efficiency of laser into x-ray energy increases with the atomic number from 2 % for Be to 40 % for Au (at λ = 0.53 μm and $3 \cdot 10^{13}$ W/cm² laser intensity). In particular, the gold spectrum was measured for a wide range of intensities ($3 \cdot 10^{11}$ to $5 \cdot 10^{15}$ W/cm²).

For discussion of the results a non LTE atomic physics model was developed, which calculates the emission from laser plasmas by taking into account the energy transport in the plasma by electrons and by radiation. The calculations are compared with the experiment.

This work was supported in part by the Commission of the European Communities in the framework of the Association Euratom/IPP.

1. Introduction

High power lasers allow the production of extremely hot and dense plasmas, which otherwise exist in nature only under stellar conditions. The intense x-radiation emitted by the laser plasma has been studied for a long time. For high-Z target material a conversion of laser energy into x-ray energy as high as 50 % has been reported /1/. Calculations have predicted even for low-Z target material (Z = 6 and 13) considerable conversion efficiencies of up to 30 % /2/. There is strong interest in the study of the emitted radiation because of its importance for the radiative energy transport in the plasma. In connection with inertial confinement fusion an efficient conversion of laser light into x-radiation may help to improve the irradiation uniformity /3/. The laser plasma as an x-ray source also allows many other applications, e.g. in x-ray lithography and radiometry or for pumping of x-ray lasers.

In this paper we describe experiments with plane targets of different material irradiated by laser from one side. The emphasis of the experimental study is on the absolute measurement of the emitted radiation in the sub-keV region, where the main emission takes place. In the first part of this paper we describe the experimental technique and present experimental results. To understand the radiation processes occuring in the plasma an atomic physics model has been developed and it is discussed in the second part.

2. Experimental results

For plasma production we used a frequency doubled Nd-laser emitting 7 J/3 ns-pulses at $\lambda = 0.53$ µm and, for a higher intensity range, an iodine-laser emitting 100 J/300 ps - pulses at $\lambda = 1.3$ µm. The laser was focussed on thin plane foils with the atomic number Z ranging from 4 (Be) to 82 (Pb).

The emitted radiation was measured by using as dispersive element a transmission grating consisting of free-standing gold bars up to 2000

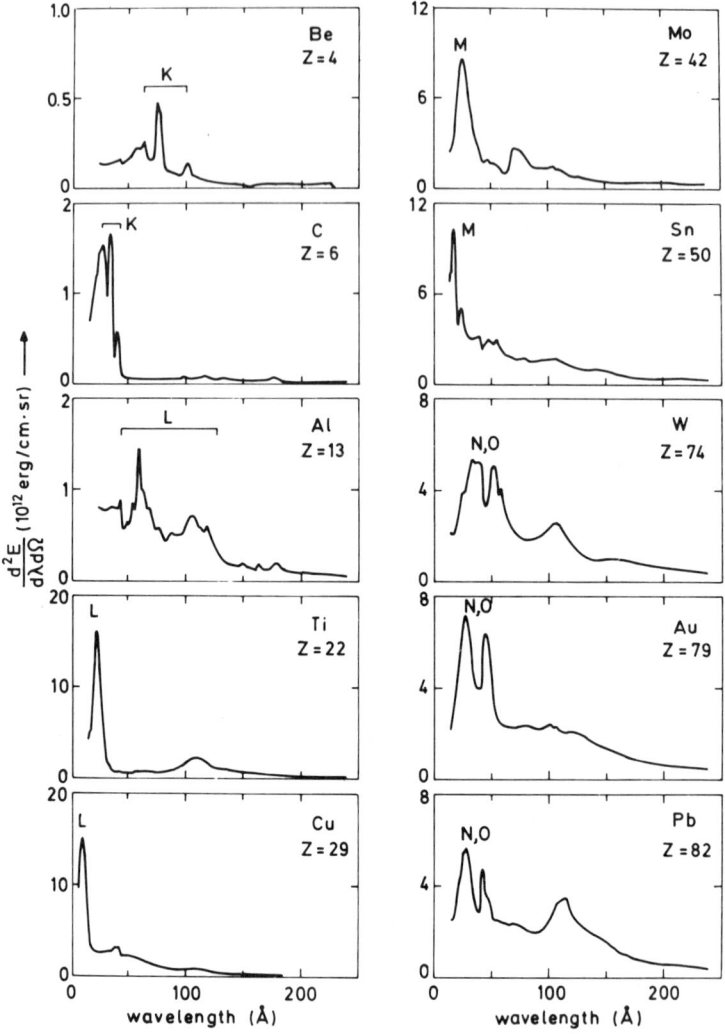

Fig. 1: Absolute measured spectra emitted by laser irradiated plane targets of different elements. Intensity 3×10^{13} W/cm^2, pulse duration 3 ns, wavelength 0.53 µm. Angle of observation was 45° to the laser axis. Note the change of the vertical scale for the different spectra.

lines per mm /4/. It was positioned under different angles to the axis of the incident laser at a distance of ~ 50 cm. To get a spectrum in the detector plane (30 to 100 cm behind the grating) the grating was bounded by a slit (50 µm to 200 µm slit width) or a pinhole (25 µm to 50 µm pinhole diameter). The latter was used to obtain spatial resolution in addition to spectral resolution. The spectra were observed in the first diffraction order. The grating efficiency is typically 10 % /4/. The wavelength resolution of the spectrometer is a few Å depending on the size of the plasma source and the geometry of the spectrometer. The accessible wavelength range was 10 Å $\leq \lambda \cong$ 250 Å. (The lower limit λ = 10 Å is due to overlapping of the zeroth order with the first order for $\lambda \leq$ 10 Å). As detector we have used routinely Kodak 101-01 x-ray film. Absolute measurements and the calibration of the x-ray film was made with a bolometer. Time resolved spectra were obtained by an x-ray streak camera /5/. For more details of the experimental techniques we refer to /6/.

Spectra measured with different target elements are shown in fig. 1. The plotted absolute values of the energy radiated per solid angle and wavelength were obtained by unfolding densitometer traces of film spectra (see fig. 3 and 4) taking into account the spectral dependence of the film sensitivity and of the grating efficiency. The structures seen in the spectra depend in a characteristic way on the target material and consist mainly of line emission and recombination continuum. With the Be and C targets individual hydrogen- und helium-like lines due to K-shell electrons are resolved. With increasing Z the emission results from electrons in shells with principal quantum numbers n > 1. One observes L-shell emission for Al, Ti and Cu, M-shell emission for Mo and Sn and finally N- and O- shell emission for the heaviest elements W, Au and Pb.

The spectra have been integrated over the wavelength to obtain the total emitted energy. We have also measured the angular dependence of the emission which allows us to perform the integration over the angle. The conversion of incident laser energy into x-ray energy increases from 2 % to about 40 % for the heaviest elements (full points in fig. 2). It agrees within ± 20 % with values obtained independently with a bolometer which collected the undispersed radiation in a range extended

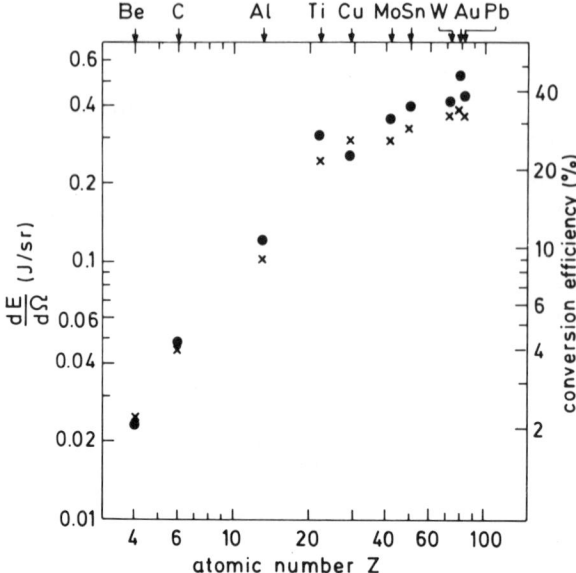

Fig. 2: X-ray energy per solid angle emitted under 45° to the laser axis as a function of the atomic number of the target material. The total conversion efficiency (right scale) was obtained by integration over the full solid angle. Full points: integration of the spectra in fig. 1 (10 Å < λ < 250 Å); crosses: independent measurement of the undispersed radiation by a bolometer (2.5 Å < λ < 250 Å).

to short wavelengths 2.5 Å ≦ λ ≦ 250 Å (crosses in fig. 2). Thus no large contributions are emitted in the region below 10 Å compared to the emission above 10 Å.

We have also recorded spectra at different laser intensities. In fig. 3 and 4 we have put together original densitometer traces of film spectra measured with a gold target in a wide laser intensity range from $3 \cdot 10^{11}$ to $5 \cdot 10^{15}$ W/cm². The shape of the spectrum changes slowly with the intensity. With increasing intensity the maximum of the emission shifts to shorter wavelengths, which is reasonable because of the increasing temperature. The detailed structure of the spectra is caused by the spectral properties of the gold ions in the plasma. The

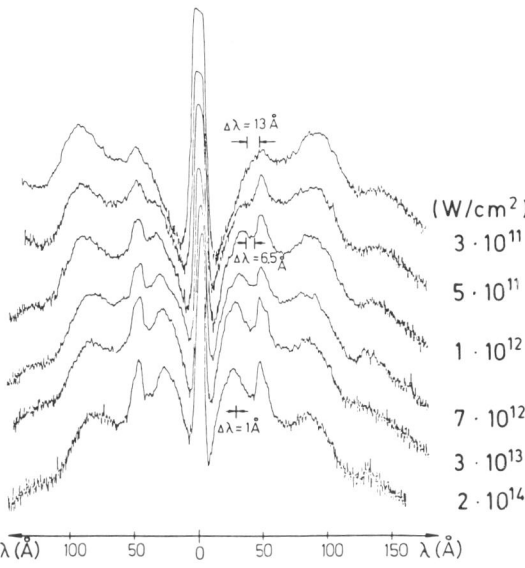

Fig. 3: Densitometertraces of gold spectra on film at different laser intensities. The strong peak at λ = 0 is the zeroth order, on both sides one sees the spectrum in first order. The fog level of each micro densitometer trace has been arbitrarily shifted (in the vertical direction) for clarity.
Irradiation conditions: 7 J/3 ns pulses at λ = 0.53 µm with different spot sizes (the plotted wavelength resolution Δλ depends on the spot size).

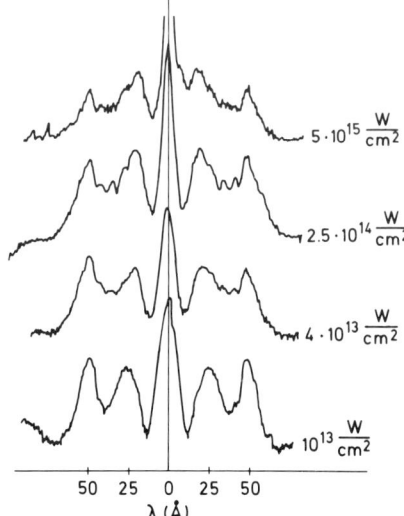

Fig. 4: As fig. 3. Irradiation conditions: 300 ps pulses at λ = 1.3 µm.

two characteristic peaks at $\lambda \cong 20\text{-}30$ Å and $\lambda \cong 50$ Å are seen in all spectra with exception of the lowest intensity ($2 \cdot 10^{11}$ W/cm² in fig. 3), in which case the peak at $\cong 20$ -30 Å is no more excited. The maximum of the 50 Å-peak remains there at all intensities whereas with increasing intensity the center of the 20-30 Å-peak shifts from 30 Å to 18 Å. Note also that the spectrum at $2.5 \cdot 10^{14}$ W/cm² (in fig. 4) shows a substructure with smaller peaks between the two main peaks.

The spectra of fig. 3 and 4 have been unfolded to determine the conversion efficiency. For the irradiation conditions of fig. 3 (3 ns pulses at $\lambda = 0.53$ µm) the conversion efficiency (x-ray-energy/incident energy) decreases slightly from 40 % at $3 \cdot 10^{13}$ W/cm² to 30 % at the lowest intensity $3 \cdot 10^{11}$ W/cm². For the conditions of fig. 4 (300 ps pulses at $\lambda = 1.3$ µm) the conversion decreases from 23 % (40 to 50 %) to 4 % (10 to 15 %) when the laser intensity increases from 10^{13} W/cm² to $5 \cdot 10^{15}$ W/cm² (the numbers in brackets are estimated values for the conversion normalized to the absorbed laser energy with absorption data taken from /7/).

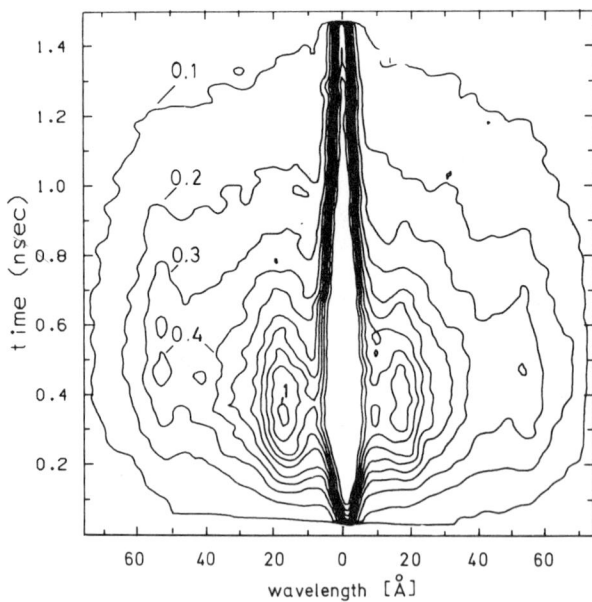

Fig. 5: Time resolved spectrum of a gold target irradiated by a 10 J/300 ps ($5 \cdot 10^{14}$ W/cm²) pulse at $\lambda = 1.3$ µm. Plotted are lines of constant intensity in the λ,t - plane (the relative intensities are indicated by the numbers 0.1 ... 1).

The emission was also studied with temporal or spatial resolution. Typically the x-ray pulse was somewhat longer than the laser pulse. An example of a time resolved spectrum is shown in fig. 5, when a gold target was irradiated by a 300 ps pulse. Plotted are lines of equal intensity at different times and wavelengths. A fast increase of the emission is followed by a slower decay. The duration (FWHM) of the x-ray pulse is 500 ps at $\lambda \cong 20$ Å and exceeds the laser pulse duration of 300 ps. For the 3 ns laser pulses the lengthening of the x-ray emission was small compared to the 3 ns time scale of the laser pulse.

Space resolved spectra measured by pinhole gratings gave the size of the x-ray emitting region, which is somewhat larger than the optical spot size as a consequence of lateral effects; e.g. with the 3 ns pulses and an optical spot size of 100 µm we measured for gold target a diameter of 150 µm (FWHM of the intensity) for the x-ray emitting region.

3. Theoretical discussion

The emitted radiation depends on the spatial distribution of the temperature and the density in the laser heated plasma layer. For this the laser plasma coupling and the energy transport in the plasma are important. At low Z the energy is transported by electrons, whereas with increasing Z the radiative transport in the dense plasma becomes dominant. Thus a complete description requires very complex models which couple the radiation processes with the hydrodynamics /1,2/.

As first step we have worked out an atomic physics model, which calculates the radiation from a plasma layer with a given spatial temperature and density profile. Although the atomic physics is not coupled with the hydrodynamics, which could be done in a later step, this treatment is useful for understanding of the experimental results. In this section we present some preliminary results.

As an appropriate atomic physics model we used a version described by Campbell et al. /8/, which was modified for our purpose. The model takes into account the non-LTE features of the laser plasma /9/ and it allows the calculation of the radiation for any element. The main assumptions are the following:

The charge state of the ions is calculated by a stationary balance between ion production by electron collisions and ion loss by collisional, radiative and dielectronic recombination. This balance equation includes the Saha- and the corona equilibrium as limiting cases at high and low densities respectively. The free-free, free-bound and bound-bound emission is then calculated by using standard formulas. For the populations of the excited states Boltzmann distribution was assumed for high densities and corona equilibrium for low densities /10/. The energy levels are calculated in a hydrogenlike approximation using Slater coefficients to take into account the screening of the Coulomb potential by the electrons in the different shells. (We found that the energies of the outer non-occupied shells could be calculated quite accurately, especially for low Z material, by replacing for non-occupied shells the usual Z_{eff} by $Z_{eff}+1$.) A crude approximation was made for the oscillator strengths, for which we used hydrogenic values multiplied with the number of the electrons in the atomic shell under consideration. This may be justified on the basis of the f-sum rule. The absorption coefficient is calculated from the emission coefficient by Kirchhoffs law assuming that strong absorption is important only in dense plasma in LTE.

We have checked this model by comparing some results with the data in the SESAME library /11/. Because the SESAME data are valid for LTE only, we have artificially suppressed for this comparison the non-LTE parts in our model. We find that our Planck opacities are in agreement with SESAME values within a factor two inspite of the crude approximations involved. An example for Eu (Z = 63), the heaviest element published in /11/, is shown in fig. 6. This gives us confidence that the trends of the emission are described quite well by our simplified model.

The intensity I_ν emitted by a plasma layer was calculated by the transport equation

$$\frac{dI_\nu}{dx} = \varepsilon_\nu(x) - \kappa_\nu(x) I_\nu(x)$$

with ε the emission- and κ the absorption coefficient. We have calculated the intensity emitted by different types of plasma layers.

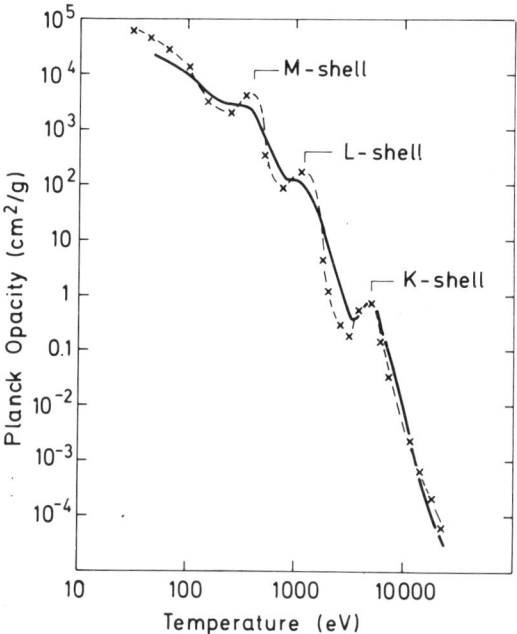

Fig. 6: Planck opacities for Eu (Z = 63) and $\rho = 0 \cdot 1$ g/cm^3 calculated by us (dashed) and according /11/ (full line). The peaks are due to the excitation of the different atomic shells.

The simplest case is a homogeneous layer with constant temperature and density. When the emission does not occur in a too wide photon energy range (i.e. when different spectral regions of the emission are not created in layers of different temperatures and densities), then a homogeneous layer may reproduce quite well the observed spectra. An example is given in fig. 7 for the measured carbon spectrum in fig. 1. We find that for a temperature kT = 300 eV, a density $\rho = 2.3 \cdot 10^{-2}$ g/cm^3 and a layer thickness $1.3 \cdot 10^{-5}$ g/cm^2 the computed and measured spectrum agree quite well. The thermal pressure in the layer corresponds to 3 Mbar, which is a reasonable value for the laser irradiation conditions of fig. 1 /12/. The layer thickness is determined so that the calculated and measured x-ray intensity are in agreement. In fig. 7a we have plotted the calculated intensity as a function of the photon energy and in 7b (for comparison with the measured spectrum) as a function of the wavelength after folding it with the experimental

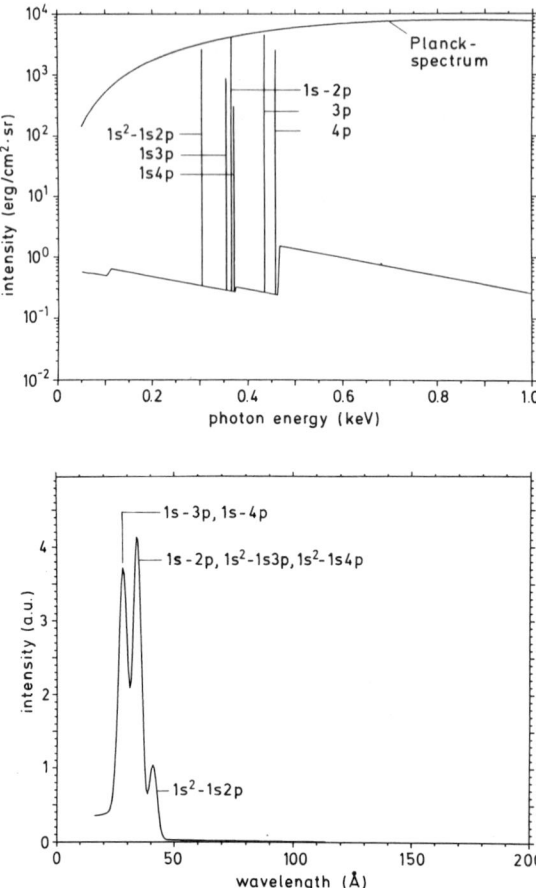

Fig. 7a,b: Caclulated spectrum emitted by a homogeneous carbon plasma (kT=300 eV, $\rho=2.3\cdot10^{-2}$g/cm^3, layer thickness= $1.3\cdot10^{-5}$g/cm^2). a: calculated spectrum versus hν; b: spectrum versus λ after smearing it with the experimental resolution for comparison with the experiment.

wavelength resolution. The fraction of the recombination continuum to the total emission is 20 %, the other 80 % are emitted in hydrogenlike and heliumlike lines. As line shape we have used a Doppler-profile with ion temperature equal to 300 eV. The strongest lines are optically thick as can be seen in fig. 7a where the corresponding Planck spectrum is shown for comparison. We note that for optically thick situations the line emission depends on the assumed line broadening mechanism.

We have also calculated the emission for more realistic inhomogeneous laser plasma layers. For low Z targets the electron energy transport dominates. For that case we have used the analytical formulas given by Manheimer et al. /13/ for the spatial density and temperature distribution in the overdense layer. We note that the typical values of the density, temperature and layer thickness of the electron heat wave driven layer are quite similar to the values used for the homogeneous layer of fig. 7. Calculated conversion efficiencies for the lightest elements Be, C and Al are presented in table 1 together with the measured data. The calculation has been done at different laser intensities. One finds that with the actual incident laser intensity (3×10^{13} W/cm²) the calculated conversion efficiency is too high. This is especially observed for Al-targets, which show for $\phi > 10^{13}$ W/cm² a large amount of radiation emitted in K-lines at 1.5-2 keV (see the numbers in brackets in table 1) in contradiction to the experiment, which showed no increase in efficiency at $\lambda < 10$ Å (see the crosses in fig. 2). With decreasing laser intensity the K-line emission decreases rapidly due to the lower temperature. The result that the calculated and measured x-ray conversion efficiencies agree better at a reduced laser intensity could be due to two-dimensional effects occuring in the experiment, which could cause cooling and which are not considered in the one dimensional plane plasma flow model /12/ used in our calculations. Indeed we observed lateral broadening of the x-ray emitting region compared with the optical spot size. For comparison table 1 contains also conversion efficiencies calculated by Duston et al. /2/ with a more detailed model for conditions similar to ours (laser intensity 10^{13} W/cm², pulse width 3 ns, $\lambda = 0.35$ μm).

Finally we have considered the emission from a high Z target (gold). The thin and hot plasma layers (typically of a thickness of a few 10^{-5} g/cm³ and a temperature of a few 100 eV) used so far to simulate the emission from low Z targets are characteristic of electron heat waves. When we calculate the emission from such thin and hot gold layer we find only a strong peak at $\lambda = 20 - 30$ Å, which is caused by electron transitions in the N-shell. In contrast the measured spectrum (see Fig. 1) shows in addition relative large contributions at longer wavelengths. To simulate the total observed gold spectrum in addition to the thin layer a thicker layer with lower temperature is required. Such profiles can be created in high Z plasmas as a consequence of the strong radiation transport in the dense plasma.

The ablative radiation heat wave alone has been studied in /14/. The real situation including electron heat conduction in the laser absorption zone is well discussed by R.F. Schmalz et al. /15/. In this model the radiation is described by a single temperature (which can be different from the matter temperature) and is transported by diffusion using LTE-Rosseland opacities. We have calculated with this model the temperature and density profiles in a gold plasma for the irradiation conditions of fig. 1. The result is shown in fig. 8. One observes clearly the two regimes with dominating electronic and radiative heat transport respectively. In a second step we have used our atomic physics model to calculate the radiation emitted by this layer. The found overall conversion efficiency was, similar as the experimental value, about 40 % (the exact calculated value depends somewhat on the

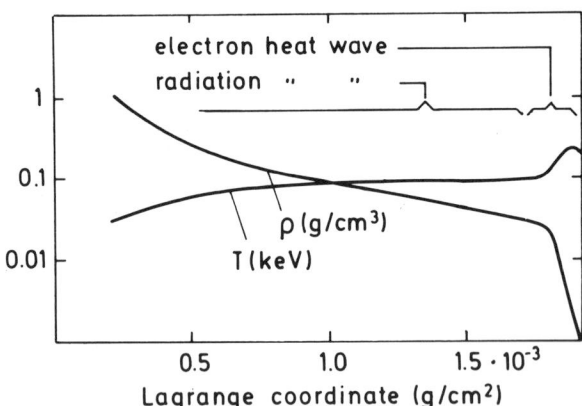

Fig. 8: Electron and density in a laser heated gold target (irradiation conditions as in fig. 1) using the MINIRA-code discribed in /15/. The laser is incident from the right side.

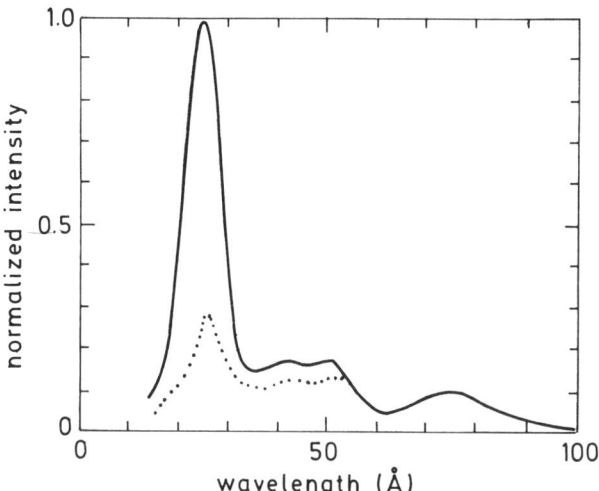

Fig. 9: Gold spectrum calculated for the profiles shown in fig. 8. Full line: spectrum emitted by the total layer; dotted line: contribution of the radiation heat wave region alone. (The spectra were folded with the experimental wavelength resolution.)

assumptions made, e.g. on the broadening of the gold lines). The computed spectrum is shown in fig. 9 (full line). It shows now emission at wavelengths above the N-peak at 25 Å, which is caused by transitions in the O-shell. The O-shell contributions are primarily radiated by the radiation heat wave region as can be seen by comparing the full and the dotted line. The dotted line is the calculated spectrum emitted by the colder radiation heat wave region alone. In contrast the N-peak is created primarily in the thin and hot electron heat wave layer.

Thus the radiation heat wave contributes to more emission at longer wavelengths as was observed experimentally. We also mention that the existence of radiative heat waves has been found in experiments with gold cavity targets /16/.

Finally we note that the calculated (fig. 9) and measured (fig. 1) gold spectra do not agree in detail (the ratio of the long wavelength emission above the N-peak to the emission in the N-peak is in the experiment larger than in the calculation). A possible reason for that could be the hydrogenic approximation used, which may be poor for gold. Gold shows strong ℓ-splitting of the energy levels for different angular momentum quantum numbers ℓ (compare the level structure of neutral gold /17/). In this case $\Delta n = 0$ transitions (not included in the hydrogenic approximation) could be important. Indication for a more complex level structure has been directly seen in the experimental spectra shown in fig. 4.

Also, we must point out that the two-step procedure used to calculate the gold spectrum is inconsistent in the sense that the hydrodynamics was calculated by LTE opacities and the emission was afterwards calculated by a non-LTE atomic physics model. An improved hydrodynamic model using non-LTE opacities and multigroup photon diffusion is being developed /18/.

4. Summary and Conclusion

The sub-keV emission from laser plasmas has been studied both experimentally and theoretically. A systematic study of the emission from targets of different elements (at $\lambda = 0.53$ µm) showed an increase of the x-ray conversion efficiency from 2 % for beryllium to 40 % for gold. The x-ray emission has been studied theoretically by a simplified non-LTE atomic physics model. The comparison between the calculation and the experiment is useful to study the energy transport in the plasma. For low Z targets an electron heat wave dominated plasma gives agreement with the experiment. At high Z the radiation heat wave propagating in the dense plasma becomes important. We note that the existence of the radiation heat wave could have practical consequences because it may help to increase the ablation pressure in the plasma /14,15/ for a more efficient acceleration of target material. Finally it turned out that a more detailed understanding of the radiative processes in high Z targets requires better approximations than the hydrogenic one used in this work.

Table 1

Comparison of measured and calculated (using electron transport dominated plasmas /13/) conversion efficiencies (in %) for low Z elements. The numbers in brackets in case of Al are the conversion efficiencies in Al-K lines.

	measured at $3 \cdot 10^{13}$ W/cm²	calculated for $5 \cdot 10^{12}$ W/cm²	10^{13} W/cm²	$2 \cdot 10^{13}$ W/cm²	Duston et al. ref. /2/
Be	2.3	2.3	2.8	3.4	
C	5.0	7.9	9.6	10.0	18.8
Al	11.0	6.0 (0.2)	12.0 (3)	31.0 (19)	30.7

References

/1/ W.C. Mead et al., Phys. Fluids 26 (1983) 2316;
T. Mochizuki et al., Phys. Rev. A, 33 (1986) 525.

/2/ D. Duston et al., Phys. Rev. A, 27 (1983) 1441 and Phys. Rev. A, 31 (1985) 3220.

/3/ R. Pakula and R. Sigel, Z. Naturforsch. 41 (1986) 463.

/4/ H. Bräuninger, P. Predehl and K.P. Beuermann, Appl. Opt., 18 (1979) 368.

/5/ R. Sigel, A.G.M. Maaswinkel and G.D. Tsakiris, 16th International Congress on High Speed Photography and Photonics, Strasbourg, France, 1984, SPIE Vol. 491, p. 814; G.D. Tsakiris, to be published.

/6/ K. Eidmann, T. Kishimoto et al., accepted by Laser and Particle Beams (1986); T. Kishimoto, MPQ Garching, Germany, Report MPQ 108 (1984).

/7/ I.B. Földes, J. Bayerl, P. Sachsenmaier and R. Sigel, MPQ Garching, FRG, Report MPQ 97 (1985).

/8/ P.M. Campbell, J.J. Kubis and D. Mitrovich, KMS-Fusion, Ann Arbor, Report KMSF-U457 (1976).

/9/ R.M. More, Lawrence Livermore National Laboratory, Report UCRL-84991 (1981).

/10/ H.R. Griem, Plasma Spectroscopy, McGraw-Hill Book Company, New York (1964).

/11/ T4-Group, Los Alamos National Laboratory, Report LALP-83-4 (1983).

/12/ A.G.M. Maaswinkel, K. Eidmann, R. Sigel and S. Witkowski, Optics Comm., 51 (1984) 255.

/13/ W.M. Manheimer, D. Colombant and J.H. Gardner, Phys. Fluids 25 (1982) 1644.

/14/ R. Pakula and R. Sigel, Phys. Fluids 28 (1985) 232.

/15/ R.F. Schmalz, J. Meyer-ter-Vehn, and R. Ramis, accepted by Phys. Rev. A (1986).

/16/ G.D. Tsakiris, P. Herrmann, R. Pakula, R. Schmalz, R. Sigel and S. Witkowski, accepted for publication in Europhysics Letters (1986).

/17/ J.A. Bearden and A.F. Burr, Rev. of Modern Phys., 19 (1967) 125.

/18/ R. Ramis, R.F. Schmalz and J. Meyer-ter-Vehn, Report MPQ 110 (1986).

X-RAY SPECTROSCOPY OF NE-LIKE AND NA-LIKE STRONTIUM IONS

J.C. GAUTHIER, J.P. GEINDRE AND P. MONIER

Institut d'Electronique Fondamentale, Bât. 220
Laboratoire associé au C.N.R.S. , Université Paris XI
91405 Orsay (France).
and Groupement de Recherches Coordonnées de
l'Intéraction Laser Matière
Ecole Polytechnique 91128 Palaiseau (France)

C. CHENAIS-POPOVICS

Laboratoire de Physique des Milieux Ionisés
and Groupement de Recherches Coordonnées de
l'Intéraction Laser Matière
Ecole Polytechnique 91128 Palaiseau (France)

J.F. WYART AND E. LUC-KOENIG

Laboratoire Aimé Cotton, Bât. 501,
Université Paris XI 91405 Orsay (France)

ABSTRACT

We present spectroscopic studies in the X-ray range of Ne-like and Na-like strontium ions as a prerequisite to detailed experiments demonstrating the feasibility of a resonant photoexcitation scheme ensuring the enhancement of gain in collisional excitation or recombination population mechanisms of lasing levels in Ne-like ions.

In addition, our work emphasizes density sensitive line ratios of X-ray transitions which may open a promising new spectroscopic diagnostic for electron density in high-Z laser plasmas.

I - INTRODUCTION

For over a decade, there has been considerable interest in producing amplification of spontaneous emission (ASE) in the soft X-ray wavelength range. The achievement of a coherent X-ray laser would open new powerful diagnostic techniques such as X-ray holography, of interest for biological imaging, and phase-contrast X-ray microscopy. In one of the possible approaches to the design of a XUV amplifier, the lasing transition pertains to a highly-stripped ion produced in a high temperature plasma and population inversion arises through either upwards collisional excitation or resonant photoexcitation mechanisms or downwards recombination or charge exchange processes.

Recently, Rosen, Matthews and co-workers (1) (2), at Lawrence Livermore Laboratory, achieved a great success in the demonstration of a soft X-ray amplifier using 3p-3s transitions in Ne-like selenium (SeXXV). The proposed mechanism for ASE was electron collisional excitation from the ground state $2p^6$ state to $2p^5 3p$ states. An alternative scheme involving recombination pumping following rapid radiative cooling of an overstripped selenium plasma was recently put forward by Apruzese and co-workers at Naval Research Laboratory (3).

However, gains to be expected (4) from the naturally favorable system involving Ne-like ions may be greatly enhanced through the use of external pumping mechanisms. Collisional-pumping by the hot electron component of laser-produced plasmas at 1.06 um wavelength (5) has already been proposed at NRL. The starting point of the present study is to combine collisional excitation and resonant photo-pumping to obtain maximum gain.

Current interest exists to find suitable spectral lines for resonant photo-excitation (6). We have focused our study on wavelength coincidence between resonance levels of Ne-like ions with Z scaling from 36 (Kr) to 42 (Mo) and H-like and He-like transitions of lighter elements (12 < Z < 17). A close matching has been found both theoretically and experimentally (7) between the 1s-3p transition of H-like aluminium (λ = 6.053 Å) and the $2s^2 2p^6$ (^1S) - $2s^2 2p^5 3d$ (^3D), J = 0 - 1 transition in Ne-like strontium (λ = 6.059 Å). The wavelength difference can be easily matched by doppler shifts of the aluminium plasma in the coronal region with typical expansion velocities of $3 \cdot 10^7$ cm/s or by self-absorption and stark broadening (8).

X-ray spectroscopy is a valuable tool to determine plasma conditions in laser-irradiated targets. However, an excited population model is needed to simulate spectral line intensities which can be compared with experiment. This comparison enables us to check the accuracy of theoretical data on wavelengths and oscillator strengths of the highly-charged ions of interest and also to determine the equilibrium balance between the excited levels. The occurence of large electron-collisional-rate coefficients for the $2p^53s$ and $2p^53d$ states from the ground state of Ne-like ions, allows us to search for density sensitive line ratios that occur at high electron densities and which can be of great use in determining the plasma parameters.

II - X-RAY SPECTROSCOPY OF RESONANCE LINES IN NE-LIKE AND NA-LIKE IONS

The spectra of highly-ionized strontium ions were generated by focusing one beam of the Groupement de Recherche Coordonnées Interaction Laser-Matière (GRECO-ILM) laser onto a solid target. The targets consist either of a 1 µm thick deposit of strontium fluoride on a silicon substrate or a massive strontium rod. In the first case, silicon resonance lines of He-like and H-like ions were used to provide wavelength standards for calibration. We used 600 ps pulses of 0.53 µm radiation and laser intensities in the 10^{14} W/cm^2 range. The X-ray spectra were recorded with a flat pentaerythritol (PET) crystal spectrograph giving a spectral range of 4.5-7 Å on KODAK SB2 films. A 30 µm slit or a knife edge (9) in front of the spectrograph were alternatively used to provide spatial resolution along the laser axis to study the dense and coronal regions of the plasma expansion. After densitometry, the recordings were corrected for film exposure non linearities and computer deconvoluted to obtain spatially-resolved spectra.

A typical spectrogram is shown in Fig.1. When compared to earlier observations (10), we note that the signal to noise ratio and the spectral resolution shown in this figure are both improved. The strongest features labelled C,H,N, and P are those from Ne-like ions which produce a simple characteristic spectrum corresponding to transitions from $2p^53d$ and $2p^53s$ configurations to the closed-shell ground state $2p^6$. They are accompanied on the long wavelength side by partially resolved satellite lines originating mostly from Na-like ions (11).

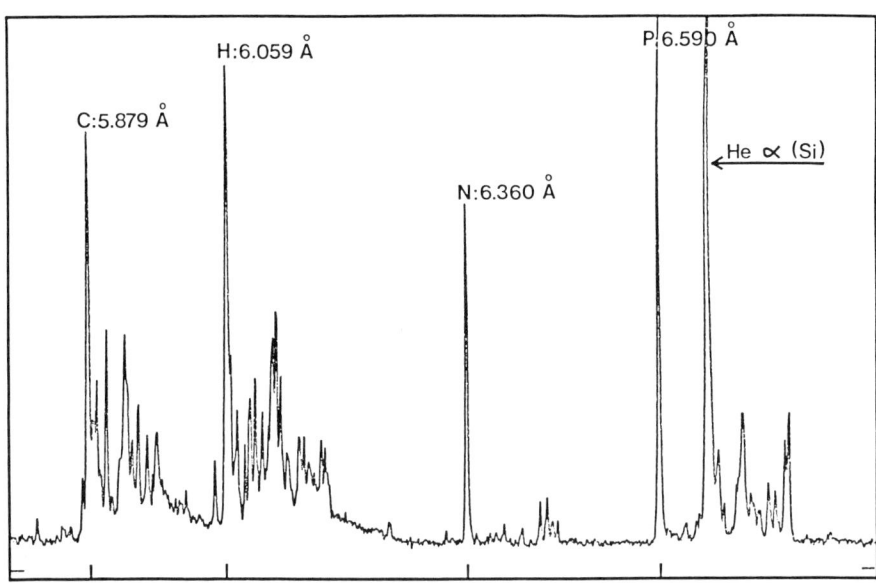

Fig. 1 : Densitogram of strontium Ne-like and Na-like transitions around 6 Å showing the strong features H,N,C and P together with silicon lines wavelengths standards.

In all our study we adjusted the laser intensity to get a minimum amount of F-like ions in the coronal region. This enables us to get an unambiguous identification of the line features from the dominant Ne-like and Na-like ionization stages. 1D Lagrangian code simulations with an average-atom model for ionization show that our observations were consistent with plasma conditions of T_e = 800 eV and n_e in the 10^{19} - 10^{21} range.

Table I gives a complete list of observed wavelengths together with a letter key which will be used to label the appropriate spectral features.

Ab-initio calculations have been made using the relativistic parametric potential method (12). The RELAC code yielded the level energies and the oscillator strenghts of several excited levels of Na-like ions pertaining to the configurations $2s^2 2p^6 3s$-$3p$-$3d$, $2s^2 2p^5 3s^2$-$3s3p$-$3s3d$-$3p^2$-$3p3d$-$3d^2$, and $2s2p^6 3s^2$-$3p^2$-$3d^2$-$3s3d$-$3s3p$-$3p3d$.

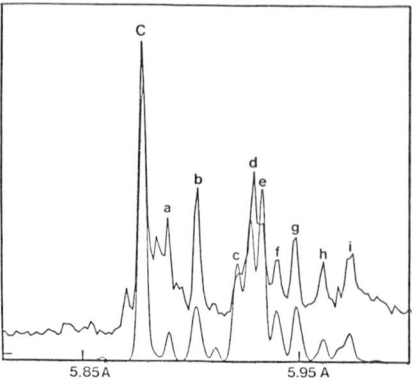 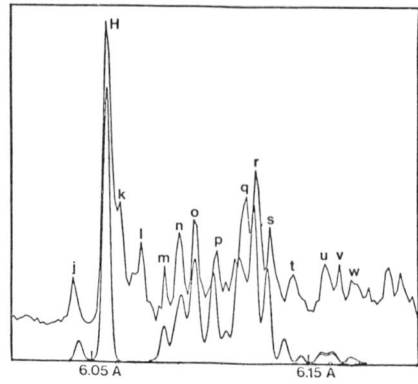

Fig. 2 : a : Na-like satellite lines in the vicinity of the C Ne-like line.
 b : Na-like satellite lines in the vicinity of the H Ne-like line.
Lines are labelled according to Table I. The population ratio of Ne/Na-like ionization stages is given by Saha equilibrium at 10^{20} cm^{-3} and 800 eV.

Synthetic spectra are reproduced in Fig. 2a for Na-like lines close to the C line and in Fig. 2b for Na-like lines close to the H line to show directly the comparison between experiment and theory. In order to take into account the finite instrumental width of our apparatus, we convoluted the theoretical line intensities with gaussian profiles of 7 mÅ width. The population ratio of ground states Na and Ne where calculated under Saha equilibrium with T_e = 800 eV and n_e = 10^{20} cm^{-3}.

The agreement is outstanding both on the wavelength position of the lines and on their relative intensities. Discrepancies in line positions are well within 2 mÅ corresponding to uncertainties in wavelength calibration. The broad feature on the long-wavelength side of H and C are satellite lines with one spectator electron in the orbitals 4ℓ or higher. For example, the line labelled l in Fig. 2b involves a 4s spectator electron in transitions from $2p^5 3d4s$ to $2p^6 4s$.

III - COLLISIONAL-RADIATIVE EXCITATION MODEL

A collisional-radiative model has been developped to compute the populations of the excited states of Ne-like strontium. Figure 3 shows the detailed energy level diagram of the $2p^6$, $2p^5 3s$, $2p^5 3p$ and $2p^5 3d$ configurations. In addition to these 27 levels (ground state plus 26 n = 3 excited states) four super-levels corresponding to the configurations with a n = 4 jumping electron were inserted in the model.

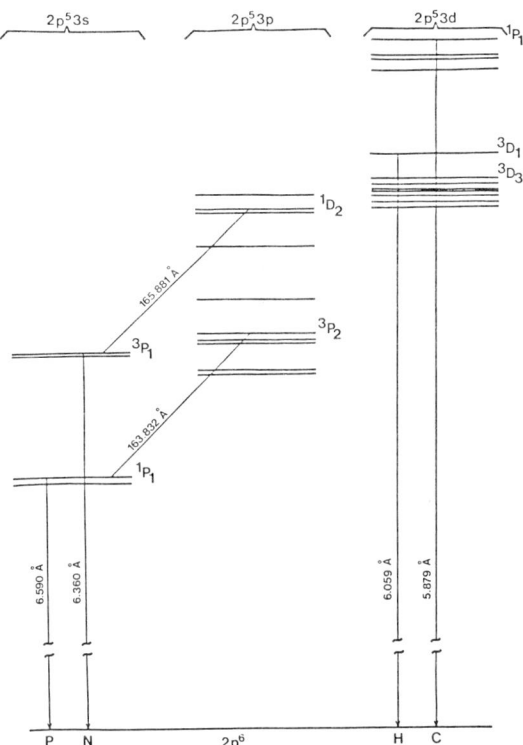

Fig. 3 : *Energy level diagram used in the collisional radiative model. Levels are labelled according to the LS coupling scheme only for convenience of refering in the discussion of Chapter III. The lasing transitions in selenium are sketched with their corresponding wavelengths in strontium.*

Energy levels and spontaneous decay rates A_{ul} have been calculated with the SUPERSTRUCTURE computer package which uses a statistical and semi-relativistic perturbational method.

Collisional excitation rate coefficients have been evaluated using an extrapolation of KrXXVII collisional strengths calculated by Feldman (13). They are given by (14) :

$$C^e_{1u}(s^{-1}) = \frac{\pi a_o^2}{g_1} \sqrt{\frac{8kT_e}{\pi m}} n_e \left(\frac{Ryd}{T_e}\right) \Omega_{1u} \, e^{-\frac{(E_u - E_1)}{kT_e}} \tag{1}$$

where g_1 is the statistical weight of the lower level, n_e the electron density, E_u and E_l the upper and lower energies of the levels. Collisional ionization has been taken into account with the Landshoff and Perez formula (15)

$$K_{ic}(S^{-1}) = 6.7 \cdot 10^{-9} \left(\frac{Ryd}{E_i}\right)^2 n_e \, e^{-\frac{E_i}{kT_e}} (kT_e)^{1/2} g \tag{2}$$

where

$$g = \frac{0.915}{\left(1 + 0.064 \frac{kT_e}{E_i}\right)^2} + \frac{0.42}{\left(1 + 0.5 \frac{kT_e}{E_i}\right)^2}$$

and we have made use of the Seaton radiative-recombination rate coefficient (16)

$$R_{ci}(S^{-1}) = 5.2 \cdot 10^{-14} n_e Z \left(\frac{E_i}{kT_e}\right)^{3/2} e^{\frac{E_i}{kT_e}} E_1\left(\frac{E_i}{kT_e}\right) \tag{3}$$

where E_1 is the exponential integral and Z the ionic charge. Energy levels with principal quantum number above 4 where also introduced, but neglecting l-splitting, by using hydrogen-like excitation and ionization rates (14). Model predictions being rather sensitive to the maximum quantum number incorporated in the calculations (17), we used a thermal band to ensure the continuity between discrete and continuum states, including the lowering of the ionization potential.

In order to evaluate the excited level populations N_i, we have solved a set of coupled equations, in steady-state

$$\left[\sum_{j<i}(\overline{A}_{ij} + C^d_{ij}) + \sum_{j>i} C^e_{ij} + K^{ionis}_{ic}\right] N_i =$$

$$\sum_{j<i} C^e_{ji} N_j + \sum_{j>i}(\overline{A}_{ji} + C^d_{ji}) N_j + (K^{recomb}_{ci} + R_{ci}) N_F \tag{4}$$

where N_1 and N_F are respectively the SrXXIX and SrXXX ground state populations as parameters of the problem. In calculating the level populations, self-absorption has been introduced using a simple escape factor approximation (18) to reduce the spontaneous decay rates.

IV - RESULTS

a) Excited state populations

The set of coupled equations has been solved for an electron temperature of 800 eV and electron densities ranging between 10^{19} an 10^{23} cm^{-3}. In these calculations, we have assumed a Ne-like fractional population $f_N = N_1/N_I$ (where N_I is the total ion density) very close to unity, in accordance to our experimental observations.

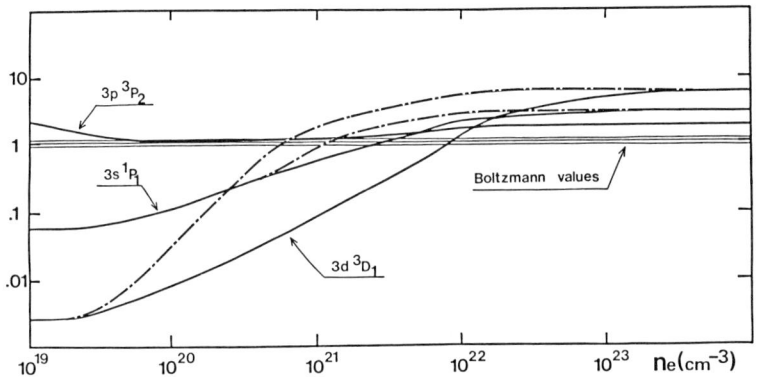

Fig. 4 : *Population ratios normalized to the $3d^3D_3$ as a function of electron density at an electron temperature of 800 eV. Solid line : optically-thin lines. Dash dotted lines : radiative transfer in a 30 µm plasma included.*

In Fig. 4, the three ratios $N'(3s^1P_1)/N'(3d^3D_3)$, $N'(3p^3P_2)/N'(3d^3D_3)$ and $N'(3d^3D_1)/N'(3d^3D_3)$ where N' stands for the level population per unit statistical weight are plotted as a function of n_e. These levels are of particular interest in that they are really representative of the population mechanisms of the three excited configurations $2p^53s$, $2p^53p$, and $2p^53d$. Full lines have been obtained assuming an optically thin plasma, and dotted lines show the modification introduced by the inclusion of self-absorption in a 30 µm thick plasma. These results are compared with population ratios calculated using the Boltzmann equation. Above $n_e = 10^{22}$ cm^{-3}, all the levels are in collisional equilibrium at value slightly higher than the Boltzmann values. This difference is due to the fact that in the model, ground states of Ne-like and F-like ions have been chosen far from Saha equilibrium at T_e = 800 eV.

Let us first examine the $3d\,^3D_1$ level population. This level, whose ground state spontaneous decay is extremely high (A = 9.5 10^{13} s^{-1}) behaves as though it was in coronal equilibrium for electron densities roughly equal to 5.10^{20} cm^{-3} : at this density, and including self-absorption effects, the ratio $N'(3d\,^3D_1)/N'(3d\,^3D_3)$ fits the value it would take if Boltzmann collisional equilibrium was reached. Here self-absorption plays an important role in bringing the levels closer to Boltzmann equilibrium (19). As a result of this effect and for n_e less than 5.10^{20} cm^{-3}, the $3d\,^3D_1$ level remains very weakly coupled to the other 3d levels.

In the same way, we have studied the population mechanisms for the $3p\,^3P_2$ level. This level is basically populated by two collisional channels : from the ground state (C = 3.1 10^{-13} cm^{-3} at 800 eV) and from some of the 3d levels essentially the $3d\,^3D_3$ level (C = 9.1 10^{-10} cm^{-3} at 800 eV. The latter means of population is clearly emphasized in Fig. 4 which shows that the $3p\,^3P_2$ level is always closely coupled to the $3d\,^3D_3$ level. Self-absorption does not affect this level since its radiative decay to the ground state is dipole-forbidden.

One may finally observe that the $3d\,^3D_1$ level does not directly populate the $3p\,^3P_2$ for two basic reasons : first the $3d\,^3D_1$ to $3p\,^3P_2$ collisional deexcitation rate coefficient (C = 6.9 10^{-12} cm^{-3}) is lower than for the $3d\,^3D_3$ to $3p\,^3P_2$ (C = 9.1 10^{-10} cm^{-3}) ; moreover the $(3d\,^3D_1 - 3p\,^3P_2) / (3d\,^3D_1 - 2p\,^1S_0)$ branching ratio is much less than unity (8.6 10^{-6} for n_e = 10^{20} cm^{-3} and T_e = 800 eV). The $3d\,^3D_1$ thus rapidly decays radiatively towards the $2p^6$ ground state.

b) Density sensitive line-ratios

The intensity ratios of several resonance line of Ne-like ions have been computed to compare to experiment. Density-sensitive line ratios in the Be-like, B-like and Ne-like sequences have already been noticed by Seely and co-workers (20) for their potential applications as an electron density diagnostic in laser-created plasmas. Here we show directly the use of such line ratios to infer the electron density.

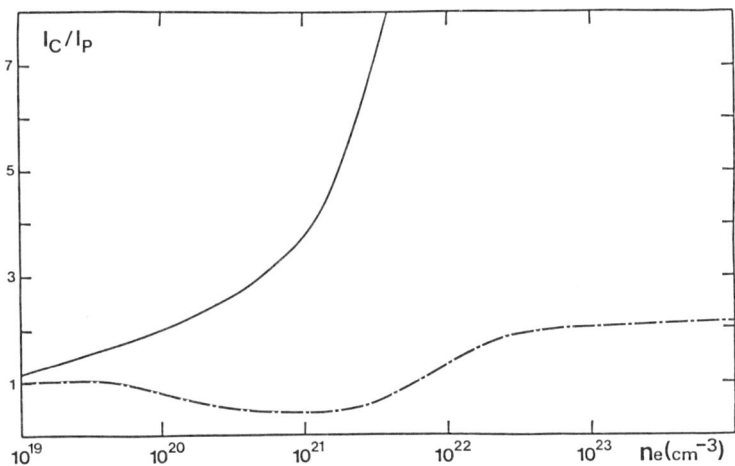

Fig. 5 : Ratio I_C/I_P as a function of electron density at $T_e = 800$ eV. Solid line : optically-thin lines. Dash-dotted line : radiative transfer in a 30 μm plasma included.

I_c/I_p has been plotted in Fig. 5 as a function of electron density. The electron temperature was 800 eV and the Ne-like fractional population was still assumed to be close to unity. The solid line curve represents the results for an optically thin plasma while the dotted curve has been obtained by solving the radiative transfer equation with an homogeneous plasma of 30 μm depth and using escape factors in the rate equations. In the first case, the ratio I_c/I_p is very strongly electron density dependent due to the coronal equilibrium way of population of the two levels $3d\,^1P_1$ and $3s\,^1P_1$ in a large range of electron densities up to 10^{22} cm^{-3} for the $3d\,^1P_1$ state whose spontaneous decay rate is extremely high. Between 10^{19} and 10^{22} cm^{-3}, I_c/I_p changes from 1.1 to 18.6 where collisional equilibrium is set up. Extrapolation of the results for KrXXVII at $T_e = 800$ eV show a reasonable agreement with the results previously obtained by Seely (20).

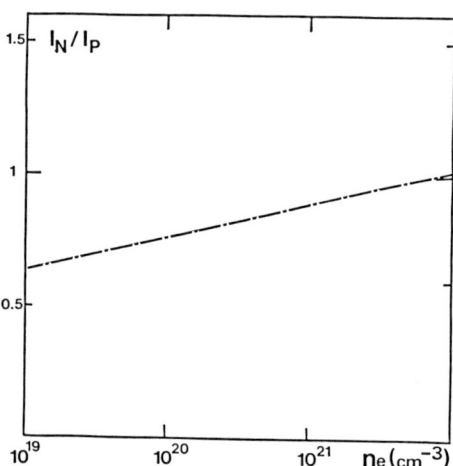

<u>Fig. 6</u> : Ratio I_N/I_p as a function of electron density at $T_e = 800$ eV. Radiative transfer in a 30 μm plasma is included.

Conversely, when opacity effects are taken into account, the I_c/I_p dependence with n_e is clearly reduced. For electron densities lower than 10^{21} cm^{-3}, line C alone is affected by photon-trapping and I_c/I_p decreases. Then, both spectral lines are optically thick at much higher electron densities.

Fig. 6 shows the ratio I_N/I_p for the same plasma conditions, but with radiation trapping. Table 2 shows the results of our spatially-resolved measurements taken from different laser shots. These time-integrated results reflect the conditions of the coronal region in the plasma expansion. From these results, the electron density can be deduced to be $4 \pm 2 \; 10^{19}$ cm^{-3} and, as noted previously from the ratio of Na-like to Ne-like lines, the electron temperature is close to 800 eV. These results are consistent with our hydrocode calculations.

V - CONCLUSION

We have made detailed measurements of the spectroscopic properties of Ne-like and Na-like strontium ions in a laser-created plasma. A comprehensive identification of several Na-like satellites with n = 3 and n = 4 of Ne-like resonance lines has been made for the first time.

We have demonstrated the potential of using X-ray line ratios from highly-charged ions of the NeI sequence for measuring electron densities in high-density plasmas.

Future work will concern identification of soft X-ray lines of the $2p^53p - 2p^53s$ and $2p^53d - 2p^53p$ transitions in Ne-like ions and the possibility of using resonant photoexcitation to enhance the gain of lasing lines from these transitions.

ACKNOWLEDGEMENTS We thank J. DUBAU et M. CORNILLE for providing the calculations of the radiative transition probabilities of Na-like and Ne-like strontium ions using SUPERSTRUCTURE and for useful discussions.

REFERENCES

1) M.D. ROSEN et al., Phys. Rev. Letters, 54, 106 (1985).

2) D.L. MATTHEWS et al., Phys. Rev. Letters, 54, 110 (1985).

3) J.P. APRUZESE et al., Phys. Rev. Letters, 55, 1877 (1985).

4) A.V. VINOGRADOV, I.I. SOBEL'MAN and E.A. YUKOV, Sov. J. Quantum Electron. 7, (1977).

5) J.P. APRUZESE and J. DAVIS, Phys. Rev. 28A, 3686 (1983).

6) P.G. BURKHALTER, G. CHARATIS and P.D. ROCKETT
J. Appl. Phys. 54, 6138 (1983).

7) P. AUDEBERT et al., GRECO ILM Internal Report (1984).

8) B.A. NORTON and N.J. PEACOCK, J. Phys. B : Atom. Mol. Phys., 8, 989 (1975).

9) Ph. ALATERRE et al., Optics Comm. 49, 140 (1984).

10) H. GORDON et al., J. Phys. B : Atom. Mol. Phys. 12, 881 (1979).

11) P.G. BURKHALTER, D.J. NAGEL and R.D. COWAN
Phys. Rev. A11, 782 (1975).

12) E. LUC-KOENIG, Physica, 62, 393 (1972).

13) U. FELDMAN, A.K. BHATIA and S. SUCKEWER, J. Appl. Phys., 54, 2188 (1983).

14) L.B. GOLDEN et al., Astrophysical Suppl. Series 45, 603 (1981).

15) R.K. LANDSHOFF and J.D. PEREZ, Phys. Rev., A13, 1619 (1976).

16) M.J. SEATON, Mont. Notices R. Astron. Soc., 119, 81 (1959).

17) J. WEISHEIT (private communication).

18) J.P. APRUZESE et al., J. Quant. Spectrosc. Radiat. Transfer, 23, 479 (1980).

19) J.P. APRUZESE, this conference p xxx .

20) U. FELDMAN, J.F. SEELY and A.K. BHATIA, J. Appl. Phys., 58, 3954 (1985).

λ_{exp}		lower relativistic configuration	J	Parity	upper relativistic configuration	J	Parity	λ_{th}	g_f
a	5.890	$3s+$	1/2	e	$:\overline{2p}-3s+3d+$	3/2	0	5.890	0.957
		$3p+$	3/2	0	$:\overline{2p}-3p+3d-$	5/2	e	5.902	1.55
b	5.903	$3p-$	1/2	0	$:\overline{2p}-3p-3d-$	3/2	e	5.900	0.575
		$3p+$	3/2	0	$:\overline{2p}-3p+3d-$	1/2	e	5.904	0.223
		$3p+$	3/2	0	$:\overline{2p}-3p+3d+$	3/2	e	5.904	0.223
	5.910	$3d-$	3/2	e	$:\overline{2p}-3d-3d+$	3/2	0	5.912	0.547
c	5.921	$3d-$	3/2	e	$:\overline{2p}-3d^2$	3/2	0	5.922	2.57
		$3d+$	5/2	e	$:\overline{2p}-3d-3d+$	3/2	0	5.921	1.85
		$3d+$	5/2	e	$:\overline{2p}-3d-3d+$	7/2	0	5.928	3.37
d	5.927	$3s+$	1/2	e	$:\overline{2p}-3s+3d-$	1/2	0	5.927	1.03
		$3p-$	1/2	0	$:\overline{2p}-3p-3d-$	1/2	e	5.924	0.967
		$3p+$	3/2	0	$:\overline{2p}-3p+3d-$	1/2	e	5.926	0.951
		$3d+$	5/2	e	$:\overline{2p}-3d-3d+$	5/2	0	5.933	3.62
		$3p+$	3/2	0	$:\overline{2p}-3p+3d-$	3/2	e	5.932	1.96
e	5.932	$3d-$	3/2	e	$:\overline{2p}-3d^2-$	1/2	0	5.934	0.611
		$3d+$	5/2	e	$:\overline{2p}-3d^2-$	3/2	0	5.931	0.413
		$3s+$	1/2	e	$:\overline{2p}-3p^2+$	1/2	0	5.934	0.247
		$3s+$	1/2	e	$:\overline{2p}-3s+3d-$	3/2	0	5.939	1.43
f	5.938	$3p+$	3/2	0	$:\overline{2p}-3p-3d+$	5/2	e	5.941	0.707
		$3d-$	3/2	e	$:\overline{2p}-3d^2+$	3/2	0	5.942	0.101
		$3p-$	1/2	0	$:\overline{2p}-3p-3d-$	3/2	e	5.948	1.65
g	5.945	$3p+$	3/2	0	$:\overline{2p}-3p+3d+$	5/2	e	5.950	0.732
		$3p+$	3/2	0	$:\overline{2p}-3p-3d-$	3/2	e	5.947	0.131
		$3d-$	3/2	e	$:\overline{2p}-3d-3d+$	3/2	0	5.958	0.187
h	5.957	$3p+$	3/2	0	$:\overline{2p}-3p+3d+$	5/2	e	5.960	0.591
		$3d-$	3/2	e	$:\overline{2p}-3d-3d+$	5/2	0	5.961	0.221
		$3d+$	5/2	e	$:\overline{2p}-3d-3d+$	5/2	0	5.971	0.401
i	5.969	$3d-$	3/2	e	$:\overline{2p}-3d^2$	5/2	0	5.973	0.874
		$3p+$	3/2	0	$:\overline{2p}-3p+3d-$	3/2	e	5.973	0.168
m	6.083	$3s+$	1/2	e	$:\overline{2p}+3s+3d+$	3/2	0	6.083	1.02
		$3d+$	5/2	e	$:\overline{2p}+3d^2$	3/2	0	6.088	0.981
		$3p+$	3/2	0	$:\overline{2p}+3p+3d+$	5/2	e	6.089	0.257
n	6.089	$3s+$	1/2	e	$:\overline{2p}+3s+3d+$	1/2	0	6.091	0.742
		$3p+$	3/2	0	$:\overline{2p}+3p+3d+$	1/2	e	6.091	0.880
		$3d-$	3/2	e	$:\overline{2s}+3s+3p+$	1/2	0	6.093	0.461
		$3d-$	3/2	e	$:\overline{2p}+3d^2$	5/2	0	6.096	1.44
		$3p+$	3/2	0	$:\overline{2p}+3p+3d+$	3/2	e	6.097	0.840
o	6.096	$3p+$	3/2	0	$:\overline{2p}+3p+3d+$	5/2	e	6.097	1.01
		$3d-$	3/2	e	$:\overline{2p}+3d-3d+$	1/2	0	6.099	0.780
		$3p-$	1/2	0	$:\overline{2p}+3p+3d+$	3/2	e	6.100	0.189

λ_{exp}		lower relativistic configuration	J	Parity	upper relativistic configuration	J	Parity	λ_{th}	g_f
p	6.105	3p–	1/2	0	$\overline{2}$p+3p+3d–	1/2	e	6.103	0.328
		3d+	5/2	e	$\overline{2}$p+3d^2+	5/2	0	6.106	2.79
		3p–	1/2	0	$\overline{2}$p+3p+3d–	3/2	e	6.108	0.376
q	6.117	3p+	3/2	0	$\overline{2}$p+3p+3d+	3/2	e	6.115	1.24
		3p–	1/2	0	$\overline{2}$p+3p–3d+	1/2	e	6.116	0.723
		3s+	1/2	e	$\overline{2}$p+3s+3d+	3/2	0	6.117	0.922
		3d–	3/2	e	$\overline{2}$p+3d–3d+	5/2	0	6.120	2.34
r	6.122	3p–	1/2	0	$\overline{2}$p+3p+3d–	3/2	e	6.122	0.432
		3d–	3/2	e	$\overline{2}$p+3d–3d+	3/2	0	6.124	1.35
		3p+	3/2	0	$\overline{2}$p+3p+3d+	5/2	e	6.124	1.61
		3d+	5/2	e	$\overline{2}$p+3d^2+	7/2	0	6.125	2.10
s	6.128	3d+	5/2	e	$\overline{2}$p+3d–3d+	5/2	0	6.130	0.735
		3s+	1/2	e	$\overline{2}$p+3p^2+	3/2	0	6.130	0.276
t	6.138	3p–	1/2	0	$\overline{2}$p+3p+3d–	3/2	e	6.137	0.401
		3s+	1/2	e	$\overline{2}$p+3s+3d–	1/2	0	6.138	0.309
		3p+	3/2	0	$\overline{2}$p+3p+3d+	1/2	e	6.139	0.085
u	6.152	3p+	3/2	0	$\overline{2}$p+3p+3d+	3/2	e	6.150	0.023
		3p+	3/2	0	$\overline{2}$p+3p+3d–	1/2	e	6.154	0.014
		3d+	5/2	e	$\overline{2}$p+3d^2+	7/2	0	6.155	0.283
v	6.158	3d+	5/2	e	$\overline{2}$p+3d–3d+	5/2	0	6.157	0.044
		3p+	3/2	0	$\overline{2}$p+3p+3d–	5/2	e	6.156	0.110
		3d–	3/2	e	$\overline{2}$p+3d^2+	5/2	0	6.159	0.041
		3p–	1/2	0	$\overline{2}$p+3p–3d+	3/2	e	6.158	0.091
		3s+	1/2	e	$\overline{2}$p+3s+3d–	3/2	0	6.160	0.015
		3d–	3/2	e	$\overline{2}$p+3d^2–	3/2	0	6.162	0.112
w	6.166	3p+	3/2	0	$\overline{2}$p+3p–3d+	1/2	e	6.167	0.019
		3p+	3/2	0	$\overline{2}$p+3p+3d+	5/2	e	6.167	0.142
		3d+	5/2	e	$\overline{2}$p+3d^2+	5/2	0	6.169	0.028
		3d–	3/2	e	$\overline{2}$p+3d–3d+	3/2	0	6.171	0.082
		3d+	5/2	e	$\overline{2}$p+3d^2–	3/2	0	6.172	0.023
		3p+	3/2	0	$\overline{2}$p+3p+3d–	3/2	e	6.173	0.033
j	6.044	3p–	1/2	0	$\overline{2}$p+3p+3d+	1/2	e	6.041	0.031
		3p–	1/2	0	$\overline{2}$p+3p+3d–	3/2	e	6.043	0.606
		3p–	1/2	0	$\overline{2}$p+3p+3d+	3/2	e	6.047	0.025

<u>TABLE I</u> : Identification of the lines labelled in Fig. 2. Experimental wavelengths (on the left) are compared to theory (on the right). gf values are also listed. In the notation, closed-shell electrons have been omitted. Code letters for the other electrons are :
$\overline{2}$p– : $2p_{1/2}^5$, $\overline{2}$p+ : $2p_{3/2}^5$, $\overline{2}$s+ : $2s_{1/2}$, 3s+ : $3s_{1/2}$, 3p– : $3p_{1/2}$, 3p+ : $3p_{3/2}$, 3d– : $3d_{3/2}$, 3d+ : $3d_{5/2}$.

TABLE II

RATIOS I_c/I_p AND I_N/I_p OBTAINED EXPERIMENTALLY IN DIFFERENT LASER SHOTS

laser shot	I_c/I_p	I_N/I_p
456		0.72
458	0.99	0.68
458		0.74
474	1.1	0.60
476	1.2	0.66
477	1.1	0.69

RADIATIVE EMISSION NEAR CRITICAL DENSITY IN LASER-PRODUCED PLASMA

P.G. Burkhalter, J.P. Apruzese, and D. Duston*

Naval Research Laboratory
Washington, D. C. 20375-5000

Abstract

Experimental and theoretical spectroscopic techniques were used to determine temperature, density, and emission profiles in laser-produced plasmas. Spatially-resolved, high resolution x-ray spectra were acquired from shots using 0.53 and 1.06 μm laser light. Plasma temperatures and densities were interpreted with a collisional-radiative equilibrium ionization model incorporating multifrequency radiative transport and a 1-D planar hydrodynamic plasma expansion model. Effects of plasma opacity on line shapes, resonance line ratios, and satellite/parent line intensities were studied as a function of tracer dot diameter.

* Strategic Defense Initiative Organization, Washington, D.C. 20301-7100

I. Introduction

This study has been designed to diagnose and interpret the temperature and density profiles near the critical density of laser-produced plasma. It is an extension of previous work aimed at using tracer dots to improve spectral resolution[1] combined with and using x-ray spectroscopy and theory to interpret plasma characteristics.[2] The use of tracer dots or micron-sized targets has been applied in other laboratories to improve x-ray spectroscopy in laser-excited studies.[3-5] The interpretation of the plasma profile characteristics is based on modeling with collisional-radiative equilibrium (CRE) code calculations[6] and atomic-radiative transfer models.[7] Previous measurements of the plasma temperature and density profiles at NRL have led to agreement with 1-D hydrocode predictions in underdense plasma regions by modeling the cylindrically-shaped plasma confinement from tracer dot targets.[2] These experimental-theoretically interpreted plasma profile measurements were an extension to the study of underdense long scalelength plasma.[8] The current study improves upon the spectral and spatial resolution of the tracer-dot experiment to characterize the plasma formation at or near the critical density regime close to the target surface.[9] Both the target diameter and laser light wavelength were varied in order to assess their effects on the plasma parameters.

Distinct Li-like satellite lines in Al XI were observed in this experimental data attributed to the excellent spectral resolution ($\lambda/\Delta\lambda=3100$ provided by the (013) plane diffraction plane in potassium acid phthalate (KAP) for the spectral region near 7.8Å,[10] with both 0.53 and 1.06 μm illumination. The classification and importance of the Li-like satellite lines has been the subject of the work of Gabriel and coworkers.[11-13] The measurement and interpretation of satellite line intensities for high density diagnostics provide direct evidence of the transient nature of laser-produced plasmas.[14,15] Jacobs[16] has noted the importance of the effect of angular-collisional momentum changes on the excitation cross sections of dielectronic satellites from doubly-excited levels. The use of satellite-to-parent line ratios[17] and satellite-to-satellite line ratios[18] provide interpretational means for plasma temperature and density predictions.

II. Experimental

The Chroma laser at KMS Fusion was focused through an f/6 spherical lens into a special experimental chamber that housed the targets and diagnostics. The beam was focused to 0.30 mm diameter at the target surface. Spectral data was collected from various diameter Al tracer dots that were 1 μm in thickness formed by vaporation through lithographically-manufactured masks onto thick plastic backing. The Nd:glass laser operated with frequency-doubled green light at 0.53 μm with irradiances of 0.5-1.0 x 10^{14} W/cm^2. Readable spectral images were collected from 50, 100, and 200 μm diameter targets.

Dual convex-curved crystal spectrographs (see Fig. 1) were utilized to collect spatially-resolved x-ray spectra from aluminum tracer dots. A potassium acid phthalate crystal was curved to a 3.2 cm radius and deployed in a cylindrical housing which served as a light-tight film holder. A special gold-electroplated slit (10 μm in width) was positioned between the 12.5

micron thick Be window on the spectrograph and the target. The spectrograph was positioned at a target-to-slit distance of 1.7 cm. The dimensions of both spectrographs were such that the spatial information in the spectral lines was recorded at a magnification of about 10:1. The pentaerythritol (PET) crystal used had a radius-of-curvature of 25.4 cm. It was positioned in the spectrograph to record the entire Al spectra from 7.5 to 5.5 Å except for the He$_\alpha$ lines near 7.8 Å. A 25-μm slit was used in front of the PET spectrograph. The film used in both spectrographs was Kodak direct exposure film (DEF) for which exposure-to-film density calibration was available.[19] The spectrographs were positioned with their entrance slits nearly orthogonal (87°) to the incident laser beam. This allowed recording of the plasma emission plume imaged in the spectral lines.

Aluminum spectrograms were processed from a high-resolution KAP spectrograph used to acquire Al tracer dot data with the Pharos laser system at the Naval Research laboratory. The spectra were from 65, 115 and 220 micron diameter tracer dots irradiated with 1×10^{14} W/cm^2 of focused 1.06 μm laser light. The spectrograms were recorded through a 10 μm slit at a magnification of 20 times. The targets consisted of aluminum dots formed by evaporation through a pinhole mask onto thin (1500 μm) plastic substrates.

The spectrograms were scanned with a PDS microdensitometer using a slit of 12 μm height and 10 μm width. The spectral lines were scanned at the target surface and at intervals of 25 μm in distance from the target surface. The digital values were stored on magnetic tape and processed by computer with a program[20] that converts film density to spectral intensity accounting for the absorption in the spectrograph window, the diffraction crystal efficiencies, and the film H and D curves.

The spectral lines were integrated to yield spectral intensities. An atomic structure code from LANL[21] was used to calculate wavelengths of satellite lines in aluminum. The spectral intensity data was interpreted by comparing with theoretical model calculations that yield line profiles and line intensity ratios as functions of plasma temperature and density based on the collisional-radiative equilibrium model that has been benchmarked at NRL for Al tracer dot studies.[2] The same model has been used to calculate satellite-to-parent line ratios in Al and other elements. A multifrequency radiative transport model[22] was used to predict Al line profiles as a function of plasma density. With the assumption that the plasma is in the form of a cylindrically-shaped plume, the theoretical plasma radiation codes can accurately account for the effect on the line emission of the plasma optical depths provided the plasma attains equilibrium over the finite regions corresponding to the plasma spatial differentiation.

III. Results

The spectral data collected for a 100 μm diameter tracer dot at an irradiance of 10^{13}W/cm^2 using 0.53 μm laser light is shown in Fig. 2. The spectrogram (acquired with the 001 diffraction plane in KAP) is shown below the two intensity traces that resulted from densitometer scans at the target surface and at a distance of 50 μm in front of the target surface. All the major He-like Al XII and H-like Al XIII lines are visible together with the weaker satellite lines. In both spectral traces, the H_α line is more intense than the He_α line. Line intensities could be measured to the n=5 Rydberg member (delta line) in both ionization stages. The moderate spectral resolution, $\lambda/\Delta\lambda=800$, provides adequate line resolution and intensity to determine the various line ratios used for diagnosing plasma parameters. The gamma lines from both stages of ionization and the He-like intercombination line $1s^2$-$1s2p$ ^3P were measurable to a distance of 125 μm from the target surface. The plasma temperature profile was determined by comparing with the calculated H_γ/He_γ line ratios. The parametric curves for the γ/γ line ratios are shown in Fig. 3 for a 100 μm tracer dot. The advantages in employing the γ line ratio are that the lines have small optical depths and are nearly equal in intensity and wavelength which minimizes experimental errors in spectral processing. The Li-like satellite lines are readable at the target surface but their components are unresolved with the 001 plane.

A portion of the spectrogram that recorded the spectrum from the (013) diffraction plane in KAP is shown in Fig. 4. The Li-like satellite groups jkl, a-d, q,r as classified by Gabriel are readable for the scan performed at the target surface. The q,r and jk lines were measured together with the He-like resonance line $1s^2$-$1s2p$ ^1P line to make temperature estimates. The temperature determined near the surface from these satellite lines had a range, 240 - 400 eV. This broad range reflects the insensitivity of the temperature-density curves predicted at near the critical densities for laser light. It can be noted that the He_α resonance line, ^1P, is opacity broadened and self-reversed at the target surface. Even though the ^3P line has low intensity at the surface, the line can be measured to a distance of 200 μm while in contrast, the Li-like satellite lines had measurable intensity only at the target surface suggesting that they are emitted from high plasma density regions.

The PET crystal data is shown in Fig. 5. The curved crystal was oriented in the spectrograph to record the He_α line and its satellites and the higher Rydberg lines. Close examination of the spectral lines revealed somewhat poor line quality for quantitative intensity measurements because of the appearance of diffraction nonuniformity from bending the PET crystal; nevertheless, line profiles could be recorded. All the lines were observed to be split into two components at the target surface. Lines near the H_β region were scanned. Fig. 6 shows the He_γ and He_δ lines splitting at the target surface while the H_β line is broad with a shoulder compared to the narrow line profiles recorded 100 μm from the target surface. The profile calculation for the He_γ line at high density (near critical) is shown in Fig. 7. The static profile calculations[22] assuming stationary ion plasma emission split the line into equal intensity wings. The measured line widths (FWHM) were 11.3 and 14.5 mÅ for He_γ and He_δ. The H_β had an opacity-broadened line width at the surface of 6.9 mÅ compared to a narrower line width of 4.5 mÅ at the distance of 100 μm from the target surface for the He_γ and He_δ lines. The calculated static opacity splitting for the He_γ line was less than the full width observed experimentally. At the 100 μm distance from the target the line width of the H_β line was 4.6 mÅ. The line width of the He_α ^1P was measured in the KAP (013) spectrum to have a value of 7.4 mÅ at the surface and a line width of 4.7 mÅ at distance of 125 microns. At the target surface, the line splitting and profile matched that calculated at an ion density of 2.5-4 10^{20} cm^{-3}.

Spectral data was collected with a curved KAP spectrograph for a 115 μm Al dot with 1.06 μm laser light. Line intensity ratios were measured from the KAP spectrogram in determining the plasma parameters. The high resolution KAP (013) spectral region showing the satellite structure and line profiles of the

He_α resonance and intercombination lines is presented as Fig. 8. The spectrum is similar to that acquired with green light except that the intensity ratio of the a-d and q,r satellites has changed. We find that the a-d satellite group is diminished in intensity.

A parameterized set of line ratios and line profiles were calculated for 50, 100, 115, and 200 μm diameter Al dots to determine the plasma parameters. The plasma densities were determined from computed values for the combined intercombination line and near lying satellite lines s and t together with the line profiles determined from the computed widths of the ^{1}P and He_γ lines. The plasma temperatures were determined from parent line ratios to distances away from the surface corresponding to densities less than 1/10 critical. At the target surface temperatures were obtained from the satellite line intensities. The plasma profiles for the 0.53 μm and 1.06 μm laser light for about the same diameter dot are shown in Figs. 9 and 10 respectively. The error bars indicate an uncertainty of a factor of 2 in determining the high density values while the temperature estimate based on the satellite line ratios have a large uncertainty because of insensitivity of the satellite-to-resonance line ratio at high densities. Within the uncertainty of the measurements, the plasma parameters are the same near the target surface. The plasma profiles are steeper with the 0.53 μm laser light.

Parameterized predictions were made for the He-and H-like ion abundances for the various size dot diameters. From the plasma temperature and density determinations, the following abundances were found for the 100 μm tracer dot. At the target surface at critical density, the plasma has a 50% He ion fraction and a 40% H ion fraction. At 1/10 critical density and at 600 eV temperature, the ratios are reversed. The He-like fractional abundance curve for a 100 μm diameter plasma is shown in Fig. 11.

IV. Discussion

The experiments and analyses detailed above and summarized in Figs. 9 and 10 point out that through intercomparison of different forms of spectral data, unique characterization of plasma properties through spectroscopy may be possible. The principal tools available are resonance line ratios, satellite line ratios, and line profiles. These data can be obtained for several tracer dot sizes simultaneously. This means that the effect of plasma opacity on the measured line ratios is present in the data. These ratios are a function of plasma size as well as temperature and density, because they are affected by optical depth. For one plasma size, a given fixed line ratio establishes a contour curve giving the allowed combinations of plasma density and temperature consistent with the ratio, such as displayed in Fig. 3. This contour is different for different tracer sizes, and in principle a unique temperature and density is established where the curves cross, assuming that the plasma temperature and density are the same for each tracer dot. The same contour crossing may be employed for different line ratios at the same tracer size, and use of the line profiles which are opacity broadened in conjunction with the above data together constitute a powerful tool for plasma diagnosis. Conceivably, such techniques may be useful when more fully refined for inferring fundamental atomic rates.

The measured plasma temperature and density profiles were compared with 1-D hydrodynamic calculations for the two laser conditions. Good agreement was found for the density profile in the 0.53 μm case between a distance of 15 to 100 µm. The theoretical profile became less steep at values greater than 100 µm while between 10 µm and the solid density for aluminum the predicted density value increased to nearly ten times critical density. There was agreement between the observed and predicted temperature profile in the 0.53 µm case. For the 1.06 µm illumination, the temperature predictions agreed at the end points, near the target surface and near the 1/10 critical density value but jumped to a high temperature of 850 eV at a distance of 100 µm, unlike what was observed. The predicted density profile agreed well with the measured plasma density profile. Within 10 µm of solid density, the predicted density was twice that determined but within the uncertainty of the measurements above the critical density. The density profiles for the two laser conditions can be characterized by the 1/10 critical density scalelengths of 100 and 250 microns for 0.53-µm and 1.06-µm laser light, respectively. For short scalelength plasma profiles, the ^3P + s,t satellites/^1P ratio for a density determination and the jkl satellites/^1P ratio as a temperature diagnostic were found to be unaffected by the computed temporal dependence of the axial plasma flow.[2]

Satellite line intensities identify the high density regions near the target surface. With the high-resolution obtainable near the Al He$_\alpha$, resonance and intercombination lines, distinct Li-like Al XI satellite intensities were measured and the satellite line patterns studied for a variety of target and laser conditions. The jkl satellite lines, which are formed predominantly by dielectric recombination-to-the integrated ^1P line, yield plasma temperatures for the high density plasma. A striking difference between the two laser light conditions was the spatial extent of the satellite lines. In the 0.53 µm experiment, the satellite intensities existed only at the target surface both for the tracer dots on thick plastic and an spectrum obtained from the tip of a 125-μm diameter Aℓ wire.

The satellite line intensity ratios were noted for the variety of experimental conditions. The line ratios q,r/jkl were found to be a nearly constant 0.30-0.35 for most of the spectral data. This is consistent with electron density regime for laser-produced plasma, namely, 10^{19}-10^{22} el/cm^3.[18] The most noticeable feature of the satellite line intensities was the variation in the a-d line complex relative to the other satellite lines particularly the q,r satellites. The a-d/q,r ratio had a value of 1.65 for the 50 μm tracer dot and a value slightly less than unity (0.95) for the 100 μm dot with the 0.53 μm laser corresponding to density estimates of $8 \bullet 10^{20}$ cm^{-3} and $4 \bullet 10^{20}$ cm^{-3}, respectively, based on the theoretical prediction as shown in Fig. 12. Also, the satellite line ratio had a value of unity for the spectrum from the 125 μm wire. The a-d/q,r line ratio decreased to a value of 0.5 for the 1.06 μm data laser excited spectra, and it was found to be independent of dot diameter.

It is known that the a-d can be formed by either dielectronic recombination or by direct inner shell excitation processes while the jkl satellites are largely formed by dielectronic recombination and are independent of density as also are the q,r satellites. The excitation rates for a-d, both by dielectronic recombination and direct excitation from the ground state vary by almost two orders of magnitude in the electron density regime of 10^{20} to 10^{22} and at a plasma temperature of 300 eV.[16] Further work is indicated that would correlate the differences in the observed a-d/q,r line ratio with differences in laser-matter interactions.

V. Conclusions

The improved spectroscopic techniques and extended theoretical calculation of line ratios and line profiles have been used to interpret the plasma temperature and density profiles near the target surface. The various line ratios were found to yield consistent plasma temperatures and densities for the tracer dot targets. The plasma profiles were more steep for the shorter wavelength laser light. The satellite groups a-d and q,r form line ratios that appear sensitive to density changes and could possible be used to indicate changes in plasma opacities.

Acknowledgements

This work was supported by the U.S. Defense Nuclear Agency. We wish to express appreciation to Dan Newman for computer processing of the spectral traces and to Bob Clark for calculating a plasma profiles with a 1-D hydrocode. The authors wish to acknowledge useful discussions with Mark Herbst, Dave Nagel and Jack Davis at NRL. One of the authors (P.G.B.) has sincere appreciation for the cooperation of the scientists at KMS Fusion, especially George Charatis and Paul Rockett and for the assistance in data collection of John Brundage, David Sullivan, and Jeff Steigman. Also P.G.B. acknowledges the expertise in target fabrication of John Kosakowski and Martin Peckerar's group at the Naval Research Laboratory. The Pharos laser system was expertly operated by Nick Nocerino under the direction of Steve Obenschain.

References

1. M.J. Herbst, P.G. Burkhalter, R.R. Whitlock, J. Grun and M. Fink, Rev. Sci. Instrum. $\underline{53}$, 1418 (1982).

2. P.G. Burkhalter, M.J. Herbst, D. Duston, J. Gardner, M. Emery, R.R. Whitlock, J. Grun, J. Apruzese and J. Davis, Phys. Fluids $\underline{26}$, 3650 (1983).

3. J.G. Gauthier, J.P. Geindre, K. Najmabadi, C. Popovics, A. Poquerusse and M. Weinfeld, J. Phys. D. $\underline{16}$, 1979 (1983).

4. P. Alaterre, C. Popovics, J.P. Geindre and J.C. Gauthier, Opt. Commun. $\underline{49}$, 140 (1984).

5. P.G. Burkhalter, D.A. Newman, C.J. Hailey, P.D. Rockett, G. Charatis, B.J. MacGowan and D.L. Matthews, Opt. Soc. Am. B $\underline{2}$, 1894 (1985).

6. D. Duston, J.E. Rogerson, J. Davis and M. Blaha, Phys. Rev. $\underline{28}$, 2968 (1983).

7. J.P. Apruzese and J. Davis, Phys. Rev. A $\underline{31}$, 2976 (1985).

8. J.H. Gardner, M.J. Herbst, F.C. Young, J.A. Stamper, S.P. Obenschain, C.K. Manka, K.J. Kearney, J. Grun, D. Duston and P.G. Burkhalter, Phys. Fluids. $\underline{29}$, 1305 (1986).

9. P.G. Burkhalter, D.J. Nagel, E. Emery, P.D. Rockett and G. Charatis, First International Laser Sci. Conf. "Techniques for Soft X-Ray Spectroscopy," Dallas Texas, Nov. 1985.

10. P.G. Burkhalter, D.B. Brown and M. Gersten, J. Appl. Phys, $\underline{52}$, 4379 (1981).

11. A.H. Gabriel and T.M. Paget, J. Phys. B $\underline{5}$, 673 (1972).

12. A.H. Gabriel, Mon. Not. R Astr. Soc. $\underline{160}$, 99 (1972).

13. C.P. Blaha, A.H. Gabriel and L.P. Presnyakov, Mon. Not. R Astr. Soc, $\underline{172}$, 359 (1975).

14. U. Feldman, G.A. Doschek, D.J. Nagel, R.D. Cowan, R.R. Whitlock, Astrophys. J. $\underline{192}$, 213 (1974).

15. N.J. Peacock, M.G. Hobby and M. Galanti, J. Phys. B. $\underline{6}$ L298 (1973).

16. V.L. Jacobs and M. Blaha, Phys. Fev. A $\underline{21}$, 525 (1980).

17. D. Duston, J. E. Rogerson, J. Davis and M. Blaha, Phys. Rev. $\underline{A28}$, 2968 (1983).

18. J.D. Kilkenny, R.W. Lee, B.L. Whitten, B.J. MacGowan and D.K. Bradley, "The Density Dependence of Dielectric Satellites of Helium-Like Transitions," Proc. of 2nd International Conference on Radiative Properties of Hot Dense Matter, Ed. J. Davis, World Scientific Publishing Co., Singapore, p. 451 (1985).

19. P.D. Rockett, C.R. Bird, C.J. Hailey, D. Sullivan, K.B. Brown and P.G. Burkhalter, Appl. Optics $\underline{24}$, 2536 (1985).

20. XTLFILM program, unpublished, R.R. Whitlock, J.W. Criss, B. Sweeney and D.A. Newman, Code 4680 NRL.

21. R.D. Cowan, J. Opt. Soc, Am. $\underline{58}$, 808 (1958).

22. J.P.Apruzese, J. Quant. Spectrosc, Radiat. Transfer. $\underline{33}$, 71(1985).

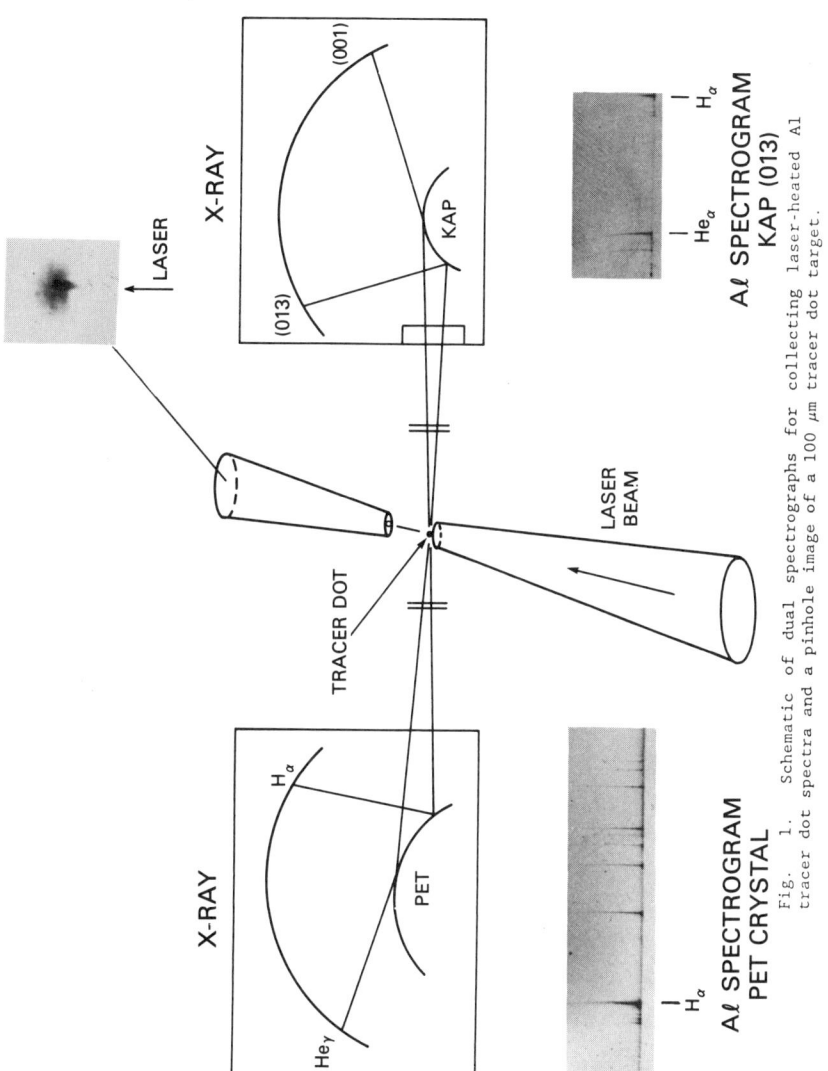

Fig. 1. Schematic of dual spectrographs for collecting laser-heated Al tracer dot spectra and a pinhole image of a 100 μm tracer dot target.

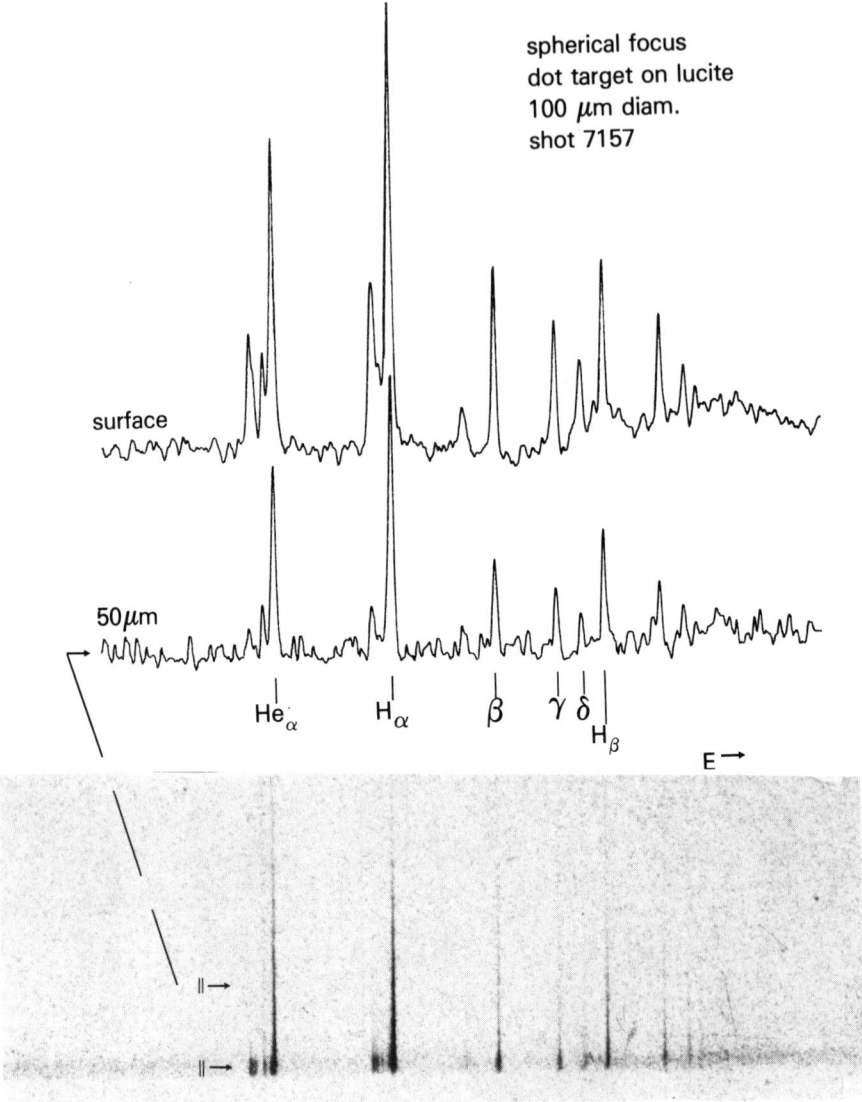

Fig. 2. Intensity traces for a spatially-resolved Al spectrum for a 100 μm diameter tracer dot scanned at the target surface and at 50 μm from the surface as indicated in the portion of spectrogram. The data was collected with the KAP (001) crystal for a laser shot with 0.53 μm light at 10^{13} W/cm^2 irradiance.

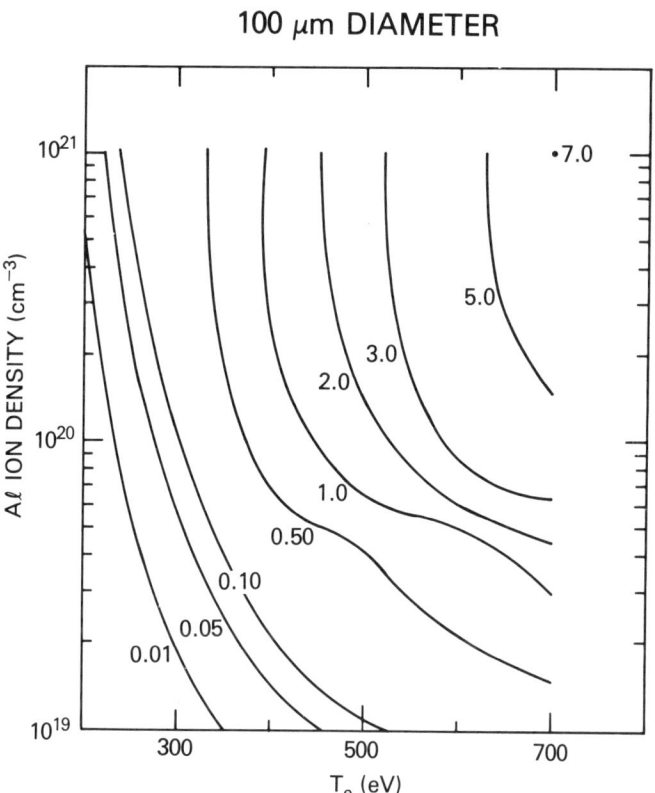

Fig. 3. Calculated H_γ/He_γ line ratio as a function of plasma temperature and density.

Fig. 4. Intensity trace at the target surface from the KAP (013) portion of the spectrograph for the same 100 μm diameter target.

spherical focus
dot target on lucite:100 μm
shot 7157
KAP (013)

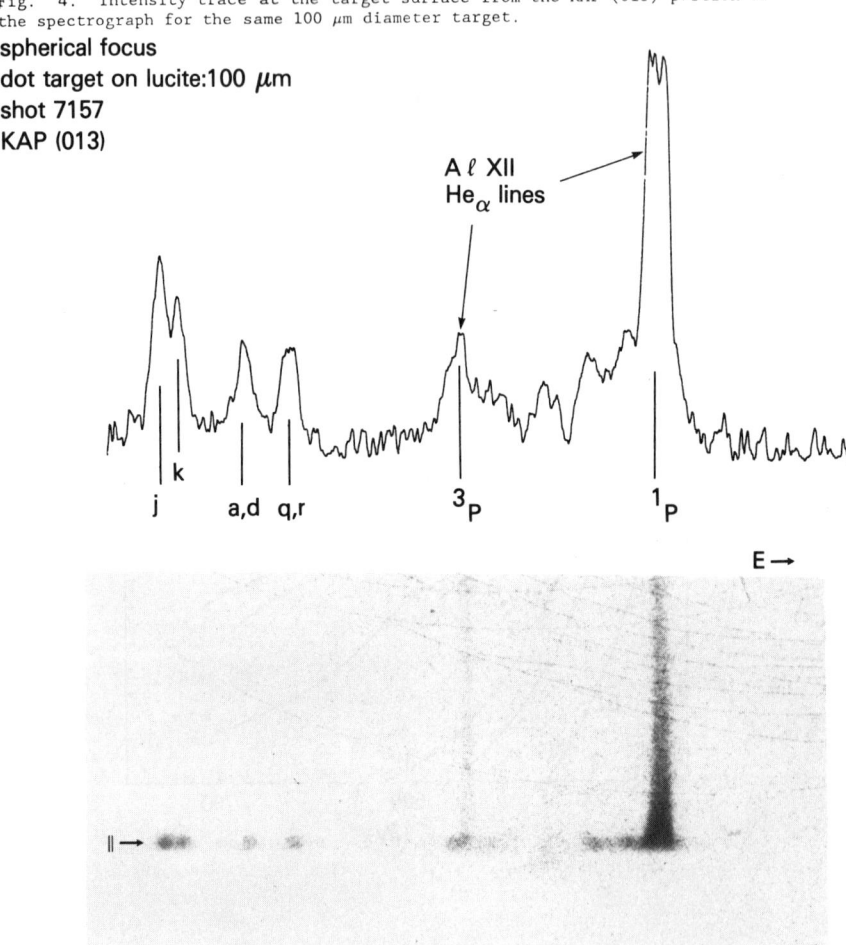

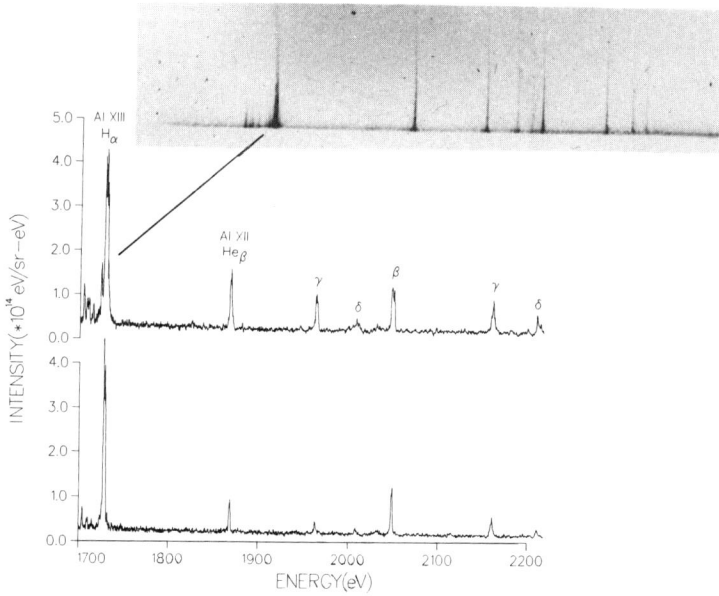

Fig. 5. Spectral traces for the same Al shot collected with a curved PET crystal and scanned at the surface and at 50 μm from the surface.

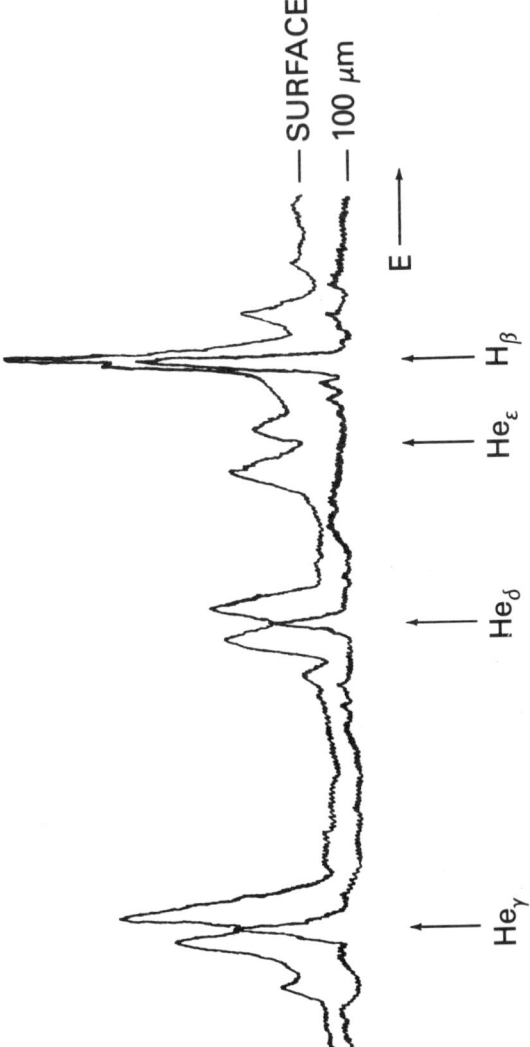

Fig. 6. Line profiles for a selected spectral region near 2 keV of the Al spectrum collected with the PET crystal with scans at the target surface and 100 μm in front.

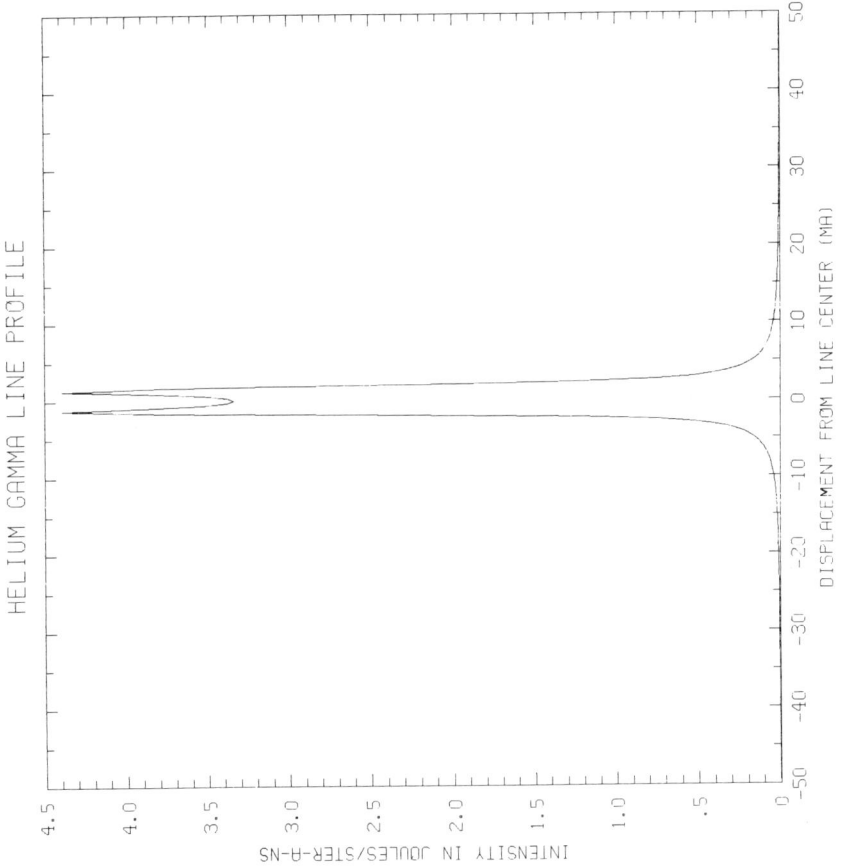

Fig. 7. The calculated self-reversed line profile of the Al He$_\gamma$ line at an ion density of $3 \cdot 10^{20}$ cm^{-3}.

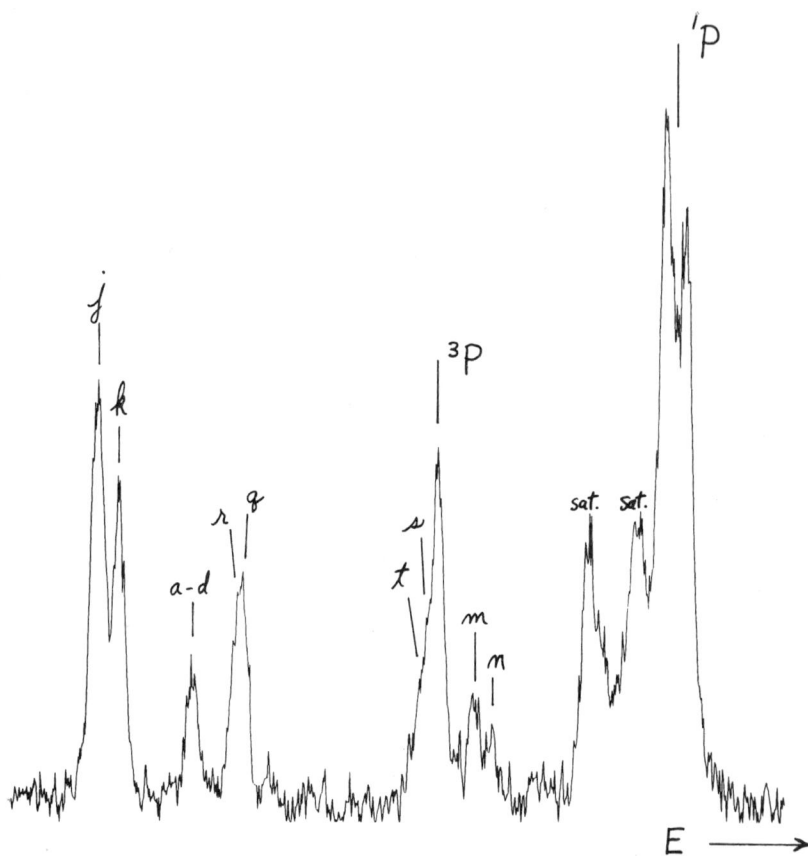

Fig. 8. Intensity trace at the target surface for a 115 μm diameter Al tracer dot irradiated with 1.06 μm laser light at 10^{14} W/cm^2. The spectrum was recorded with the KAP (013) crystal spectrograph.

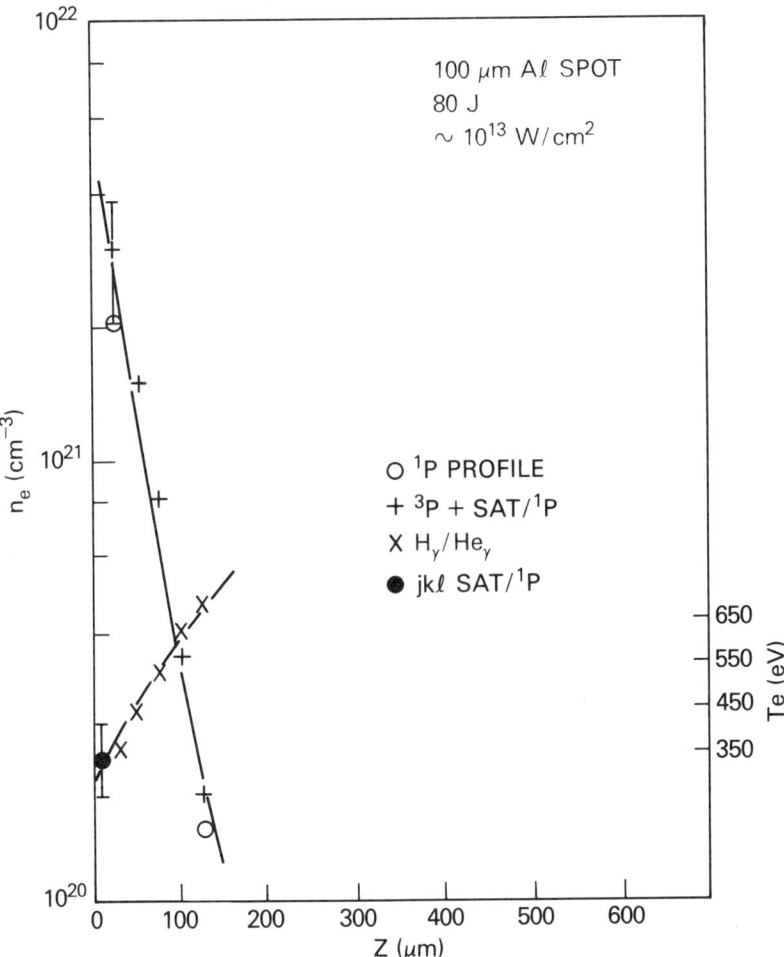

Fig. 9. Plasma temperature and density profiles for a 100 μm diameter Al dot irradiated with 0.53 μm laser light.

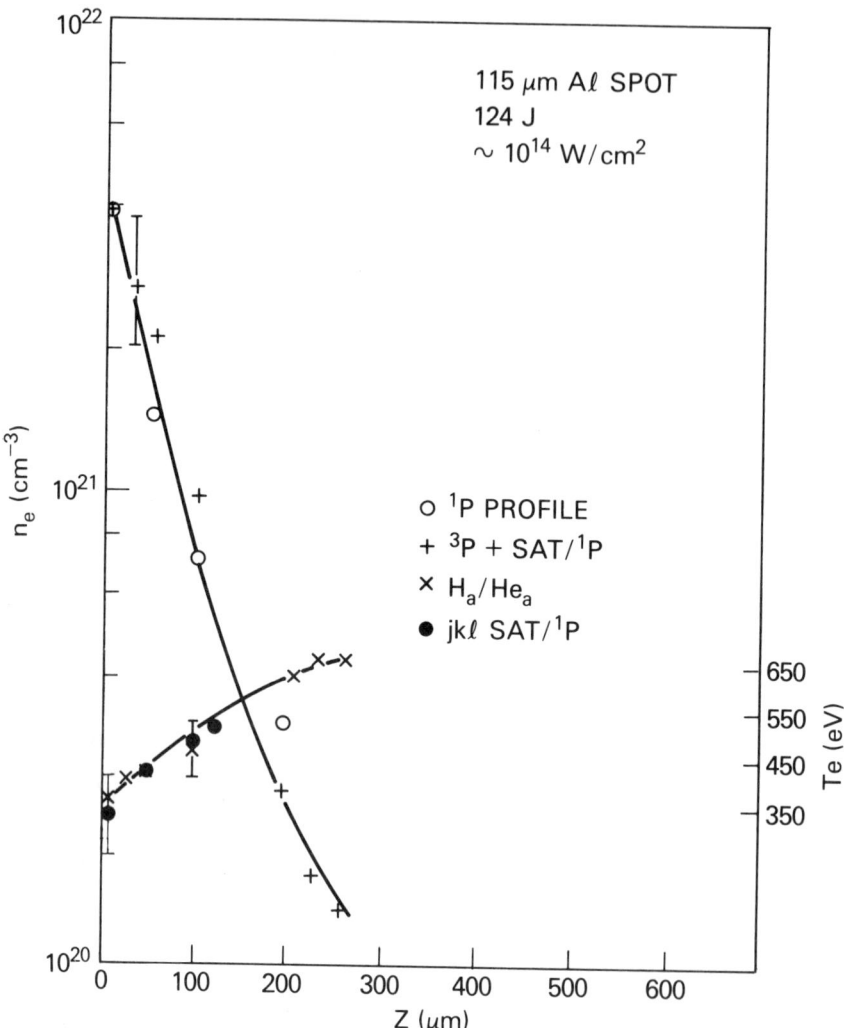

Fig. 10. Plasma temperature and density profiles for a 115 μm diameter Al dot irradiated with 1.06 μm laser light.

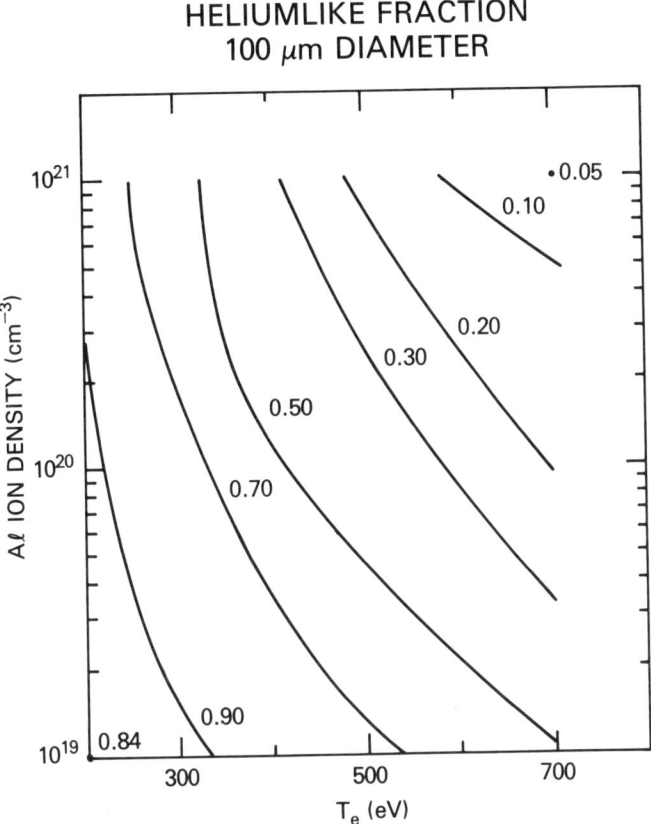

Fig.11. Fractional abundance curves for He-like Al XII (100 μm tracer diameter) as a function of plasma temperature and density.

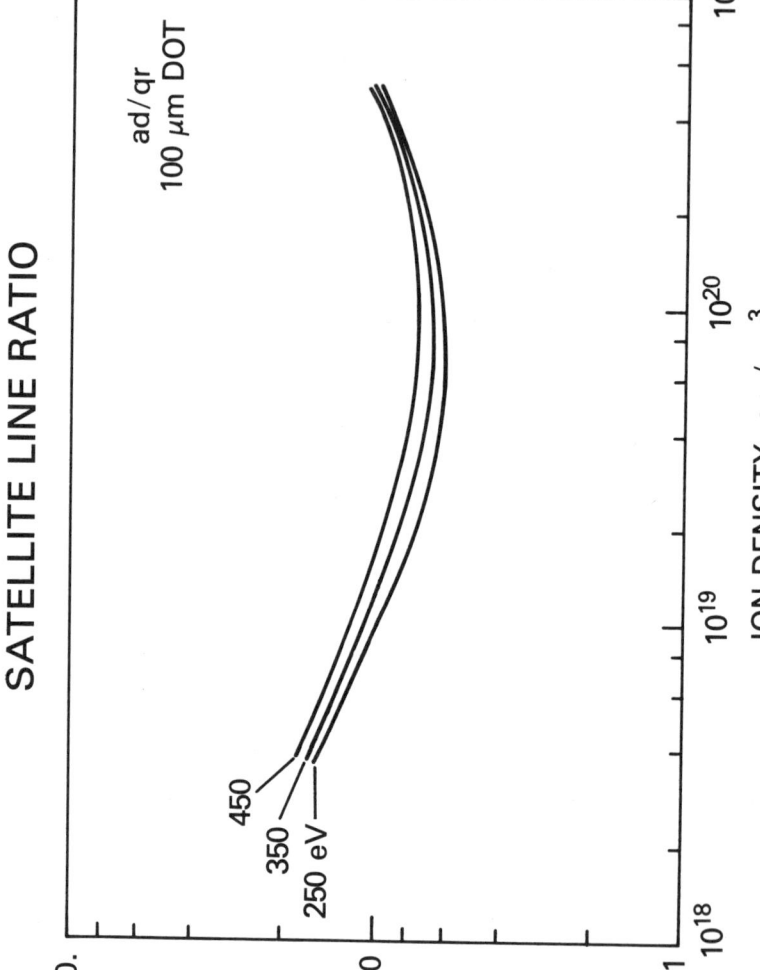

Fig. 12. Satellite line ratios (a-d/q,r) as a function of plasma temperature and density.

Argon Puff Gas Soft X-ray Laser

J. DAVIS, J. P. APRUZESE, C. AGRITELLIS[*] and P. KEPPLE

Plasma Physics Divsion
Plasma Radiation Branch
Naval Research Laboratory
Washington, D. C. 20375

Abstract

A non-LTE dynamic argon pinch model has been applied to investigate the feasibility of creating conditions favorable to gain coefficients in excess of unity during the implosion of a cylindrical annular argon gas puff plasma. Preliminary results obtained from the SIMPLODE model suggest that it is possible to achieve measureable gain in the 3p-3s lasing transition at 434 Å in neonlike argon.

I. Introduction

The successful demonstration of soft x-ray lasers[1-5] using as the lasing medium the plasma generated from high power laser-target interaction has provided considerable impetus to a number of alternative techniques for producing an x-ray laser. One such scenario involves the application of pulse power drivers to implode gas puff plasmas. Of the many variations in the target design there are three basic concepts currently under investigation. They are: (1) a single cylindrical annular gas puff, (2) a single cylindrical annular gas puff imploding on a central core plasma, and (3) similar to (1) and (2) except that the imploded plasmas are not the lasing medium but a source of intense x-ray emission for photopumping a target material. Scenarios (2) and (3) are discussed by the authors elsewhere and by T. Hussey in these proceedings. Since the single gas puff plasma is central to all three scenarios, it is essential that we understand fully its radiative properties and behavior. Therefore, the focus of this investigation will be confined to the dynamics of a single gas puff plasma. The intent here is to concentrate on the radiation kinetics while characterizing the implosion dynamics of the gas puff plasma by means of the simplest plausible description, i.e., a dynamic pinch model. Obviously, such a model ignores much of the observed pathological behavior of these plasmas, but it does provide a starting point for determining whether it is at all feasible to produce temperature and density conditions favorable to x-ray lasing. The influence of the development and growth of plasma instabilities and their impact on the uniformity and homogeneity of the evolving plasma is a critical issue and is currently under investigation. Preliminary results based on 2-D Magneto-Radiation-Hydrodynamic simulations and experiments,[7] employing plasma erosion switches, suggest a significant reduction in the role of instabilities and provides encouragement for further pursuit of this design.

The treatment presented here is based on a simple model, hereafter referred to as SIMPLODE, with emphasis on the atomic kinetics and radiation dynamics self-consistently coupled to a dynamic pinch model in order to obtain information on the plasma environment as it evolves under the influence of the current discharge. This provides us with a self-consistent picture of temperature, density, size, and level population during the plasma's evolution. The influence of radiation transport on the level inversion and gain calculations is taken into account using Sobolev's method.[8] In either a stationary medium or one moving without velocity gradients, the final emission of optically thick line photons or continuum from the medium generally occurs from unit optical depth from the boundary. Photons originating deep in the interior of the medium undergo successive scatterings on their flight to the surface before escaping. In this instance the emitted radiation reflects nonlocal conditions. On the other hand, if the medium has a velocity gradient then radiation originating deep in the interior of the plasma can escape the medium directly because of the Doppler effect and this radiation reflects the local interior conditions rather than conditions at other points in the medium and, in particular, the boundary. This is the essence of Sobolev's approximation. A more thorough discussion of the effect of radiation transport on the gain of the lasing transitions is presented by J. Apruzese in these proceedings. Also, since the experiments on the GAMBLE II facility at NRL are being done with an argon gas puff plasma, the theory and analysis presented here is for argon.

So far the discussion has been general focusing on the plasma and its properties. However, in particular, we will investigate the feasibility of creating a population inversion in the 3p levels of neonlike argon due to electron impact collisional excitation from the $2p^6$ ground state and estimate the gain coefficients in the n=3, Δn=0 lasing transitions.

II. Physical Model

The simplest description of an imploding Z-pinch plasma is probably the Bennett Pinch equilibrium model. This model is based on an equilibrium balance between the fluid and magnetic pressures in combination with an equilibrium balance between the sources and sinks of energy. The classical Bennet pinch ignores radiation and hence ignores excitation/ionization energy (chemical potential) and energy lost by radiation (radiation cooling). An obvious extension to the Bennett pinch equilibrium model is the inclusion of the flow parameters describing the temporal evolution of the plasma. That is, maintain the simple philosophy of the Bennett pinch but allow the plasma to radiate and evolve in time. In essence, this is our philosophy - a radiating dynamic Bennett pinch; SIMPLODE. Also, like Shearer,[9] who included a radiation cooling term in the form of Bremsstrahlung losses from a pure hydrogen plasma, we include radiation cooling but in a much more extensive fashion. The radiation cooling term in our model includes contributions from free-free, free-bound, and bound-bound transitions and is determined from a non-LTE collisional-radiative model of the level dynamics.[10]

The SIMPLODE model describes the radial implosion of a cylindrical annular gas puff plasma of uniform density carrying a uniform current in the axial direction. Only radial motion is considered, i.e., there is no axial structure and the plasma is always uniform in this direction. The radial motion is determined from the force equation, vis.

$$m \frac{d^2 r}{dt^2} = (P - \frac{I^2}{2\pi r^2 c^2}) A \qquad (1)$$

where r is the radial distance measured from the symmetry axis, P is the fluid pressure, A is the area over which the force is exerted, i.e., $2\pi r \ell$, and $I^2/2\pi r^2 c^2$ is the magnetic pressure, i.e., $B^2/8\pi$. The thermal energy, $E_{th} = 3/2 (1+Z) n_i kT + \Sigma \mu_p$, varies in time as

$$\frac{dE_{th}}{dt} = -\frac{P}{N}(\frac{\dot{V}}{V}) + \frac{zI^2}{A_{curr.}} \ell - P_{rad} V \qquad (2)$$

where $\Sigma \mu_p$ is the sum of ionization energies and is loosely referred to as chemical potential. $A_{curr} = \pi(R_B^2 - R_A)^2$ where $R_B(R_A)$ represents the outer (inner) radius of the annular plasma, $\dot{V} = 2\pi \ell r \dot{r}$, $N = (m/m_i)/\pi r^2 \ell$ where m is the plasma mass, m_i is the atomic mass, $n_i = m/m_i$, and the thermal pressure is $N(1+Z)kT$. Finally, ℓ is the length of plasma, Z is the charge state, z is the classical resistivity, and P_{rad} is the power radiated per unit volume V. The remaining sysmbols have their usual meaning. The first term on the RHS of Eq. (2) represents the work done in compressing the plasma, the second expression is the joule heating source term, and the third term is the power radiated, i.e., radiative cooling.

The collisional-radiative model describing the level dynamics contains an extensive number of levels in the K- and L-shells with particular emphasis on the neonlike ionization stage which contains 42 excited states in j-j representation and 385 lines. The line profile functions are represented by Voigt functions and include natural and Doppler broadening. The transport of radiation employs the Sobolev escape model.

The "x-ray" lasing scheme considered here takes advantage of the large monopole excitation rate from the ground state to the excited 3p state of the neonlike ionization stage. Although we are focusing on a collisional excitation scheme, all other scenarios, such as recombination lasers, can be investigated with this model. However, by controlling the implosion we hope to avoid burning through the neonlike stage and instead create conditions for producing a stable abundance of neonlike argon. A simplified energy level diagram for neonlike argon is shown in Fig. 1. A population inversion can occur in the 3d and/or 3p levels leading to gain and lasing in the 3d-3p and 3p-3s transitions. This is indicated specifically for the 3p-3s transition with the label L(434Å).

III. Results and Discussion

Preliminary estimates for the neonlike fractional abundance and gain coefficient in a stationary environment were obtained from the collisional-radiative model for prescribed values of temperature, density, velocity, and size. These estimates provide a measure of the parameter space over which gain can be expected as well as providing guidance for the SIMPLODE simulations. In Fig. 2 the neonlike ground state fractional abundance is presented as a function of temperature for several ion densities in the absence of opacity effects, i.e., the optically thin case. For temperatures from about 30 to 70 eV, for densities typical of imploding gas puff plasmas, a significant amount of neonlike argon prevails. The gain coefficient for the 3p-3s transition is shown in Fig. 3, for an optically thin plasma. The gain for a Doppler broadened line is given by

$$\alpha = 10^{-16} f_{osc.} \lambda(A)(M/T)^{1/2} (N_2 - \frac{g_1}{g_2} N_1) \text{ cm}^{-1} . \qquad (3)$$

where $f_{osc.}$ is the absorption oscillator strength of the line, λ is the wavelength in angstroms, M is the atomic mass of the radiating ion, T is the temperature of the plasma in eV, N_2 (N_1) is the upper (lower) level population, and g_i is the statistical weight of level i. The gain coefficient for a Doppler broadened line is directly proportional to the wavelength of the lasing line and the difference between the upper and lower level population densities, and inversely proportional to the square root of the temperature. For an optically thin plasma, gain coefficients greater than unity exist for all three densities over a broad temperture range with a peak gain coefficient of about 30 cm^{-1} at 60 eV for an ion density of $1.5 \times 10^{19} \text{cm}^{-3}$. For a fixed density and increasing temperature the number of neonlike ions decreases due to increased ionization causing burnthrough; for decreasing temperature the plasma becomes too cold to support the existence of neonlike ions, and the collisional excitation rates of the 3p neonlike levels drop drastically. Similarly, for a fixed temperature and increasing density the fractional abundance of neonlike ions decreases due to the increase in the ionization rate causing burnthrough; another way of viewing this situation is to note that as the density increases for a fixed temperature the ionization balance tends toward LTE which causes a given ionization state to appear at a lower temperature than a plasma in collisional-radiative equilibrium.

In Figs. 4 and 5 we have presented results for a optically thick plasma with a diameter of 0.9mm and imploding with a peak velocity of 2×10^7 cm/sec. (These parameters are typical of argon implosions on the GAMBLE II facility.) In this case the neonlike ion fraction starts falling off more rapidly at lower temperature than in the optically thin case. The differences are greatest for the higher densities due to the combined effects of collisions and the radiation field.

The combined effects of velocity and opacity can but be understood by first considering them separately and then as a composite. In Fig. 6 the results of several calculations are presented for varied conditions while maintaining the total ion density fixed at 5×10^{17} cm^{-3}. The optically thin result is included for reference and comparison purposes. The influence of opacity on the production of neonlike ions is shown on the v=0, diameter = 0.9mm curve. In comparison with the optically thin case, it is seen that an increase in plasma size manifests itself with an increase in opacity which maintains the radiation field in the plasma thereby making it easier to achieve a given degree of ionization, in this instance the neonlike stage, at a lower electron temperature than the optically thin case. Therefore, for a fixed electron temperature opacity effects will enhance the plasmas' degree of ionization above that achieved in a purely collisional optically thin plasma. Hence, the rapid decrease in neonlike ions when the plasma is opaque. With increasing temperature collisional ionization will reduce the number of neonlike ions in both cases. The influence of velocity on the neonlike ion fraction can be understood in the following way. In plasmas where the directed motional velocity exceeds the thermal velocity the escape probability is enhanced. In essence, when the line quanta impinge on a region of plasma where the local velocity has shifted the apparent frequency of the line away from the line center, where absorption is greatest, the quanta are able to escape the entire plasma. Therefore, the effects of velocity tend to reduce the effective opacity, associated with a stationary plasma, making the plasma effectively thin. Even though the velocity tends to mitigate the effects of opacity, it does not entirely remove opacity in the regimes considered, but the neonlike fraction does increase in the direction of the optically thin result. Finally, to explore the effects of mixing elements we have included the results of calculations where 10^{19} neon ions per cm^3 are mixed with 5×10^{17} argon ions per cm^3. These results are shown also in Fig. 6 with the consequence of further reducing the number of neonlike argon ions due to increased collisional effects induced by the higher density.

The gain calculations for the conditions depicted in Fig. 6 are shown in Fig. 7. Note the inclusion of an additional gain curve for a velocity of 1×10^7 cm/sec and a diameter of d=0 .9mm. For the conditions described in Fig. 7, gain coefficients can be achieved in excess of unity in the 3p-3s lasing transition at 434Å for peak implosion velocities in excess of 1×10^7 cm/sec. It is clear from these single plasma calculations that for prescribed values of temperature, density, velocity an size, that are representative of gas puff implosions, it should be possible to create a population inversion and achieve gain coefficients in excess of unity.

We will now investigate whether these conclusions will prevail in a dynamic environment such as that generated by the SIMPLODE model. For illustrative purposes, we present the results of calculations for a reference case of a 4 cm long, 35 µgm/cm argon gas puff plasma distributed uniformly between the outer and inner radius of 1.55 and 0.95 cm, respectively. The driving current waveform typical of the GAMBLE II generator, with the Plasma Erosion Opening Switch, is shown in Fig. 8. Peak current of 8.75×10^5 amps is attained in about 50 nsec, decaying to about 6×10^5 amps in another 70 nsec and then falling precipitously to zero in 100 nsec. The temporal behavior of the radiis is shown in Fig. 9. The inner radius collapses and stagnates on axis while the outer radius continues inward until the back pressure is sufficient to impede the forward motion and bounces outward. The final pinch radius is about 1/10 the initial radius and is in good agreement with the bulk of experimental data accumulated over the years from a variety of generators and plasma loads. The variation of velocity as a function of time, shown in Fig. 10, reaches a peak value of 1.1×10^7 cm/sec in about 160 nsec which is well after peak current. This behavior is characteristic of the GAMBLE II generator and is especially true of driving currents with sharp risetimes. The implosion phenomenology suggests that the plasma heats up and percolates for a time and then eventually coasts inward. The temporal variations of temperature and density are shown in Figs. 11 and 12, respectively. The ion density peaks at the pinch and reaches a value of 3×10^{18} ions/cm^3 while the peak temperature occurs some 15 nsec earlier and reaches a value of about 165 ev. The total radiative yield which essentially comes from the L-shell, is roughly 1 kilojoule and exhibits a pulse duration of about 25 nsec as shown in Fig. 13. This result is in reasonably good agreement with the experimental observations from GAMBLE II.[7]

A sample of the gain coefficient for the 434 Å line is shown as a function of time in Fig. 14 for the illustrative case represented in Figs. 8-12. The gain coefficient was greater than unity for a long time reaching a peak value of about 4 cm^{-1} late in the implosion. The values obtained for the gain coefficient are probably reasonable around peak compression but become less reliable after the bounce because of the lack of an adequate physics description of this late stage. However, it is encouraging that conditions prevail for producing measurable gains over a 4 cm length of plasma assuming, of course, stability of the column. Work is currently in progress using more sophisticated models to assess the validity of our findings here.

Finally, a series of simulations were performed to determine the plasma parameters at maximum gain as a function of M/ℓ for fixed $\Delta r=0.60$ corresponding to $R_B=1.55$ cm and $R_A=0.95$ cm. The ion density and temperature, at maximum gain, are shown as a function of M/ℓ in Figs. 15 and 16, respectively. The gain coefficient as a function of M/ℓ is shown in Fig. 17. For $\Delta r=0.60$, the gain coefficient peaks at $M/\ell=60\mu gm/cm$ and has a value of about 4 cm^{-1}.

Summary

It has been theoretically demonstrated that it is possible to create conditions favorable to population inversion and gain coefficients in excess of unity for a variety of conditions by imploding a cylindrical annular argon gas puff plasma. Gain coefficients of 4 cm^{-1} have been calculated for the 3p-3s lasing transition in neonlike argon at 434 Å.

Acknowledgments

This work was supported in part by the Strategic Defense Initiative Organization through the Defense Nuclear Agency and by the Office of Naval Research. We would like to thank Drs. F. Young and S. Stephanakis for making the experimental results available prior to publication.

Figure Captions

Fig. 1   Abbreviated energy level diagram for neonlike argon.

Fig. 2   Neonlike ion fraction as a function of temperature. Plasma assumed optically thin.

Fig. 3   Gain coefficient as a function of temperature for the 3p-3s transition at 434 Å. Plasma is assumed optically thin.

Fig. 4   Neonlike ion fraction as a function of temperature - Opaque case.

Fig. 5   Gain coefficient as a function of temperature - Opaque case.

Fig. 6   Neonlike ion fraction as a function of temperature - Mixed velocity and opacity case.

Fig. 7   Gain coefficient as a function of temperature - Mixed velocity and opacity case.

Fig. 8   Current waveform as a function of time.

Fig. 9   Inner and outer shell radius as a function of time.

Fig. 10 Implosion velocity as a function of time.

Fig. 11   Temperature as a function of time.

Fig. 12   Ion density as a function of time.

Fig. 13   Total radiative yield (joules) as a function of time.

Fig. 14   Gain coefficient at 434 Å as a function of time.

Fig. 15   Ion density at maximum gain as a function of mass per unit length.

Fig. 16   Temperature at maximum gain as a function of mass per unit length.

Fig. 17   Gain coefficient at 434 Å as a function of mass per unit length.

References

* Permanent address Science Applications International Inc., McLean, Va.

1. D. Jacoby, G. J. Pert, L. D. Shorrock, and G. L. Tallents, J. Phys. B 15, 3557 (1982).

2. M. D. Rosen, et al., Phys. Rev. Lett. 54, 106 (1985).

3. D. L. Matthews, et al., Phys. Rev. Lett. 54, 110 (1985).

4. S. Suckewer, et al., Phys. Rev. Lett. 55, 1753 (1985).

5. J. F. Seely, et al., Optics Commun. 54, 289 (1985).

6. J. P. Apruzese and J. Davis, Phys. Rev. A 31, 2976 (1985).

7. F. C. Young, S. J. Stephanakis, et al., Bull. APS 30, 1389 (1985).

8. V. V. Sobolev, Sov. Astron. 1, 678 (1957).

9. J. Shearer, Phys. Fluids 19, 1426 (1976).

10. J. Davis and K. G. Whitney, JAP 47, 1426 (1975) and D. Duston and J. Davis, Phys. Rev. A 21, 1664 (1980).

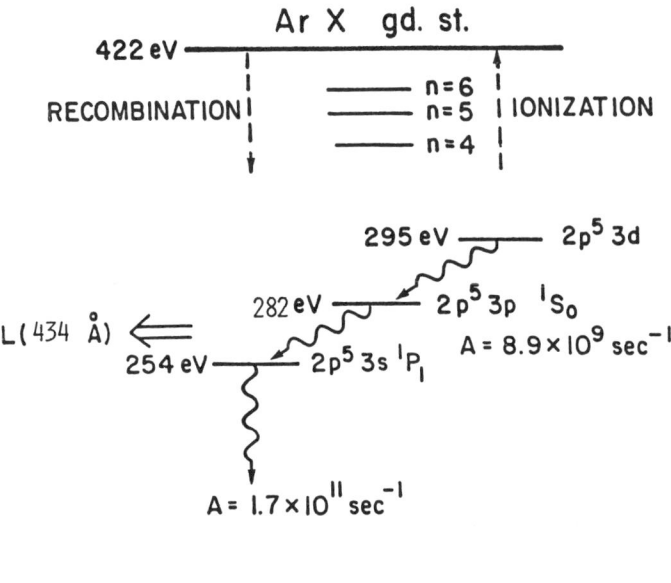

Figure 1

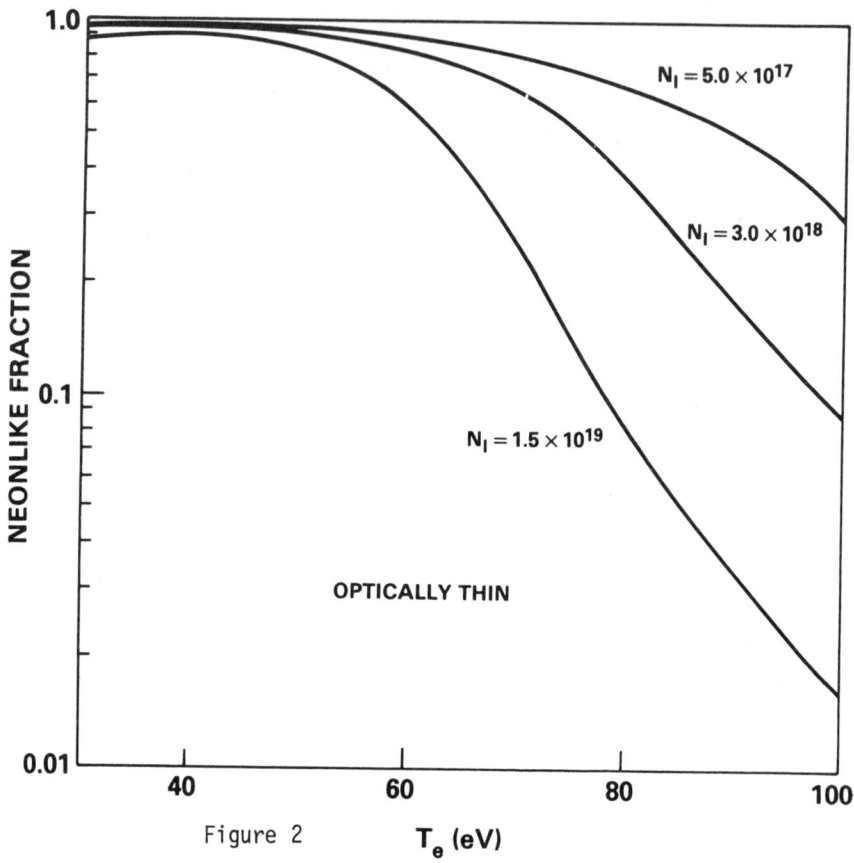

Figure 2

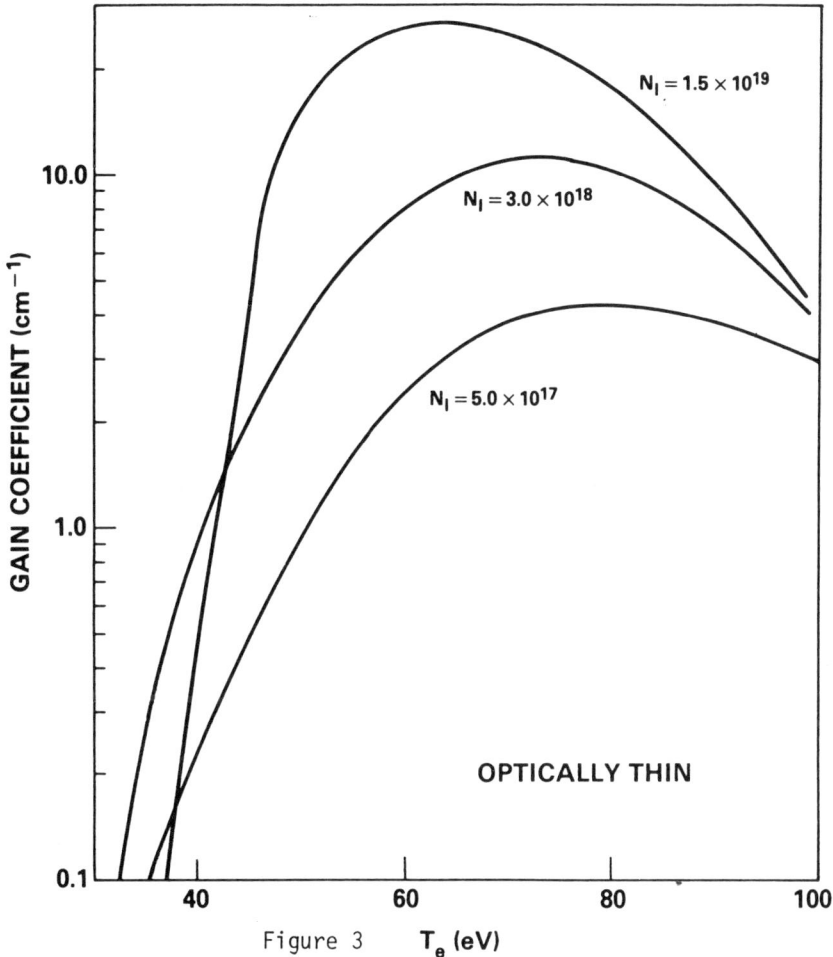
Figure 3

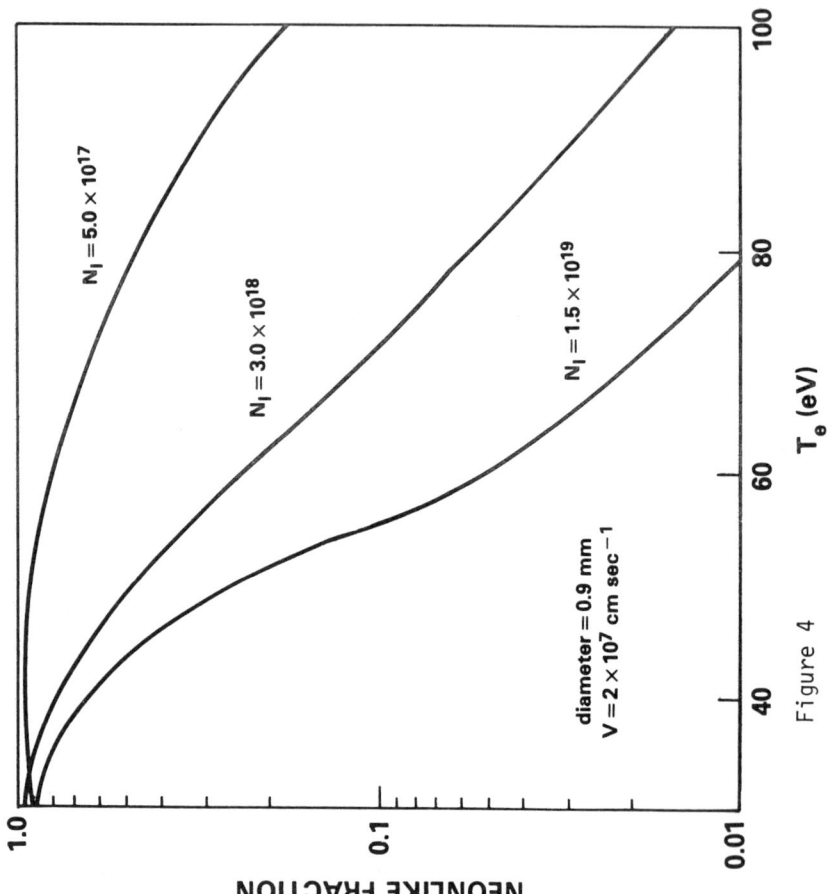

Figure 4

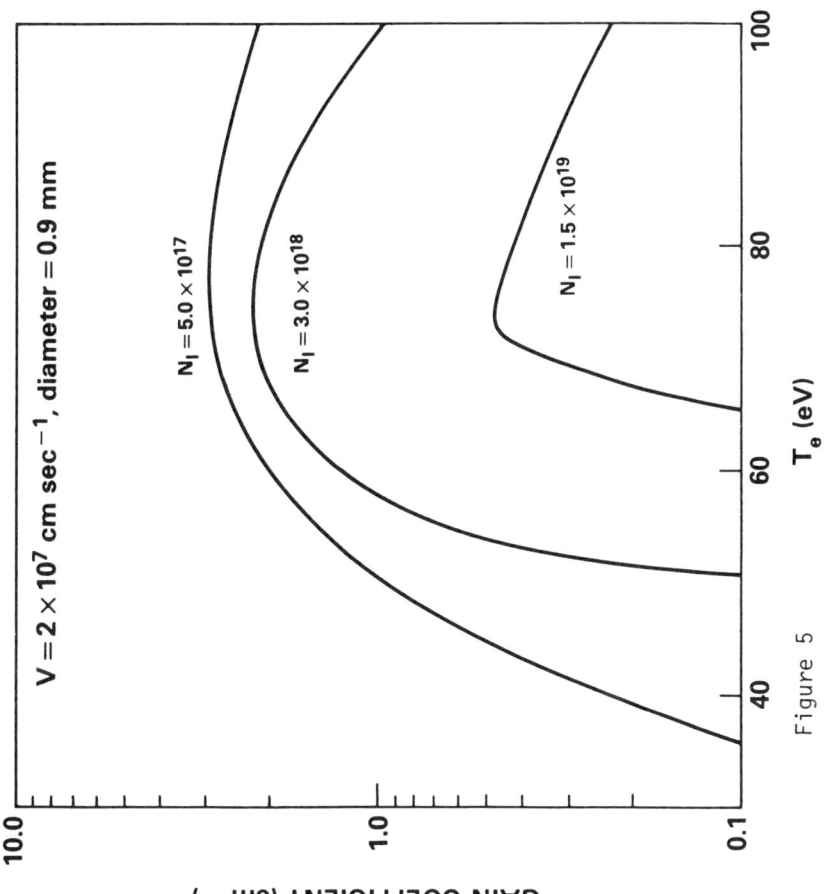

Figure 5

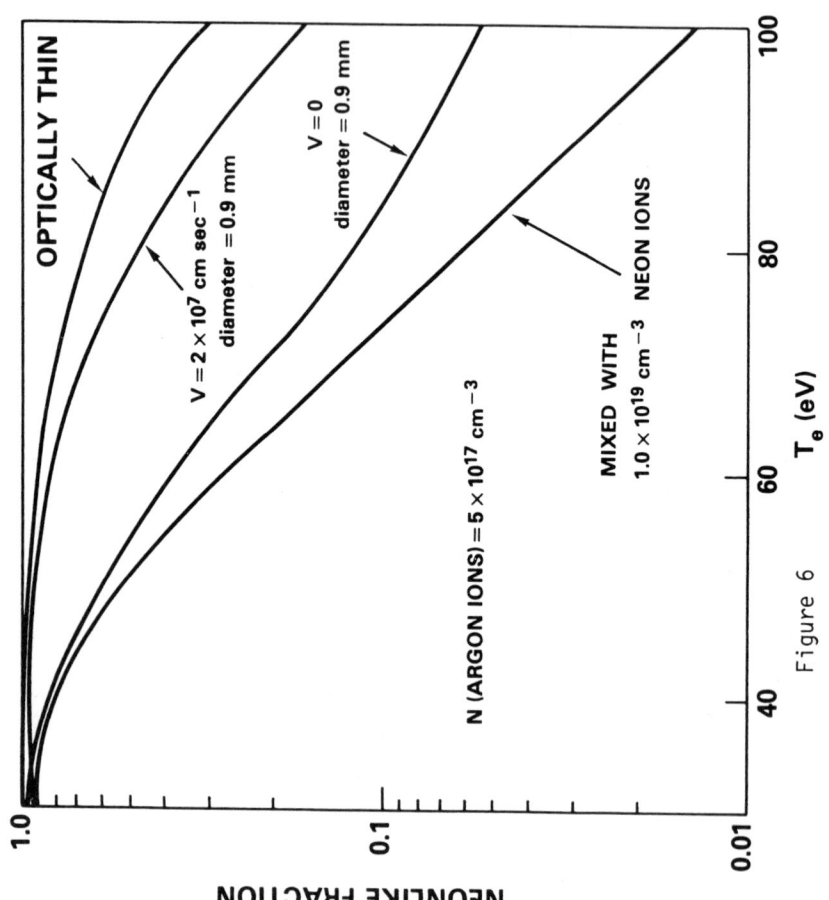

Figure 6

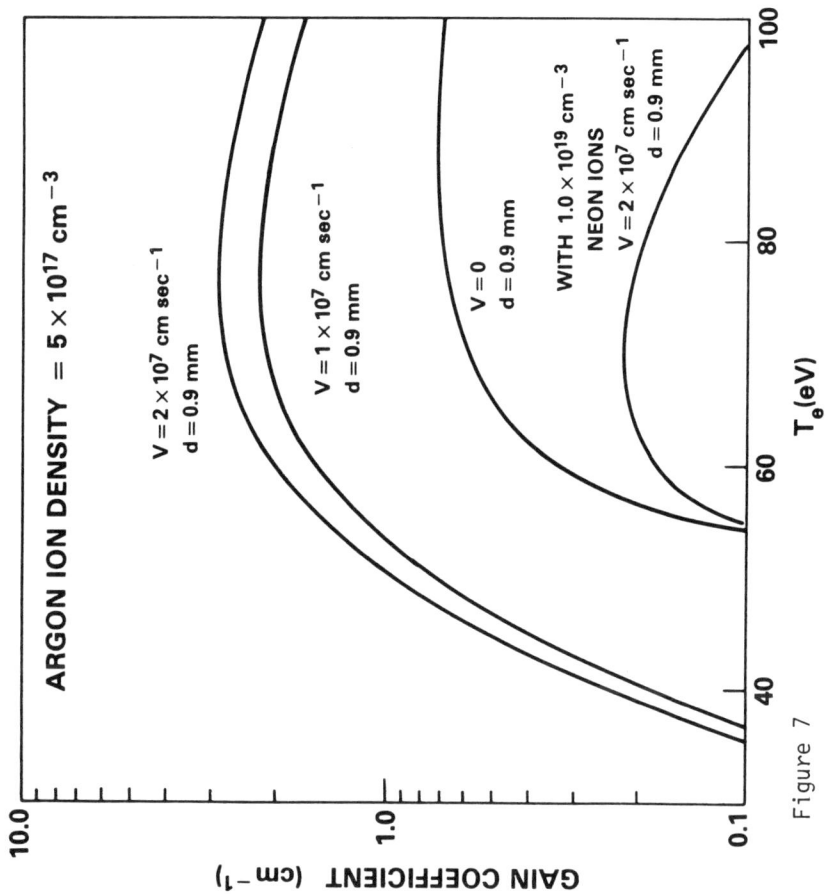

Figure 7

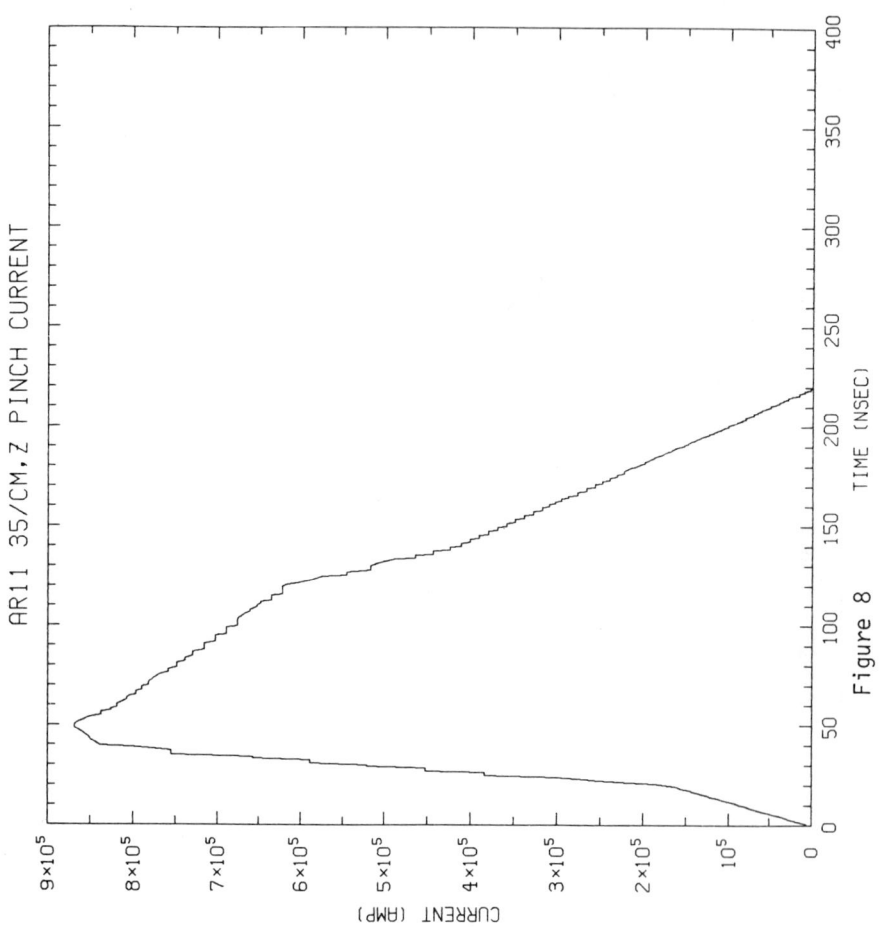

Figure 8

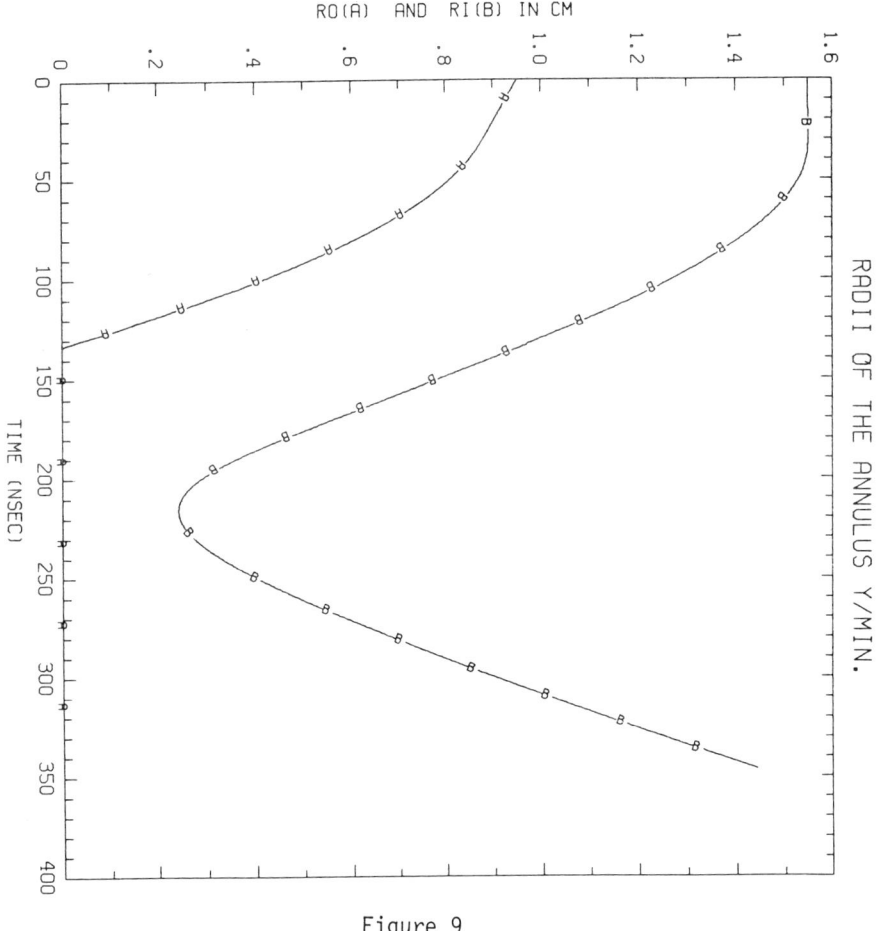

Figure 9

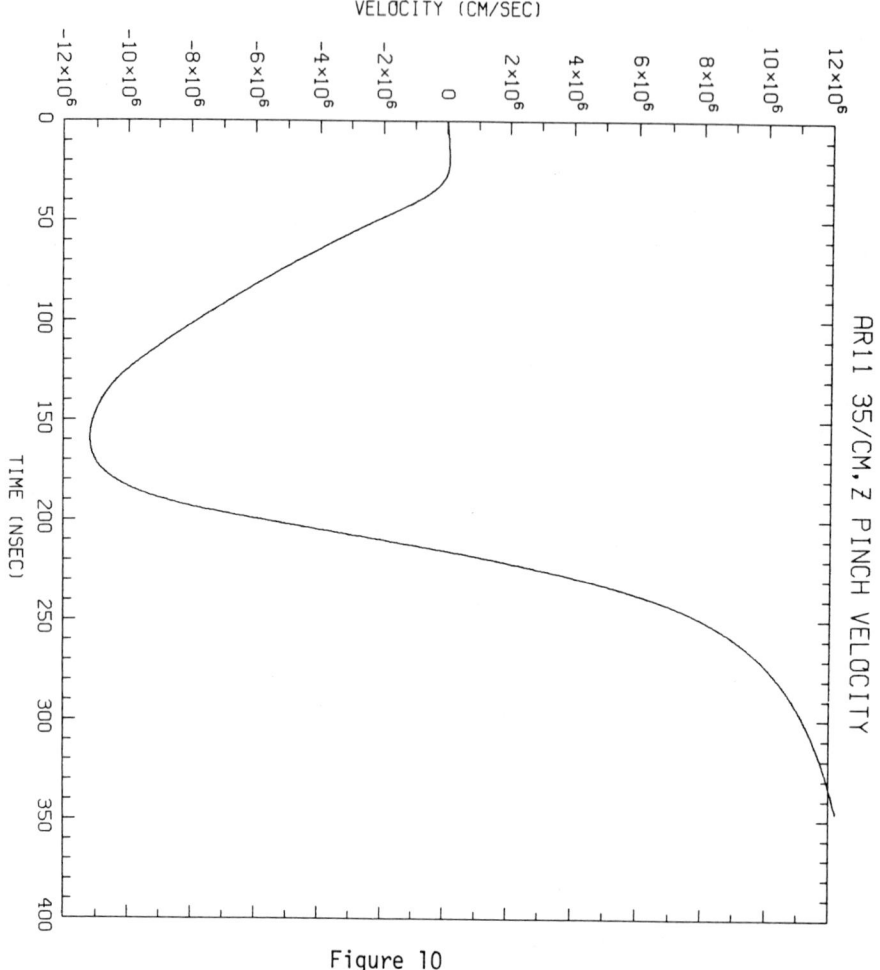

Figure 10

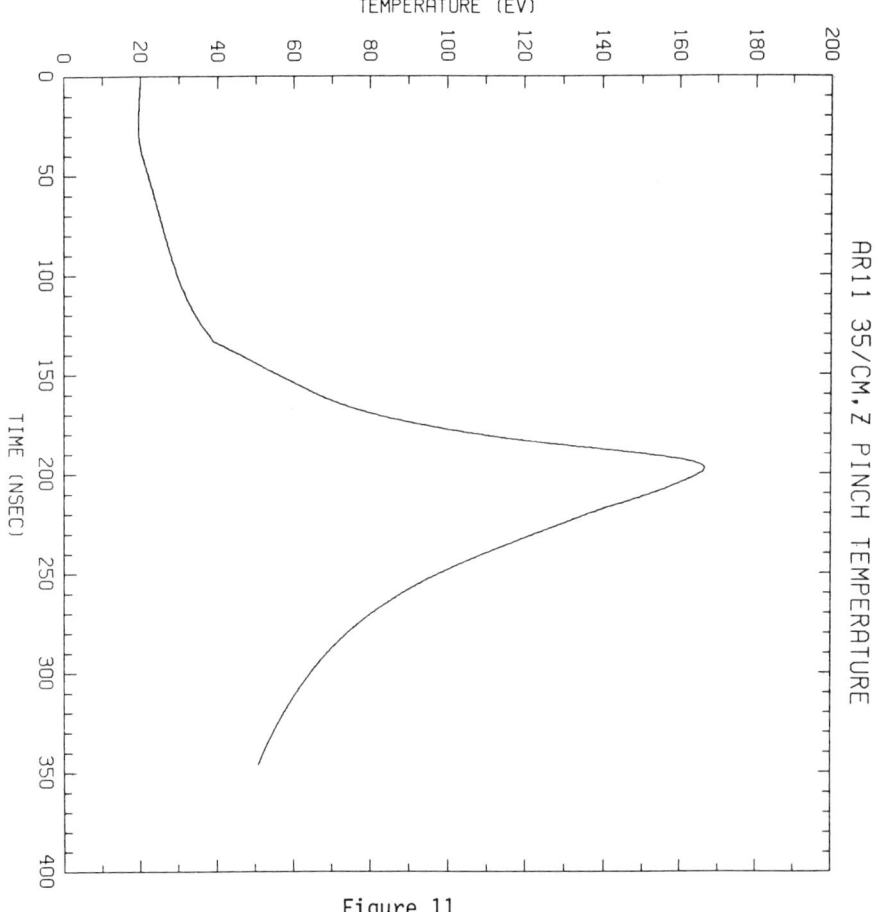

Figure 11

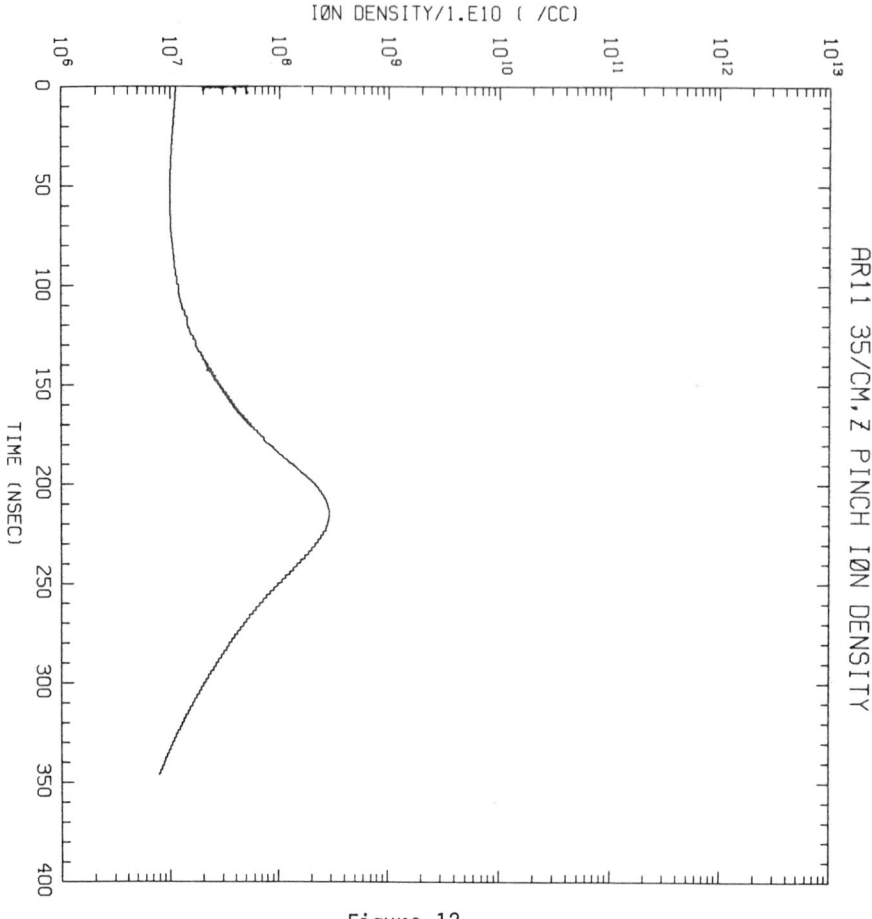

Figure 12

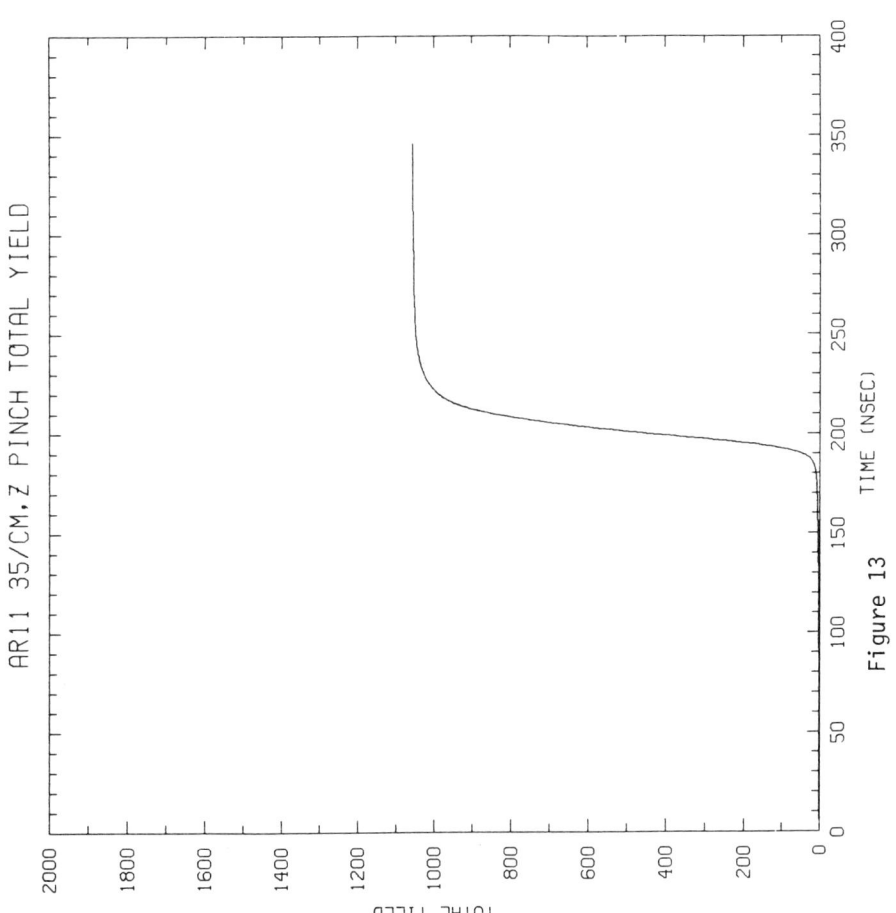

Figure 13

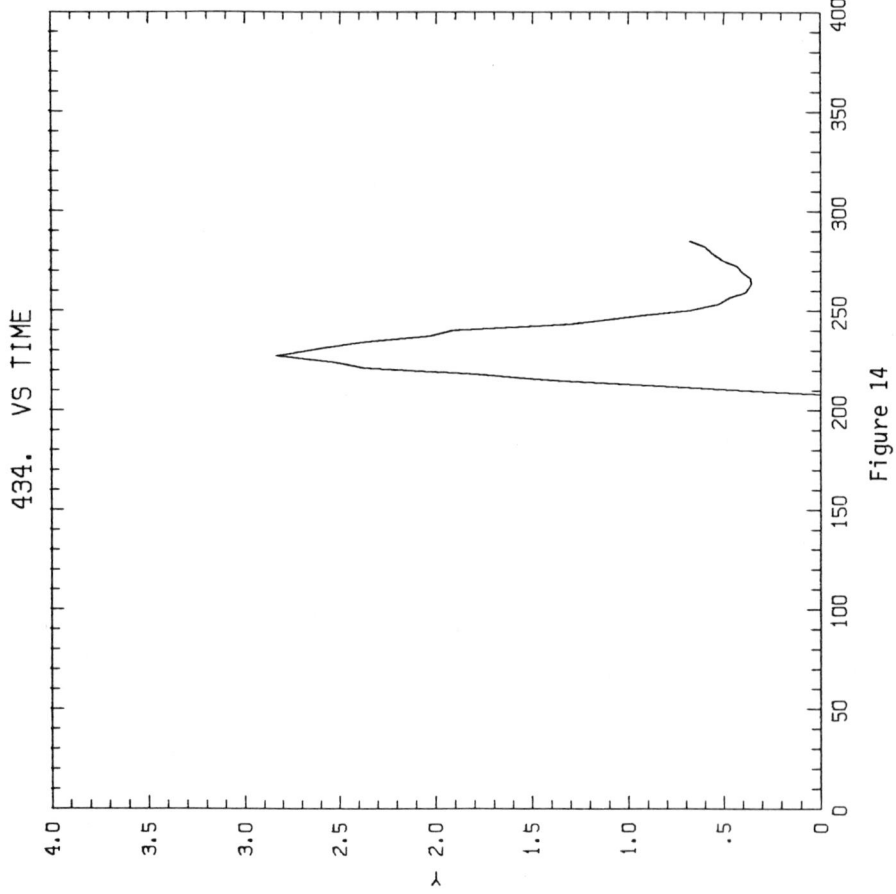

Figure 14

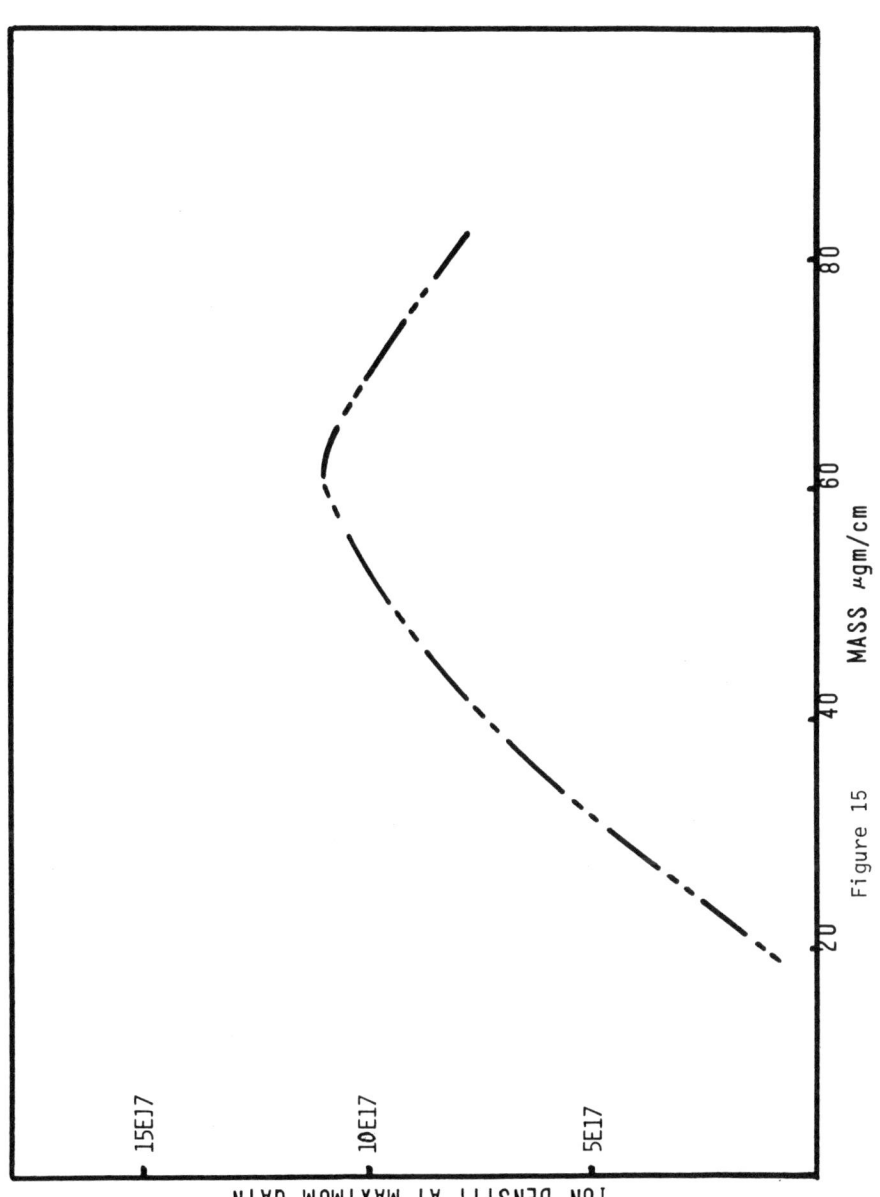

Figure 15

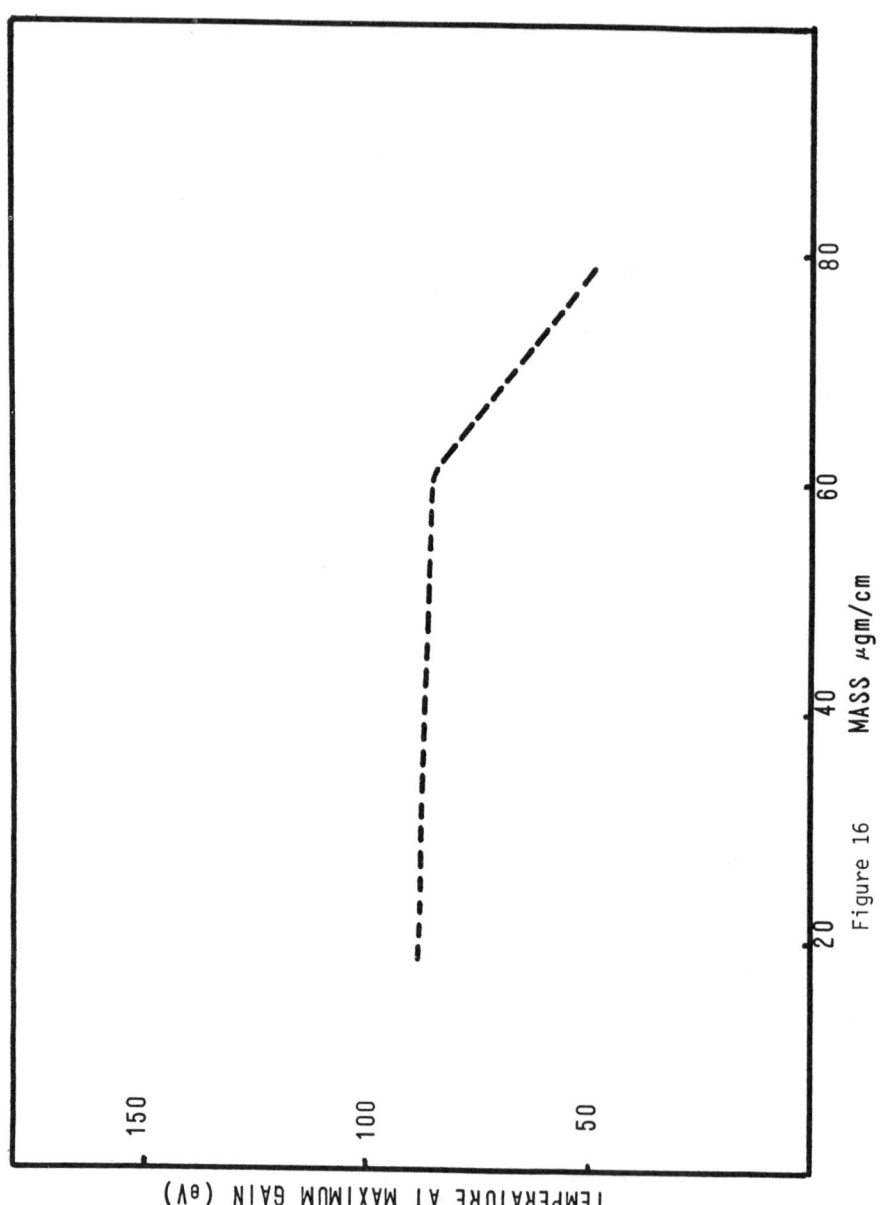

Figure 16

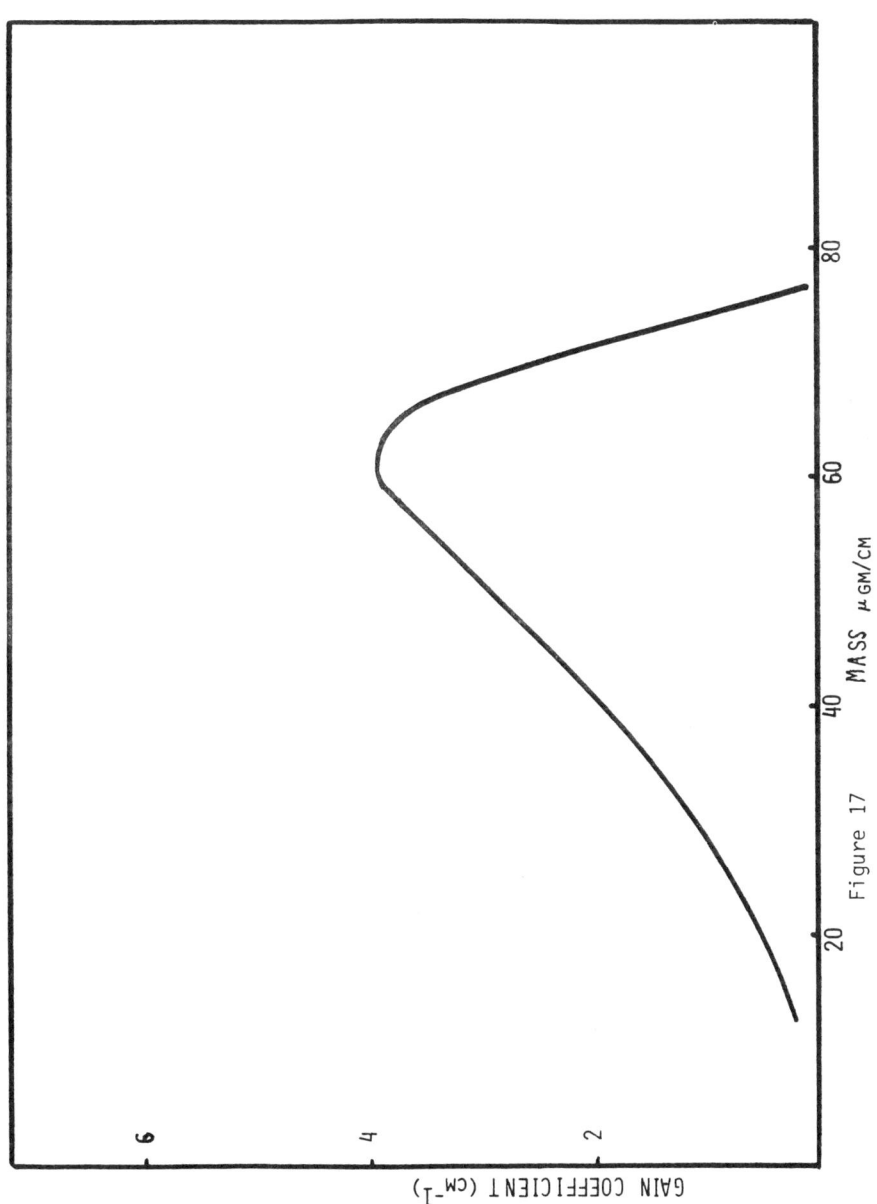

Figure 17

COLLECTIVE VECTOR METHOD FOR CALCULATION OF E1 MOMENTS IN ATOMIC TRANSITION ARRAYS

S. D. Bloom and A. Goldberg,

Lawrence Livermore National Laboratory
and
University of California-Davis, Department of Applied Science
Livermore, California 94550

ABSTRACT

The CV (collective vector) method for calculating E1 moments for a transition array is described and applied in two cases, herein denoted Z26A and Z26B, pertaining to two different configurations of the M-shell. The basic idea of the method is to create a CV from each of the parent ("initial state") state-vectors of the transition array by application of the E1 operator. The moments of each of these CV's, referred to the parent energy, are then the rigorous moments for that parent, requiring no state decomposition of the manifold of daughter state-vectors. Since, in cases of practical interest, the daughter manifold can be orders of magnitude larger in size than the parent manifold, this makes possible the calculation of many moments higher than the second in situations hitherto unattainable via standard methods. The combination of the moments of all the parents, with proper statistical weighting, then yields the transition array moments from which the transition strength distribution can be derived by various procedures. We describe two of these procedures: (1) The well-known GC (Gram-Charlier) expansion in terms of Hermite polynomials, (2) The Lanczos algorithm or Stieltjes imaging method, also called herein the delta expansion. Application is made in the cases of Z26A (50 lines) and Z26B (5523 lines) and the relative merits and shortcomings of the two procedures are discussed.

I. Introduction

The use of collective state-vectors in getting transition strength distributions in many fermion systems is a standard technique in, for example, the RPA (random phase approximation) or the TDA (Tamm-Dancoff approximation). Here we describe a method where such state-vectors can, in principle, be used to obtain moments of arbitrary multipolarity to any order without making any approximations in the character of either the excitation or the correlations in the "ground" or parent state vector $|P\rangle$, as we shall term it here. As in any other method there are important limitations to this method. However our emphasis here is on the calculation of E1 (electric dipole) moments in atomic transition arrays where, as we shall show, these limitations are sufficiently loose so as to permit the study of complex atoms where ordinary diagonalization is impractical or other methods are limited to the evaluation of at most the second or, in some cases, the third moment.[1,2,3] For obvious reasons we call this the CV (collective vector) method. More general discussions of the method, in the context of nuclear applications, can be found in Refs. 4 and 5.

The definition of the CV is simple since, as usual, it requires only $|P\rangle$, the parent state vector and the E1 one-body operator, denoted by (E1). The action of (E1) on $|P\rangle$ produces $|CE1;P\rangle$, the collective E1 or daughter state vector,

$$|CE1;P\rangle \equiv (E1)|P\rangle \qquad (1)$$

where it is to be noted that in general $|CE1;P\rangle$ is <u>not</u> an eigenvector of the Hamiltonian, as $|P\rangle$ is. Note also that the model space for the daughter is much larger than for the parent by virtue of the extra particle/hole excitation produced by the E1 excitation (see below). The <u>total</u> strength for the transition, sometimes called the EUS, i.e. the energy unweighted sum, is given by,

$$(N_p)^2 = \langle CE1;P|CE1;P\rangle \tag{2}$$

where the aptness of the symbol $(N_p)^2$ will become clear shortly. The centroidal moments $\mu_{cn}(P)$ pertaining to each parent $|P\rangle$ are given by,

$$\mu_{cn}(P) = \langle CE1;P|(H-\langle H\rangle)^n|CE1;P\rangle/(N_p)^2; \qquad n>1 \tag{3a}$$

$$\langle H\rangle \equiv \langle CE1;P|H|CE1;P\rangle \tag{3b}$$

where we now see that the total strength $(N_p)^2$ is the normalizing factor for the moments.

Besides the centroidal moments $\mu_{cn}(P)$ we need the difference in energy between the parent state $|P\rangle$ and the centroid of $|CE1;P\rangle$ for which we use the symbol $\langle H1\rangle$,

$$\langle H1\rangle = [\langle CE1;P|H|CE1;P\rangle/(N_p)^2] - \langle P|H|P\rangle \tag{4}$$

In principle we now have all that is needed to get the E1 strength distribution but we have not yet considered the spin projection dependence explicitly. We will not give in all detail the equations with this feature included but give only the full expression for the E1 operator, which is very simple and will suffice to make the major arguments we need,

$$(E1) = \sum_\mu (E1)\mu \tag{5}$$

where μ is the spin projection for the E1 operator. With Eq. (5) we can easily derive from Eq. (3) the familiar result for the n'th centroidal moment in factorized form,

$$\mu_n(P) = \sum_{J'M'} \langle J'M'|(E1)|JM\rangle^2 \langle J'|(H-\langle H\rangle)^n|J\rangle/(N_p)^2; \quad n>1 \tag{6}$$

where (JM, J'M') refer to (parent, daughter) spins and spin projections. Note that this result is independent of M due to Eq. (5) and the scalarity of H, which is to say μ_n is, as it must be, a scalar. Because of this a further simplification of Eq. (6) can be effected by the use of reduced matrix elements yielding,

$$\mu_n(P) = \sum \langle J'||E1||J\rangle^2 \langle J'|(H-\langle H\rangle^n)|J\rangle/(N_p)^2 \tag{7a}$$

$$(N_p)^2 = \sum \langle J'||E1||J\rangle^2 \tag{7b}$$

where all explicit spin-projection dependence has been eliminated. Note that there has been a slight change in the definition of the normalization $(N_p)^2$ (by a factor of $(2J+1)$) to conform to the use of reduced matrix elements.

It is Eqs. (7 a, b) and their practical evaluation via Eq. (3), which forms the basis for all the results to be described herein. The ingredients of the actual computation are the collective state vector $|CE1;P>$, the parent state vector $|P>$, and, of course, some means of calculating the various matrix elements required. In our case we have used the system of codes called VLADIMIR at LLNL to make all the required state-vectors as well as the Hamiltonian matrix elements. We will not discuss the VLADIMIR codes here except to note that they are couched entirely in terms of Fock space representation (2nd quantization) which, for the purposes involved here, is particularly useful since this makes it possible to deal with state-vectors and operators on a separate footing. [Two examples of the utility of this feature are in the implementation of Eqs. (1) and (5).] Further details may be found in Refs. 5 and 6.

II. Calculating the transition array

An atomic transition array generally has many parent states and we wish to utilize the set of moments $\mu_n(P)$ to describe the entire array. As we have seen only the state-vectors of the parent configuration are explicitly required, not those of the daughter. This makes for considerable computer economy, since, for example, in the two cases to be illustrated the parent configurations have 19 and 180 parent states, while the daughter configurations have respectively 128 and 3449 states.

The most common method for calculation of the transition array from the moments is the Gram-Charlier expansion.[7] This gives the strength function $S_p(\omega)$ arising from a single parent $|P>$ as a series in Hermite polynomials, H_n,

$$S_p(\omega) = (2/\pi)^{1/2} \sum_{n=0}^{\infty} c_n(P) \exp\left\{-(\omega - <H1>)^2/2\mu_2(P)\right\} H_n\left(\frac{(\omega - <H1>)}{[\mu_2(P)]^{1/2}}\right)$$

with $<H1>$ given in Eq. (4) and $\mu_2(P)$ being the second centroidal moment of the array arising from $|P>$. This expansion is exact if carried to infinite order, but as a practical matter is truncated at some value $n = N$, with the coefficients $c_0, c_1...c_N$ chosen so that the truncated expansion gives the first N+1 centroidal moments correctly ($\mu_0 = 1$, $\mu_1 = 0$). The complete transition array T(E1) is given by summing $S_p(\omega)$ over all P, i.e. the whole multiplet.

This procedure will be illustrated in the next section, where a deficiency associated with the GC method will become apparent. It arises from the fact that the finite expansion allows negative values for $S_p(\omega)$. These negative excursions [$S_p(\omega)$ must of course be positive definite] can become quite significant, as will be seen.

A second procedure for synthesizing the transition strength distribution is the Stieltjes Imaging or delta expansion method. Here we use the Lanczos algorithm to implement the delta expansion. This has two main advantages: (1) it is more convenient computationally, about twice as fast as a straightforward evaluation of the moments, and (2) it yields a finite width to be associated with each line. The Lanczos procedure constructs a truncated set of basis vectors in the daughter model space, and, from these, generates a set of approximate eigenvectors $|e.v.>$ of the daughter configuration. The energies of the approximate eigenvectors fix the delta expansion line energies, while the line intensities are the square of the projections of

|CE1:P> on these vectors. Finally, a computational width (as distinguished from a physical width) can be assigned to each line, to wit the energy fluctuation of that eigenvector, i.e.,

$$\Gamma^2 = <e.v.|H^2|e.v.> - <e.v.|H|e.v.>^2 \tag{9}$$

Detailed discussions of the Lanczos method can be found in Refs. 6 and 10.

III. Results

In this section we give results for two calculations, labeled Z26A and Z26B, transition arrays for two parent configurations of the M-shell. These are model problems and the results are presented only for comparison between the calculational methods. These results are not intended to represent an actual physical system. The transition arrays correspond to

Z26A: $(3d)^3 \rightarrow (3d)^2(4p)$
Z26B: $(3p)^5(3d)^4 \rightarrow (3p)^5(3d)^3(4f)$.

We discuss the Z26A case first. Here there are 4 parent energy levels and 19 distinct parent states in total J and energy. In both parent and daughter J takes on 6 distinct values, ranging from J = 1/2 to J = 11/2. This high degree of degeneracy results from the pure LS coupling used in both examples Z26A and Z26B. In this first example there are 50 transitions, distinct in energy, shown in Fig. 1a (with the exception of a few extremely weak lines).

This is an absorption spectrum save for a few lines at energies about -3eV. The arbitrary width of w = 0.06 eV was assigned to all lines for the sake of easier display, and also for consistency with the width used in the delta expansion (see below). Figure 1b exhibits the same spectrum, this time for positive transition energies only, with width w = 0.25 eV. This is a much smoother representation and, as will now be seen, is much more appropriate for comparison with the GC (Gram-Charlier) expansion results. In Fig. 2a we show the GC expansion results for 6 moments per parent, so that this represents a superposition of 19 GC expansions, each up to the 6th moment. We note the negative excursion at $\approx$ 12 eV, not too severe in this case. We have limited ourselves to the positive (absorption) spectrum only in this figure and also in Fig. 2b, which is an overlay of the GC result and the microscopic result with w = 0.25 eV (Fig. 1b). There is a fair correspondence between the broadened microscopic spectrum and the GC expansion except for such features as the peak at $\approx$ 8.5 eV. To obtain a better feeling as to how the GC method tries to fit simple spectra, we show in Fig. 3 the 6 moment result for one value of J, i.e., J = 11/2, which has only one parent and 3 lines (shown there with their proper strengths). For such spectra the GC method is completely inappropriate, but nonetheless this shows that the GC method would be equally inappropriate in cases where complex intermediate structure is of physical interest.

A more suitable method for studying structure in transition arrays is the delta expansion, the results for which are shown in Fig. 4. The delta expansion spectrum using 12 iterations per parent (corresponding to 23 moments per parent) is shown in Fig. 4a, while Fig. 4b shows that for 6 iterations per parent (11 moments). In both each line was assigned a width either given by Eq. (8) or 0.06 eV, whichever was larger. The 12 iteration result is virtually indistinguishable from the microscopic result (Fig. 1a). All the weaker lines in the 6 iteration delta expansion spectrum have widths generated by the method itself, and comparison of this with the microscopic spectrum demonstrates nicely how the Lanczos method smoothes over the discrete structure when the algorithm has not reached complete convergence, as it had in the 12 iteration calculation.

Table 1 gives the overall moments for the entire Z26A array as calculated from the microscopic spectrum, the GC expansion, and the delta expansion. The GC moments depart substantially from the exact values by the 10th moment getting rapidly worse after this, despite the fact that the GC method generates 6 x 19 = 114 moments worth of information. (The slight deviations up to the 6th overall moment are due to neglect of the weak negative energy transitions.) The delta expansion moments agree within ≈25% with the microscopic values up to the 12th moment for the 6 iteration calculation with full widths for each line (see Sect. II). (For the 12 iteration result the agreement is marred only by the missing peaks in the microscopic spectrum.) In this case there are 19 x 11 = 209 moments worth of information, corresponding roughly to the 6 moment GC computation in that the computation times required are about the same, the delta expansion method being about a factor of 2 faster.

The delta expansion result at 6 iterations per parent with full widths for each line (see Sect. II) yields moments at significant variance with the microscopic values beyond the 12th moment (not shown in Table 1) due to the distortion introduced by the intrinsic widths. However, Fig. 5 shows that the loss of some moment information is rather well compensated by the gain of the correct smoothed structural definition in both the weak and strong transition regions.

The Z26B case has 50 parent energy levels and 180 distinct parent states in both J and energy. J takes on 8 distinct values in the parent configuration (up to $J = 15/2$) and 9 for the daughter (up to $J = 17/2$). There are 5523 distinct transitions, shown in Fig. 6a with $w = 0.10$ eV and in Fig. 6b with $w = 0.25$ eV. As might be expected, no individual transitions are seen in Fig. 6b, but considerable "intermediate" structure remains. Figure 7 exhibits the overlay of Fig. 6b with the GC result using 6 moments per parent. The wings of the strength distribution are fitted quite well (except for the usual negative dips), but the center portion has a severe negative excursion, a result of the GC method trying to reproduce the double-humped structure of this particular spectrum. We show delta expansion results here only for the set of parent states with $J = 15/2$. There are 2 such, with about 150 daughters and about 100 lines. Figure 8a gives the Lanczos results with 12 iterations per parent, and with each line assigned a width of 0.2 eV. Assigning a small constant width ($\lesssim 0.2$ eV) leaves the moments unaltered, and it is to be emphasized that the first 24 moments obtained this way are exact, i.e. as essentially as if calculated at zero width. Figure 8b exhibits the same delta expansion with the computational widths, Eq. (8), assigned to each line, and Fig. 8c overlays these two. In Fig. 8d, we show the overlay of Fig. 8a, and the 6 iteration delta expansion result with full width. As above the major effect of using the full widths is to smooth the resultant spectrum,

without greatly altering its general shape. One does not really need the full
12 iterations to obtain a substantially correct strength distribution. Table
II gives the values for the moments for the J = 15/2 parent manifold, for
these various computations. Beyond the 6th moment, those obtained from the
full width spectra for both 6 and 12 iterations diverge from the correct
moments. This, as before, is a distortion due to the broadening procedure.
Nonetheless, as before, the shape of the spectrum is much better retained by
the delta expansion method than by the GC method.

IV. Conclusion

Although the second test case (Z26B) is not complete, a number of firm
conclusions may be drawn. First the general procedure described here forms a
practical method for characterizing atomic transition arrays, and is
especially applicable when the array contains more than 5000 lines
(essentially the limiting number for use of purely microscopic methods). It
is to be emphasized that although intermediate coupling and/or configuration
interaction effects have not been considered in these test cases, such effects
can be included with no additional work and with no increase in computer time.

Secondly, the use of a Gram-Charlier type of description of the array
seems to be useful only for the gross overall features determined from the
first few moments. Sixth (or higher) order Gram-Charlier expansions result in
unphysical negative strength distributions. In any case, the GC expansion as
a practical matter appears incapable of giving any structure in the transition
array. The Lanczos delta-expansion on the other hand is positive definite
everywhere and can be used to give detailed structure in the array,

unambiguously giving strength distributions quite similar to the microscopically computed array. This clearly seems the preferred method when 4 or more moments are known.

One difficulty of our method is the fact that as of now it is executed for each of the parent states. Test case Z26B has 180 such parent states, and interesting arrays could have many more. We intend then to develop a Monte Carlo-Lanczos procedure to simulate the states of the parent configuration. That is, we choose a random starting vector and then use the Lanczos method to construct approximate eigenstates within the parent manifold. This procedure will considerably shorten the computation and allow application to much larger parent manifolds, for example configurations in the N-shell.

ACKNOWLEDGMENTS

The authors are grateful to K. Reed for providing the microscopic strength distribution for test case Z26A. We also must thank S. M. Grimes, J. D. Anderson, and B. Rozsnyai for a number of enlightening conversations. Work performed under the auspices of the U.S. Department of Energy by Lawrence Livermore National Laboratory under contract #W-7405-Eng-48.

Table I Overall moments up to the 12th moment, for case Z26A, as computed from microscopic spectrum, etc. The overall moments are the weighed sums of the moments of each parent (see Sect. I). The units are arbitrary. The exponents are given in parentheses. See Sect. III.

$$\mu_n = \sum_P (2J_p+1)\, \mu_n(P)$$

n	Microscopic	GC $N_{mom} = 6$	Lanczos (full widths) $N_{it} = 6$	$N_{it} = 12$
2	7.65	7.73	7.77	7.74
3	-1.55 (1)	-1.59 (1)	-1.64 (1)	-1.62 (1)
4	2.59 (2)	2.62 (2)	2.70 (2)	2.66 (2)
5	-1.29 (3)	-1.26 (3)	-1.38 (3)	1.34 (3)
6	1.80 (4)	1.68 (4)	1.90 (4)	1.85 (4)
7	-1.34 (5)	-1.16 (5)	-1.44 (5)	-1.36 (5)
8	1.76 (6)	1.61 (6)	1.89 (6)	1.78 (6)
9	-1.62 (7)	-1.61 (7)	1.76 (7)	-1.61 (7)
10	2.04 (8)	2.57 (8)	2.23 (8)	2.03 (8)
11	-2.08 (9)	-3.63 (9)	-2.32 (9)	-2.05 (9)
12	2.55 (10)	6.34 (10)	2.86 (10)	2.50 (10)

Table II Moments for case Z26B for the parent state manifold with J = 15/2. There are 2 parent states and approximately 150 daughter states that are connected to these parents by E1 transitions. See Sect. III.

n	Lanczos			GC
	N_{it} = 12	N_{it} = 12	N_{it} = 6	N_{mom} = 6
	ω=0.2 ev (correct)	full width	full width	
2	7.46	8.25	8.57	7.44
3	-2.38 (1)	-2.42 (1)	-1.75 (1)	-2.34 (1)
4	4.06 (2)	4.55 (2)	5.00 (2)	4.04 (2)
5	-3.85 (3)	-4.05 (3)	-3.62 (3)	-3.81 (3)
6	4.78 (4)	5.52 (4)	5.71 (4)	4.73 (4)
7	-5.60 (5)	-6.58 (5)	-5.93 (5)	-4.21 (5)
8	6.91 (6)	9.02 (6)	8.41 (6)	5.76 (6)
9	-8.65 (7)	-1.26 (8)	-1.04 (8)	-6.04 (7)
10	1.13 (9)	1.94 (9)	1.54 (9)	7.97 (8)
11	-1.54 (10)	-3.28 (10)	-2.31 (10)	-1.06 (10)
12	2.28 (11)	6.19 (11)	4.11 (11)	1.29 (11)

References

1. A. L. Merts, "Moment Methods in Many-Fermion Systems", edited by B. J. Dalton, S. M. Grimes, J. P. Vary, and S. A. Williams, p. 81, 1979.

2. C. Bauche-Arnoult, J. Bauche, and M. Klapisch, Phys. Rev. A $\underline{20}$, 2424 (1979).

3. J. Bauche, et al, Phys. Rev. A, $\underline{28}$, 829 (1983).

4. S. D. Bloom and R. F. Hausman, Jr., see Ref. 1, p. 151, 1979.

5. S. D. Bloom, Progress in Particle and Nuclear Physics, $\underline{11}$, 505 (1983).

6. R. F. Hausman, Jr., "A Vector Method for Large-Scale Configuration Interaction Problems", UCRL-52178, 1976; unpublished.

7. M. Kendall and A. Stuart, The Advanced Theory of Statistics, MacMillen (New York) 1977, pp 168-175.

8. P. W. Langhoff, see Ref. 1, p. 191, 1979.

9. R. R. Whitehead, see Ref. 1, p. 235, 1979.

10. R. R. Whitehead et al, Advances in Nuclear Physics; Editors, M. Baranger and E. Vogt, Plenum, N.Y., 1977, p. 123.

FIGURE CAPTIONS

Fig. 1-a: Transition array calculated microscopically for Z26A with a width w = 0.06 eV, for 48 lines (unnormalized). This transition array is an absorption spectrum except for the two negative energy lines at ~ -3eV. See Sect. III.

Fig. 1-b: Same as Fig. 1-a except that w = 0.25 eV. and the negative energy part of the spectrum is not displayed. Also this distribution is normalized to unity. See Sect. III.

Fig. 2-a: The Gram-Charlier (GC) expansion for the Z26A transition array with 6 moments per parent (19 parents). See Sect. III for a fuller discussion.

Fig. 2-b: Overlay of Figs. 2-a and 1-b. Note that the GC representation has significant contributions from the off-scale (unphysical) energy region.

Fig. 3: Case Z26A: Overlay of the 6 moment GC expansion and the 3 microscopic strengths originating from the single parent with J = 11/2. See Sect. III for a fuller discussion.

Fig. 4-a: Delta expansion result for the transition array for Z26A with 12 iterations per parent, with minimum width w = 0.06 eV. See Sect. III.

Fig. 4-b: Same as Fig. 4-a, but with 6 iterations per parent.

Fig. 5:   Overlay of case Z26A microscopic spectrum, Fig. 1-a, and delta expansion, Fig. 4-b. See Sect. III.

Fig. 6-a: Transition array from the microscopic spectrum for test case Z26B, with each line given a width $w = 0.1$ eV. This emission spectrum contains 5523 lines, mainly unresolved, even at 0.1 eV width.

Fig. 6-b: Same as Fig. 6-a, but with width $w = 0.25$ eV. There are no unresolved lines using this width.

Fig. 7:   Overlay of GC results for Z26B transition array with 6 moments per parent (180 parents) with the microscopic spectrum of Fig. 6-b. The extreme negative excursion at the center of the spectrum arises from the attempt of the GC method to reproduce the double peaked structure of the true spectrum. See Sect. III.

Fig. 8-a: Delta expansion result for the Z26B transition array for the two parent states with $J = 15/2$. The computation used 12 iterations per parent, and each line was given a width $w = 0.2$ eV. See Sect. III.

Fig. 8-b: Same as Fig. 8-a, except each line was given its full intrinsic width as discussed in Sect. III.

Fig. 8-c: Overlay of Figs. 8-a and 8-b.

Fig. 8-d: Overlay of Fig. 8-a and the delta expansion with full width but with 6 (rather than 12) iterations per parent.

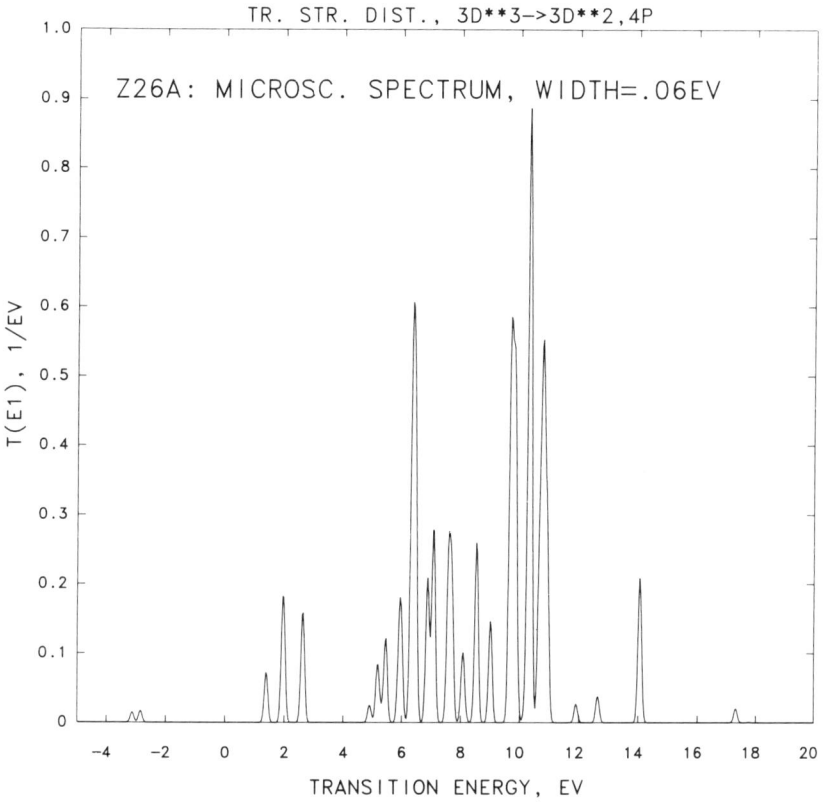

Figure 1a

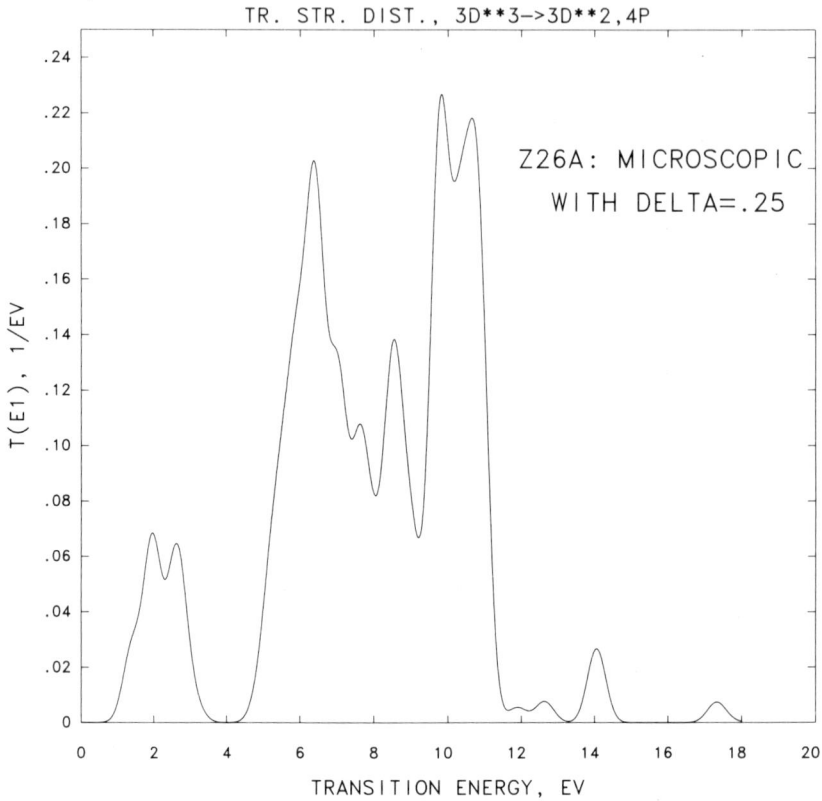

Figure 1b

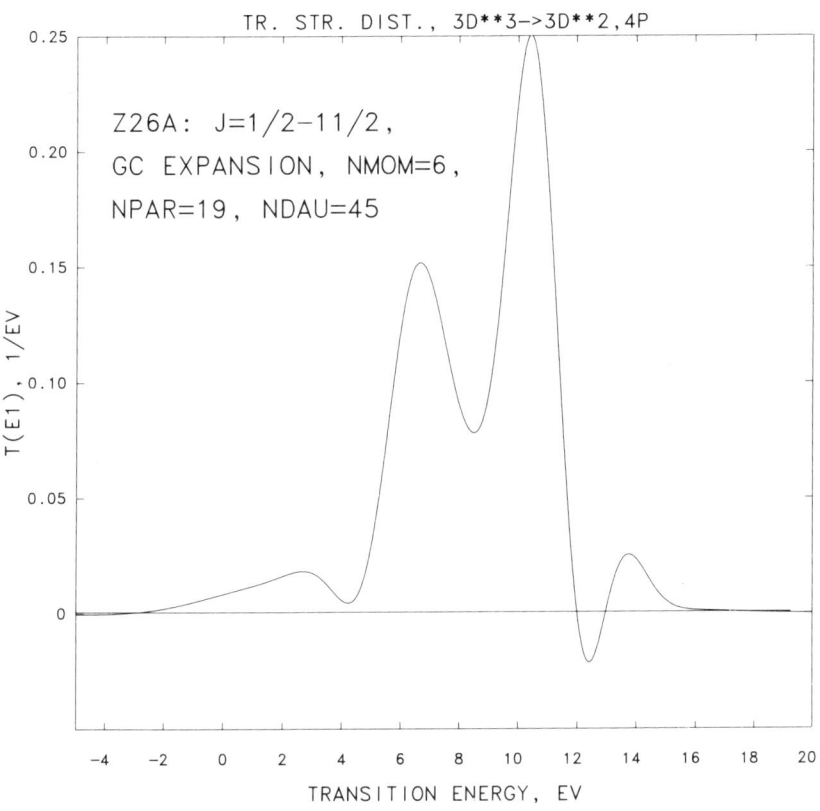

Figure 2a

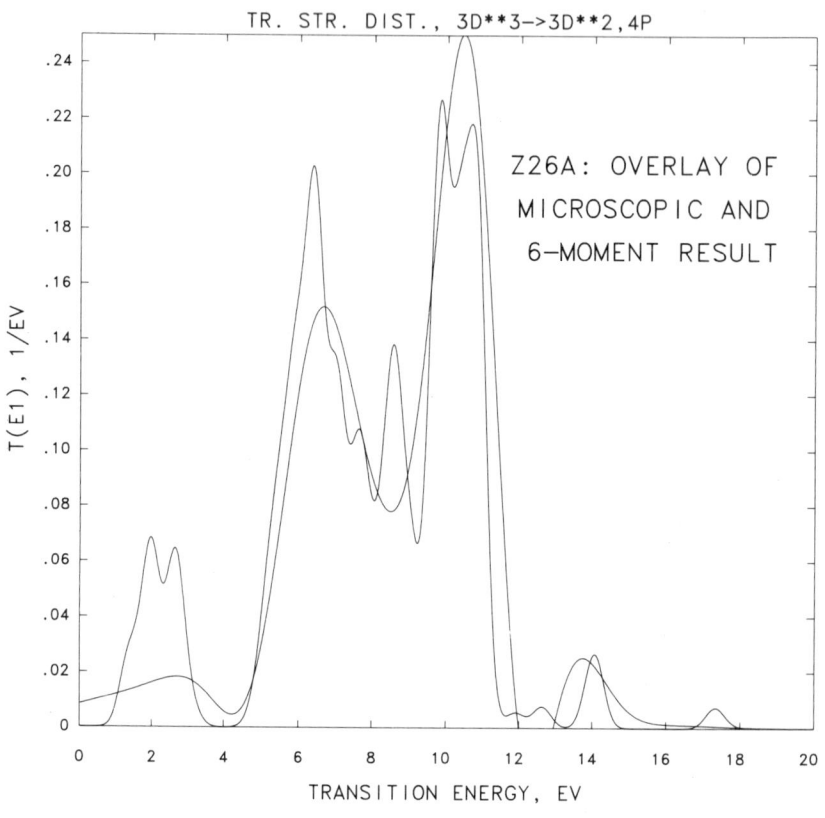

Figure 2b

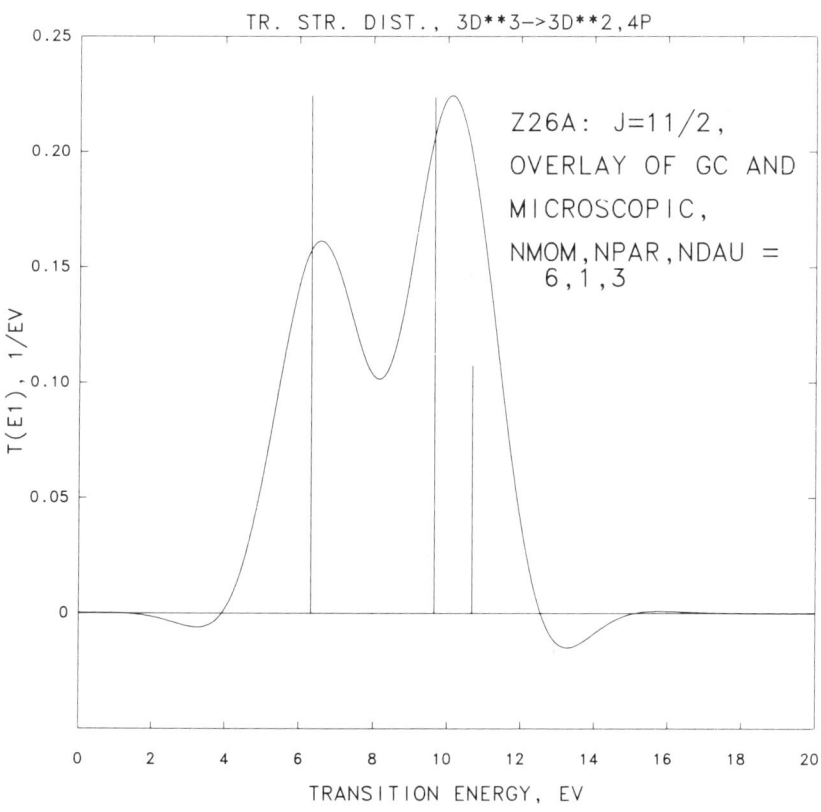

Figure 3

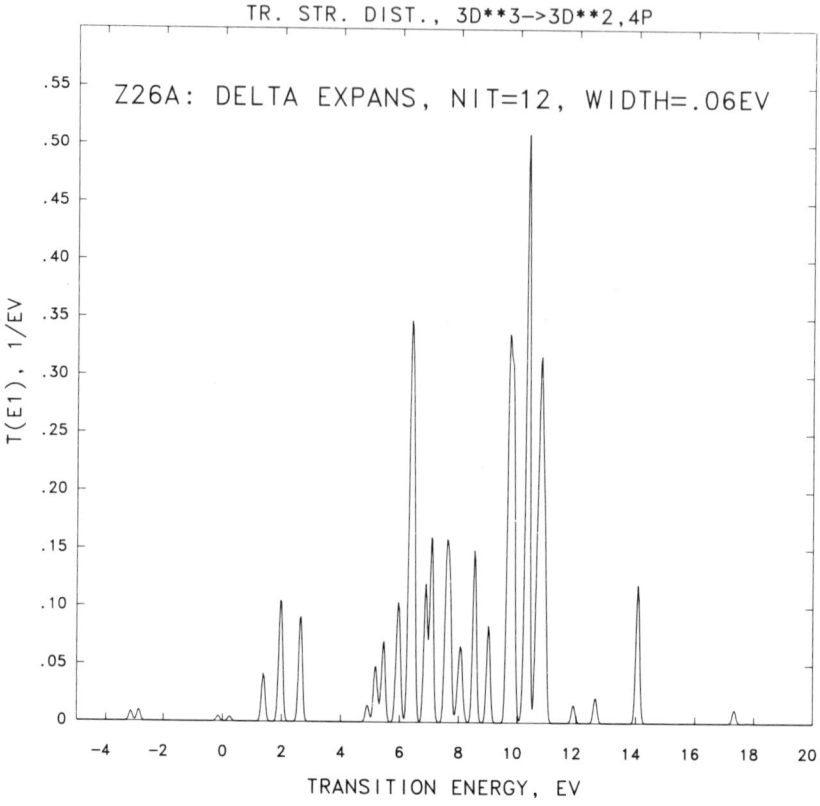

Figure 4a

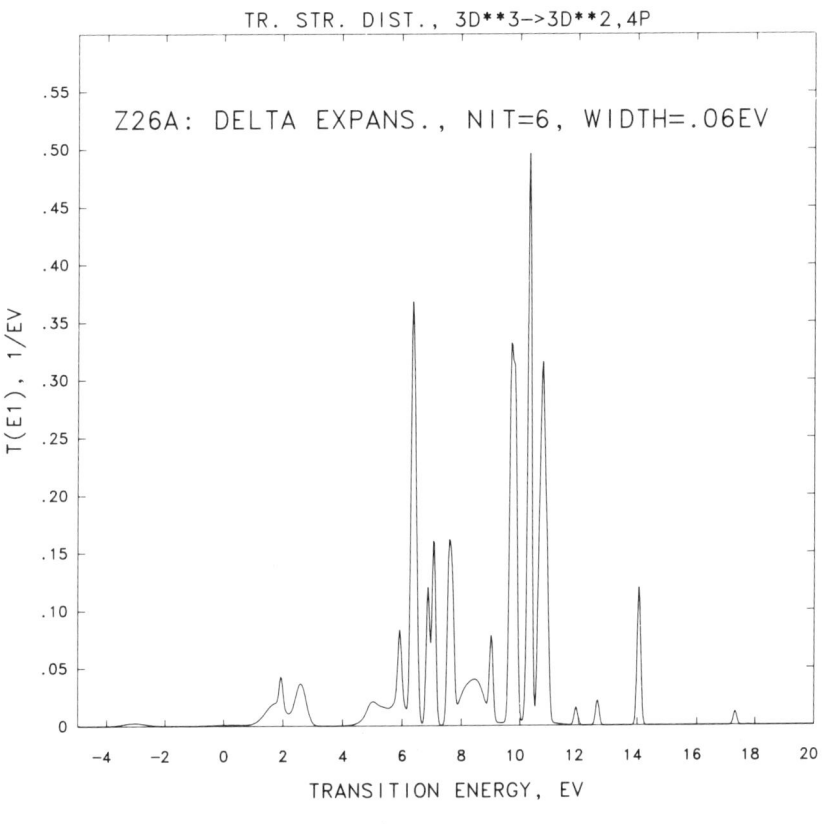

Figure 4b

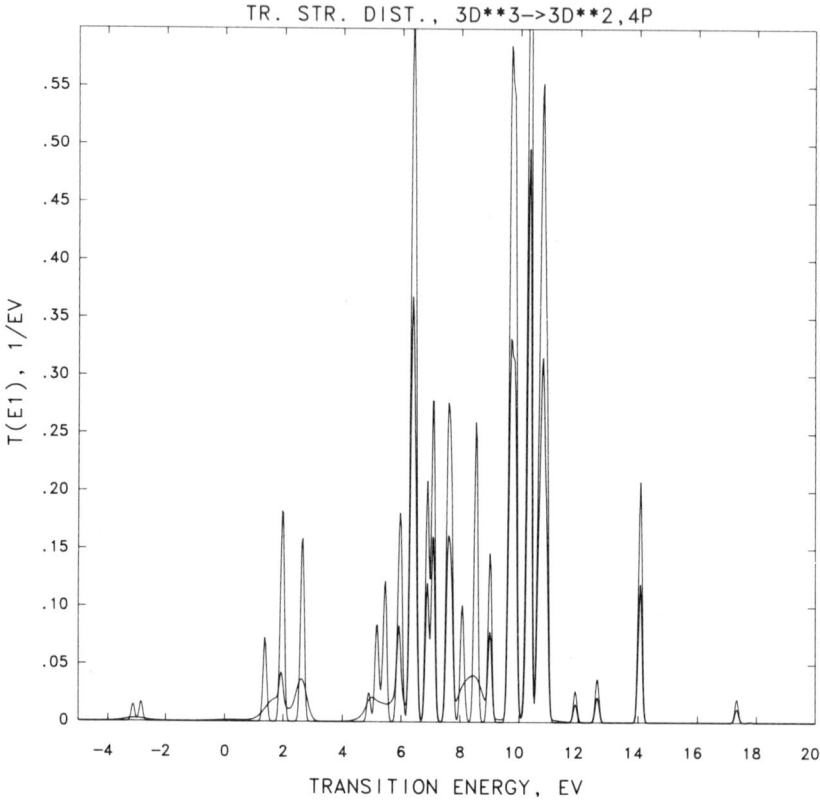

Figure 5

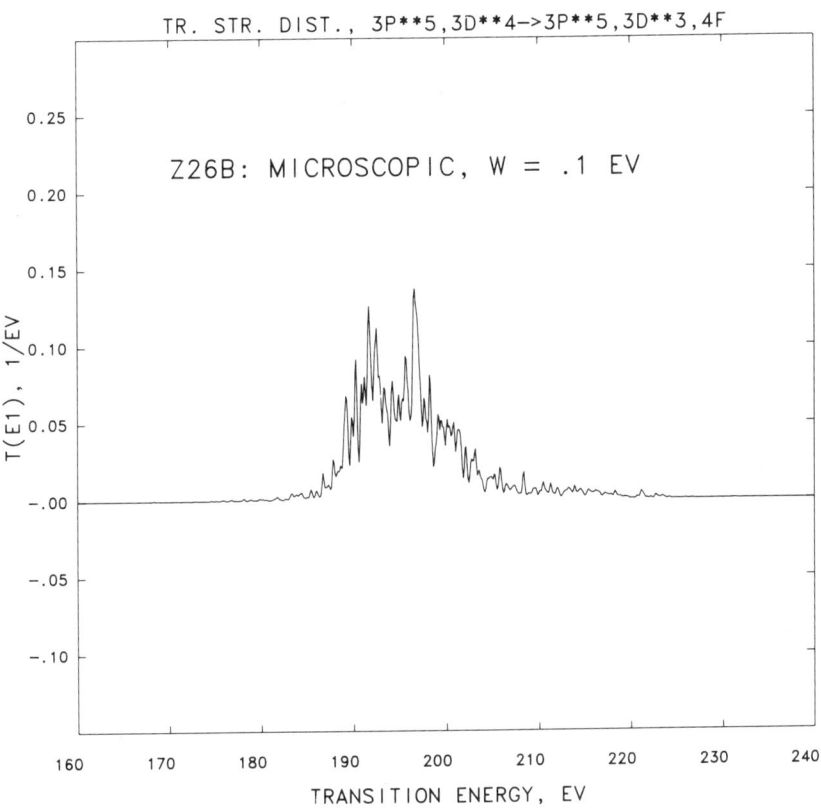

Figure 6a

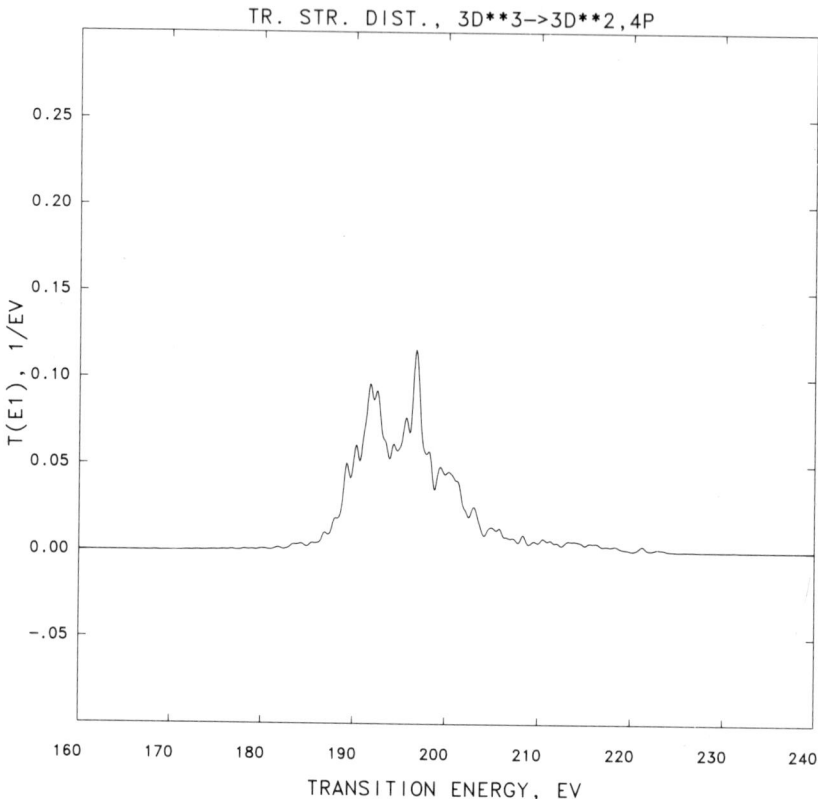

Figure 6b

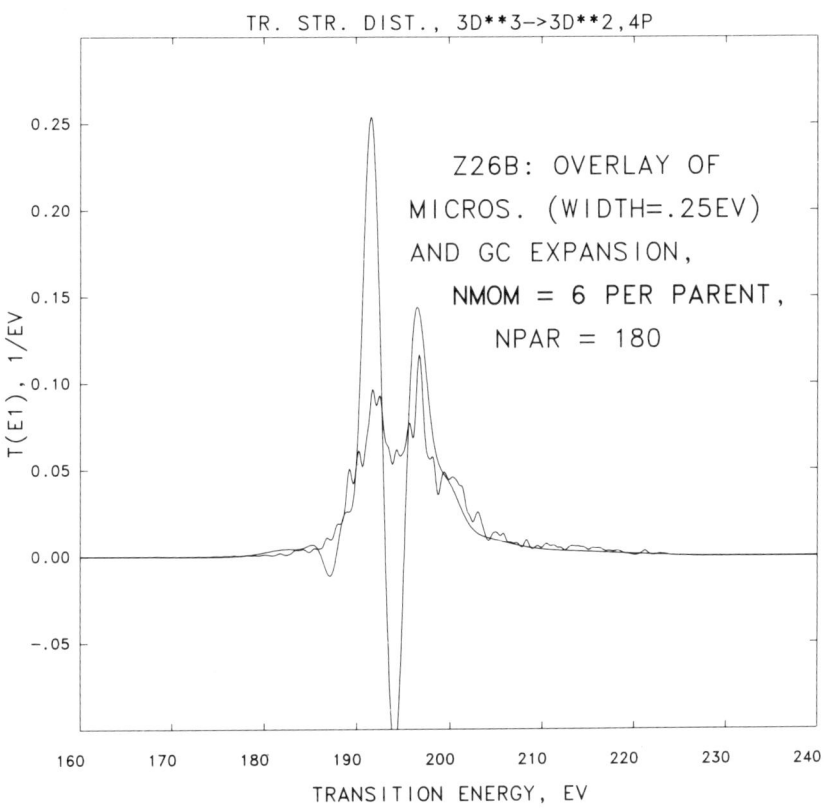

Figure 7

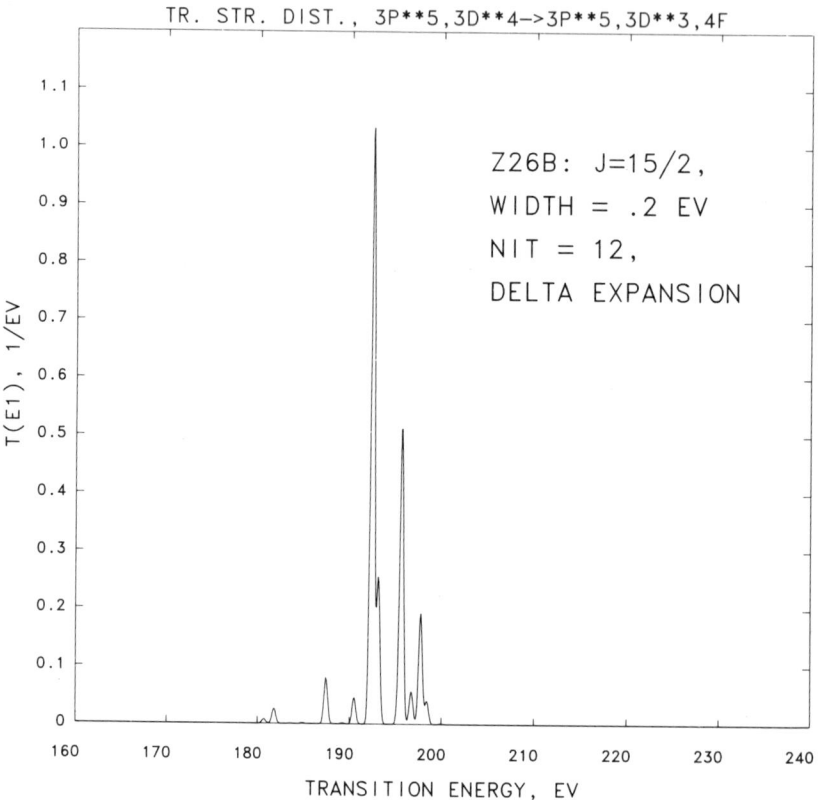

Figure 8a

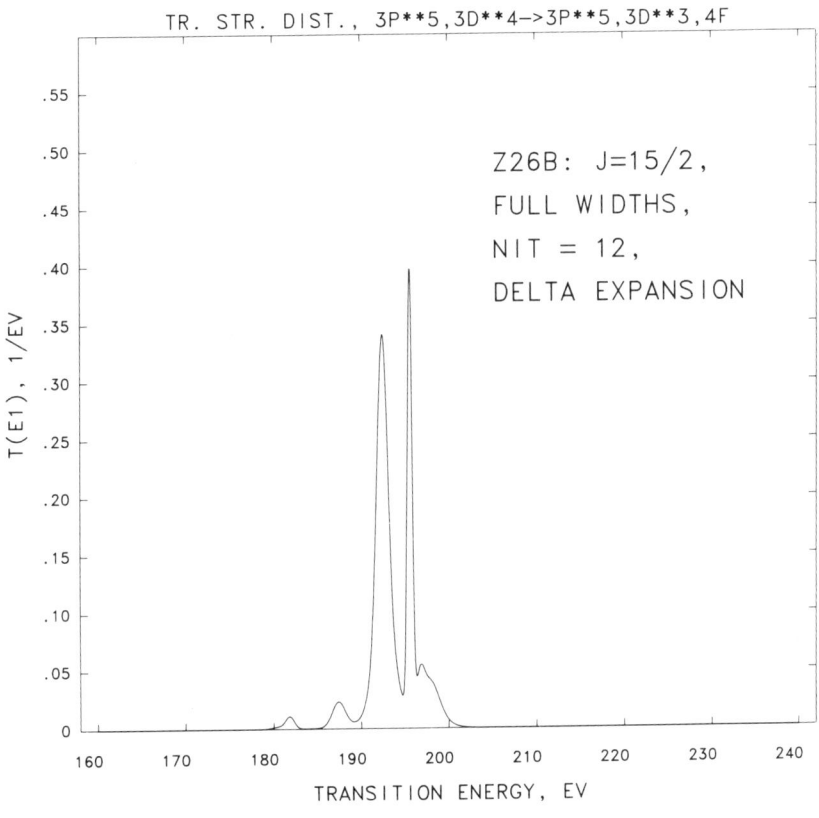

Figure 8b

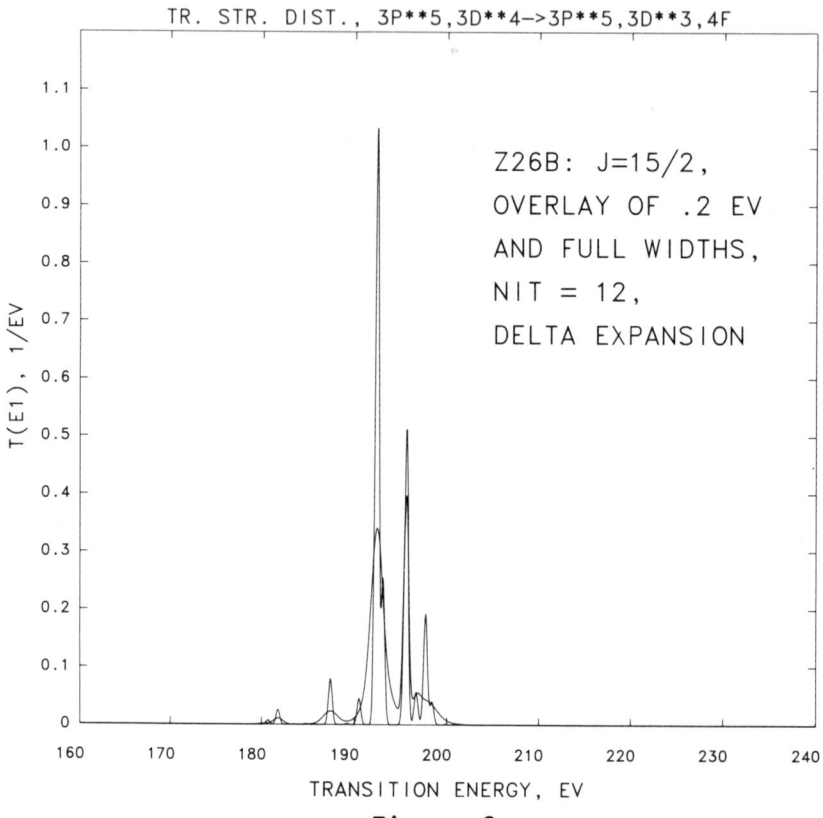

Figure 8c

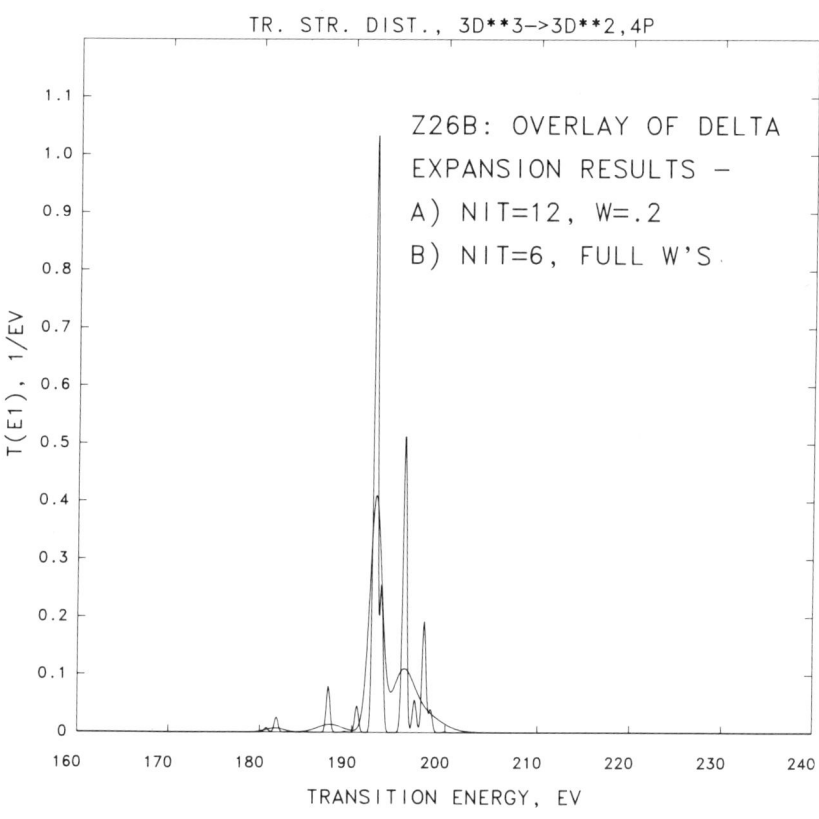

Figure 8d

Photoionization and Photorecombination Cross Sections of Non-Hydrogenic States in Plasmas

Balazs F. Rozsnyai

University of California, Lawrence Livermore National Laboratory
Livermore, California 94550

Verne L. Jacobs
E. O. Hulburt Center for Space Research
Naval Research Laboratory, Washington, D.C. 20375

Abstract

We use Hartree-Slater self-consistent-field wave functions to compute photoionization and photorecombination cross sections and rates for partially ionized iron atoms, both in their isolated state and in a plasma environment. The calculated cross sections are fitted to analytic expressions so that the photoionization and recombination rates can be given in closed forms. Our analytic representation satisfy the requirement that the oscillator-strength sums S_0, S_1 and S_2 must converge.

I. Introduction

Photoionization and photorecombination processes have been of interest in astrophysical problems. In recent years high temperature plasmas have been produced in the laboratory, and there has been a renewed interest in the theory of radiative transitions in hot plasmas. Analytic forms of the cross sections for photoionization and photorecombination are known only for hydrogen-like levels.[1,2,3] For non-hydrogenic levels these cross sections must be evaluated numerically. Non-hydrogenic levels occur in the case of partially ionized atoms and also in plasmas where the plasma screening yields a short range potential for the one-electron states. The purpose of this report is to present a procedure for the computation of the relevant parameters which are needed for analytic fits to the photoionization and recombination rates. We also investigate the effect of the plasma environment, which may alter these rates from their isolated ion values. The basis of our calculation is a temperature and density dependent Hartree-Slater self-consistent field model,[4] which provides the one-electron wave functions needed for the computation of the dipole integrals. We calculate the photoabsorption cross sections for each electron level from the bound-free threshold to a limiting value of the photon energy. Beyond this limit we use an analytic fit which satisfies the requirements that the logarithmic derivative of the cross section must be continuous and that the oscillator-strength sums must converge. In Section II we discuss theory, and in Section III we present some numerical results for iron ions.

II. Theory

In a Planckian (black body) radiation field of temperature kT the ionization rate from an electronic level nl of an ion or atom is given by

$$I_{nl} = \frac{8\pi}{h^3 c^2} \int_{\omega_{nl}}^{\infty} \frac{\omega^2}{e^{\omega/kT}-1} \sigma_i(n\ell,\omega) \, d\omega (\sec^{-1}) \quad , \qquad (1)$$

where n and l stand for the principal and angular momentum quantum numbers, respectively, $\omega_{n\ell}$ and ω are the ionization and photon energies, and $\sigma_i(n\ell,\omega)$ designates the photoionization cross section. In the case of local thermodynamic equilibrium (LTE) the induced recombination processes can be accounted for by multiplying the integrand in (1) by $1 - e^{-\omega/kT}$. Consequently, the net ionization rate is given by

$$I_{n\ell} = \frac{8\pi}{h^3 c^2} \int_{\omega_{n\ell}}^{\infty} \omega^2 e^{-\omega/kT} \sigma_i(n\ell,\omega) \, d\omega \quad (\text{sec}^{-1}) \quad . \tag{2}$$

If the ion is surrounded by free electrons, then the recombination rate into the subshell nl is given by

$$R_{n\ell} = \int_0^{\infty} \sigma_r(n\ell,\varepsilon) \, n(\varepsilon) \, v(\varepsilon) \, d\varepsilon \quad , \tag{3}$$

where $\sigma_r(n\ell,\varepsilon)$ denotes the recombination cross section, ε is the energy of the incident free electron, $v(\varepsilon)$ is the electron velocity, and $n(\varepsilon)$ represents the number of electrons with energies between ε and $\varepsilon+d\varepsilon$ per cc. The relation between the photoabsorption and photorecombination cross sections is given by

$$\sigma_r(n\ell,\varepsilon) = \frac{\omega^2}{2mc^2} \frac{G^{N+1}(n\ell)}{G^N(n\ell)} \sigma_i(n\ell,\omega) \quad , \tag{3a}$$

where $G^{N+1}(n\ell)$ and $G^N(n\ell)$ designate the statistical weights of an ion with N+1 and N electrons in the nl sub-shell, respectively, and the electron and photon energies are related by

$$\varepsilon = \omega - \omega_{n\ell} \quad .$$

If the free electrons have a Maxwellian distribution, then

$$n(\epsilon) = \frac{2\pi\rho_e}{(\pi kT)^{3/2}} \epsilon^{1/2} e^{-\epsilon/kt} \quad , \tag{3b}$$

where ρ_e is the total free electron density. Using $v(\epsilon) = [\frac{2\epsilon}{\omega}]^{1/2}$ the recombination rate is given by

$$R_{n\ell} = \frac{4\pi}{c^2} \frac{\rho_e}{(2m\pi kT)^{3/2}} e^{\omega_{n\ell}/kT} \frac{G^{N+1}(n\ell)}{G^N(n\ell)} \int_{\omega_{n\ell}}^{\infty} e^{-\omega/kT} \omega^2 \sigma_i(n\ell,\omega) \, d\omega \tag{4}$$

We note that the integral in Eq. (4) is the same as that in Eq. (2). It is customary to compute the recombination coefficient, which is the recombination rate divided by the electron density.

It is instructive to consider the recombination rate corresponding to Kramer's photoionization cross section

$$\sigma_i^K(n\ell,\omega) = \frac{2^6}{3\sqrt{3}} \pi a_0^2 \frac{e^2}{\hbar c} \frac{n}{Z^2} \left(\frac{\omega_{n\ell}}{\omega}\right)^3 \quad , \tag{5}$$

where a_0 stands for the Bohr radius. This yields for the recombination coefficient into an empty $n\ell$ shell

$$a_{n\ell}^K = \frac{2^8 \pi^2}{3\sqrt{3}c^2} a_0^2 \frac{e^2}{\hbar c} \frac{n}{Z^2} \omega_{n\ell}^3 \frac{2(2\ell+1)}{(2\pi mkT)^{3/2}} e^{\frac{\omega_{n\ell}}{kT}} \quad , \tag{6}$$

$$\times E_1(\frac{\omega_{n\ell}}{kT})$$

where E_1 is the exponential integral

$$E_1(x) = \int_x^\infty \frac{e^{-t}}{t} \, dt \quad .$$

In the case of hydrogenic degeneracy the recombination coefficients given by Eq. (6) can be summed with respect to ℓ for each principal quantum number. This results in the replacement of the term $2(2\ell+1)$ by $2n^2$, which gives the more familiar form of Kramer's recombination rates.

In order to obtain an analytic representation for the integral occurring in Eqs. (2) and (4), we fit the numerically evaluated photoionization cross section to the forms

$$\sigma_i(n\ell,\omega) = \frac{A_o}{\omega^p} \qquad \text{if } \omega_{n\ell} \leq \omega \leq \omega_{max} \tag{7a}$$

and

$$\sigma_i(n\ell,\omega) = \frac{B_o}{(\omega+\alpha)^{\frac{2\ell+7}{2}}} \qquad \text{if } \omega_{max} \leq \omega \leq \infty \tag{7b}$$

We first evaluate the photoabsorption cross sections numerically within the range $\omega_{n\ell} \leq \omega \leq \omega_{max}$, and we determine the parameters A_o and p by least square fitting. Next we determine the parameters B_o and α in such a way that the logarithmic derivative of the cross section is continuous at ω_{max}. We note that the analytic form given by Eq. (7b) insures that the oscillator sums S_o, S_1 and S_2 will converge.

Using Eqs. (7a) and (7b), the integral

$$F_{n1} = \int_{\omega_{n\ell}}^\infty \omega^2 \sigma_i(n\ell,\omega) e^{-\omega/kT} \, d\omega$$

can be given in a closed form, namely

$$F_{nl} = F^1_{nl} + F^2_{nl} \quad , \tag{8}$$

where

$$F^1_{nl} = A_0(kT)^{1-P} \left\{ \left(\frac{\omega_{n\ell}}{kT}\right)^{-\frac{P}{2}} \exp[-\omega_{n\ell}/kT] \, W_{-\frac{P}{2}, \frac{1-P}{2}}\left(\frac{\omega_{n\ell}}{kT}\right) \right. \tag{8a}$$

$$\left. - \left(\frac{\omega_{max}}{kT}\right)^{-\frac{P}{2}} \exp[-\omega_{n\ell}/kT] \, W_{-P, \frac{1-P}{2}}\left(\frac{\omega_{max}}{kT}\right) \right\}$$

and

$$F^2_{nl} = B_0 \exp(\alpha/kT) \left\{ I_{\frac{2\ell+3}{2}}(Z_0, kT) - 2\alpha \, I_{\frac{2\ell+5}{2}}(Z_0, kT) \right. \tag{8b}$$

$$\left. + \alpha^2 \, I_{\frac{2\ell+7}{2}}(Z_0, kT) \right\} \quad .$$

In Eq. (8a) W designates the Whittaker's function, and Eq. (8b) the I-s are determined by the recursion relation

$$I_{\frac{2n+1}{2}}(Z_0, kT) = \frac{2\exp(-Z/kT)}{(2n-1)Z_0^{\frac{2n-1}{2}}} - \frac{2}{(2n-1)kT} \, I_{\frac{2n-1}{2}}(Z_0, kT) \quad ,$$

where

$$Z_0 = \alpha + W_{max}$$

and

$$I_{1/2}(Z_0, kT) = \sqrt{\pi kT} \left\{ 1 - \text{ERFC}\left(\sqrt{\frac{Z_0}{kT}}\right) \right\} \qquad (9)$$

III Numerical Results

We illustrate our method by presenting numerical results for the FeXVII and FeIX ions. We performed the calculations for these ions in their isolated state and in a plasma environment having a temperature and density which is appropriate for these ionization states. Tables I-IV list the parameters of the analytic fits to the cross sections given by Eqs. (7a) and (7b). The principal and angular momentum quantum numbers for each shell are given in columns 1 and 2, respectively; column 3 contains the threshold energy, columns 4-7 give the parameters for the analytic fit, and in column 8 we tabulate the maximum photon energy for which the cross sections were evaluated numerically. All quantities in Tables I-IV are given in atomic units and the cross sections were normalized to one electron occupancy of the shells.

We present some detailed graphical illustrations for the isolated FeXVII ion in Figs. 1-3 and for the same ion in the plasma environment corresponding to that of Table II in Figs. 4-6. Figure 1 shows the photoionization cross sections of the occupied levels and some of the initially empty levels, all normalized to one electron occupancy. The values of the integral occurring in Eqs. (20 and (4) are shown in Fig. 2. Here we also present the results obtained from an analytic fit to these integrals, using the representation

$$F_{n\ell}(kT) = F_i \left\{ 1 - \exp[-(kT\beta)^{1/2}] \right\} \qquad (10)$$

and which are shown by the dotted curves. In Eq. (10) R_i is the value of the integral in the limit of infinitely high temperatures and β is a fitting parameter. We show the recombination rates for the initially empty levels in Fig. 3. In Fig. 3 we also show the recombination rates corresponding to Kramer's cross section by the dotted lines labeled by K-s. To compare our results with other published data, we computed the sum of the photoionization cross sections for the occupied levels of each ion. Our data for the occupied levels are in agreement with those of Reilman and Manson.[5]

In Figs. 4-6 we show the same quantities as in Figs. 1-3, but for the case when the FeXVII ion is in a plasma environment corresponding to that of Table II. A comparison of Figs. 1 and 4 reveals that the plasma environment considerably changes the threshold energies and photoionization profiles of the highly excited levels. The reduced threshold energies or continuum lowerings are mainly responsible for the altered recombination rates of the upper levels in a plasma. However, the recombination rates are to a good approximation inversely proportional to n^3. Consequently, the plasma effect is not very important for those states which provide the dominant contribution to the total recombination rates. A close inspection of Tables III and IV shows that similar plasma effects occur for the FeIX ion.

In summary, we conclude that the Hartree-Slater self-consistent-field wave functions give photoionization and recombination rates which differ considerably from the prediction of Kramer's classical formula, and in a dense plasma the effect of the plasma environment on the highly excited levels has to be taken into account.

Acknowledgment

Work performed under the auspices of the U.S. Department of Energy by Lawrence Livermore National Laboratory under contract #W-7405-Eng-48.

Table I. Analytic fits for the photoionization cross sections of the occupied nl shells of an isolated FeXVII ion and of the excited nl shells of an isolated FeXVI ion.

n	l	ω_{nl}	A_o	p	B_o	α	ω_{max}
1	0	2.802e+02	1.935e+03	2.680e+00	1.233e+06	2.571e+02	8.400e+02
2	0	4.935e+01	4.122e+00	2.049e+00	3.773e+04	1.047e+02	1.479e+02
2	1	4.560e+01	1.507e+02	2.900e+00	2.850e+06	7.544e+01	1.367e+02
3	0	1.805e+01	5.837e-01	1.901e+00	2.213e+03	3.831e+01	4.557e+01
3	1	1.692e+01	1.907e+00	2.235e+00	2.534e+05	4.616e+01	4.557e+01
3	2	1.536e+01	1.751e+01	3.145e+00	3.043e+06	3.411e+01	4.557e+01
4	0	9.021e+00	1.796e-01	1.781e+00	2.755e+02	1.741e+01	1.804e+01
4	1	8.566e+00	2.971e-01	1.915e+00	2.456e+04	2.436e+01	1.804e+01
4	2	7.951e+00	1.192e+00	2.549e+00	4.174e+05	2.089e+01	1.804e+01
4	3	7.568e+00	4.116e+00	3.549e+00	1.096e+06	1.512e+01	1.819e+01
5	0	5.264e+00	1.104e-01	1.838e+00	1.663e+02	1.903e+01	2.104e+01
5	1	5.040e+00	1.497e-01	1.917e+00	1.624e+04	2.713e+01	2.014e+01
5	2	4.739e+00	3.608e-01	2.370e+00	3.687e+05	2.502e+01	1.894e+01
5	3	4.558e+00	7.375e-01	3.035e+00	7.233e+05	1.467e+01	1.285e+01
5	4	4.450e+00	1.433e+00	3.989e+00	5.299e+05	8.804e+00	1.000e+01
6	0	3.334e+00	6.793e-02	1.833e+00	4.895e+01	1.212e+01	1.333e+01
6	1	3.208e+00	8.274e-02	1.866e+00	3.604e+03	1.810e+01	1.282e+01
6	2	3.039e+00	1.520e-01	2.193e+00	9.229e+04	1.832e+01	1.214e+01
6	3	2.938e+00	2.446e-01	2.726e+00	3.415e+05	1.316e+01	9.506e+00
6	4	2.881e+00	3.521e-01	3.386e+00	5.983e+05	9.321e+00	7.673e+00
6	5	2.857e+00	3.376e-01	4.186e+00	3.834e+05	6.465e+00	6.272e+00
7	0	2.211e+00	4.491e-02	1.817e+00	1.742e+01	8.184e+00	8.839e+00
7	1	2.133e+00	5.420e-02	1.848e+00	8.754e+02	1.224e+01	8.526e+00
7	2	2.029e+00	8.616e-02	2.094e+00	2.182e+04	1.319e+01	8.109e+00
7	3	1.967e+00	1.169e-01	2.641e+00	5.565e+05	1.723e+01	1.179e+01
7	4	1.933e+00	1.258e-01	2.948e+00	3.664e+05	8.723e+00	5.650e+00
7	5	1.917e+00	1.282e-01	3.551e+00	5.062e+05	6.699e+00	4.807e+00
7	6	1.914e+00	1.081e-01	4.598e+00	9.553e+04	4.267e+00	4.002e+00
8	0	1.501e+00	3.121e-02	1.701e+00	1.292e+01	7.400e+00	6.994e+00
8	1	1.449e+00	3.463e-02	1.578e+00	1.136e+03	1.295e+01	6.994e+00

Table I Continued

n	l	ω_{nl}	A_o	p	B_o	α	ω_{max}
8	2	1.380e+00	4.907e-02	1.734e+00	4.263e+04	1.519e+01	6.994e+00
8	3	1.340e+00	5.911e-02	2.268e+00	3.746e+05	1.499e+01	8.032e+00
8	4	1.317e+00	5.248e-02	2.257e+00	2.242e+07	1.834e+01	7.896e+00
8	5	1.307e+00	3.582e-02	1.349e+00	1.256e+09	1.773e+01	3.344e+00
8	6	1.304e+00	3.843e-02	2.865e+00	4.944e+06	6.943e+00	2.998e+00
8	7	1.304e+00	1.203e-02	1.915e+00	3.788e+09	1.222e+01	2.726e+00
9	0	1.023e+00	2.274e-02	1.146e+00	1.103e+02	1.437e+01	6.994e+00
9	1	9.870e-01	2.678e-02	1.385e+00	2.304e+03	1.573e+01	6.994e+00
9	2	9.391e-01	3.690e-02	1.357e+00	2.574e+05	2.136e+01	6.994e+00
9	3	9.110e-01	3.939e-02	1.657e+00	3.511e+06	2.045e+01	6.994e+00
9	4	8.953e-01	3.820e-02	2.026e+00	2.937e+07	1.889e+01	6.994e+00
9	5	8.880e-01	3.853e-02	1.303e+00	5.236e+08	1.543e+01	2.794e+00
9	6	8.858e-01	2.952e-02	2.651e+00	1.283e+06	5.735e+00	2.220e+00
9	7	8.855e-01	1.673e-02	6.286e+00	6.311e+01	1.318e+00	1.966e+00
9	8	8.854e-01	3.529e-03	1.402e+01	8.544e-05	3.188e-01	1.775e+00
10	0	6.858e-01	2.201e-02	1.813e+00	2.280e+00	3.720e+00	3.997e+00
10	1	6.600e-01	2.338e-02	1.807e+00	5.912e+01	5.954e+00	3.997e+00
10	2	6.253e-01	2.832e-02	1.759e+00	2.168e+04	1.488e+01	6.993e+00
10	3	6.050e-01	3.451e-02	1.432e+00	2.096e+05	1.109e+01	3.131e+00
10	4	5.937e-01	3.830e-02	1.348e+00	3.187e+06	1.092e+01	2.392e+00
10	5	5.883e-01	2.485e-02	2.364e+00	8.905e+04	5.159e+00	1.988e+00
10	6	5.867e-01	1.079e-02	4.218e+00	4.237e+02	2.154e+00	1.720e+00
10	7	5.864e-01	2.086e-03	7.509e+00	2.466e-01	6.054e-01	1.520e+00
10	8	5.864e-01	4.234e-05	1.522e+01	5.017e-07	3.387e-01	1.386e+00
11	0	4.394e-01	2.437e-02	1.272e+00	1.848e+01	7.003e+00	3.996e+00
11	1	4.201e-01	2.464e-02	1.355e+00	4.263e+02	9.276e+00	3.996e+00
11	2	3.942e-01	2.189e-02	1.833e+00	1.800e+02	4.498e+00	2.248e+00
11	3	3.791e-01	2.333e-02	1.815e+00	3.012e+03	5.421e+00	2.100e+00
11	4	3.707e-01	2.246e-02	1.829e+00	3.971e+04	6.062e+00	1.956e+00

Table I Continued

n	l	ω_{nl}	A_o	p	B_o	α	ω_{max}
11	5	3.666e-01	8.879e-03	3.023e+00	4.799e+02	2.663e+00	1.470e+00
11	6	3.653e-01	6.988e-03	3.359e+00	1.444e+03	2.686e+00	1.469e+00
11	7	3.651e-01	8.272e-04	5.762e+00	1.357e+00	1.038e+00	1.262e+00
11	8	3.651e-01	8.797e-05	8.178e+00	5.298e-03	4.285e-01	1.055e+00
12	0	2.538e-01	2.420e-02	1.158e+00	2.976e+01	8.079e+00	3.996e+00
12	1	2.390e-01	2.661e-02	1.145e+00	1.316e+03	1.171e+01	3.996e+00
12	2	2.192e-01	4.977e-03	2.548e+00	6.838e+00	3.194e+00	2.757e+00
12	3	2.076e-01	2.396e-02	1.510e+00	2.132e+04	7.685e+00	2.326e+00
12	4	2.011e-01	2.837e-02	1.395e+00	9.974e+05	9.544e+00	2.180e+00
12	5	1.980e-01	1.216e-02	2.022e+00	2.152e+04	4.486e+00	1.400e+00
12	6	1.970e-01	4.901e-03	2.749e+00	2.036e+03	2.915e+00	1.187e+00
13	0	1.105e-01	1.143e-02	2.019e+00	5.613e-02	5.852e-01	7.979e-01
13	1	9.888e-02	6.022e-03	2.272e+00	5.282e-02	6.536e-01	6.666e-01
13	2	8.238e-02	8.856e-03	2.170e+00	5.159e-01	1.119e+00	7.296e-01

Table II Same as Table I. But in a plasma of kT = 140 eV and .1 g/cc matter density, which corresponds to a Debye-Huckel screening length of 3.08 a_0.

n	l	ω_{nl}	A_0	p	B_0	α	ω_{max}
1	0	2.785e+02	2.970e+03	2.759e+00	9.978e+05	2.242e+02	8.350e+02
2	0	4.803e+01	4.164e+00	2.039e+00	3.915e+04	1.031e+02	1.440e+02
2	1	4.419e+01	1.284e+02	2.864e+00	2.900e+06	7.565e+01	1.325e+02
3	0	1.606e+01	4.170e-01	1.817e+00	2.431e+03	4.092e+01	4.417e+01
3	1	1.490e+01	1.467e+00	2.180e+00	2.504e+05	4.699e+01	4.416e+01
3	2	1.335e+01	1.520e+01	3.161e+06	2.258e+06	3.269e+01	4.416e+01
4	0	7.347e+00	1.371e-01	1.690e+00	2.668e+02	1.719e+01	1.605e+01
4	1	6.890e+00	3.645e-01	2.000e+00	1.443e+04	2.005e+01	1.605e+01
4	2	6.296e+00	1.878e+00	2.731e+00	1.917e+05	1.627e+01	1.605e+01
4	3	5.782e+00	7.584e+00	3.820e+00	3.653e+05	1.080e+01	1.540e+01
5	0	3.671e+00	7.848e-02	1.715e+00	1.153e+02	1.528e+01	1.467e+01
5	1	3.446e+00	1.221e-01	1.824e+00	7.930e+03	2.021e+01	1.377e+01
5	2	3.154e+00	3.473e-01	2.337e+00	1.165e+05	1.706e+01	1.261e+01
5	3	2.913e+00	6.486e-01	3.000e+00	1.890e+05	1.011e+01	8.668e+00
5	4	2.846e+00	1.519e+00	4.086e+00	7.515e+04	5.219e+00	6.248e+00
6	0	1.803e+00	4.285e-02	1.611e+00	2.701e+01	8.449e+00	7.204e+00
6	1	1.677e+00	6.893e-02	1.754e+00	9.992e+02	1.095e+01	6.994e+00
6	2	1.513e+00	1.206e-01	2.065e+00	2.108e+04	1.164e+01	6.994e+00
6	3	1.379e+00	1.257e-01	2.454e+00	3.645e+05	1.363e+01	8.268e+00
6	4	1.330e+00	2.046e-01	3.089e+00	6.340e+04	5.553e+00	3.889e+00
6	5	1.311e+00	2.184e-01	4.003e+00	3.642e+04	3.923e+00	3.492e+00
7	0	7.554e-01	2.424e-02	1.519e+00	2.118e+01	9.117e+00	6.994e+00
7	1	6.807e-01	3.671e-02	1.585e+00	2.282e+02	7.351e+00	3.997e+00
7	2	5.812e-01	4.500e-02	1.809e+00	2.677e+04	1.427e+01	6.993e+00
7	3	4.983e-01	5.008e-02	1.898e+00	2.869e+04	7.599e+00	3.134e+00
7	4	4.596e-01	5.585e-02	2.215e+00	4.241e+04	5.478e+00	2.295e+00
7	5	4.370e-01	5.247e-02	2.776e+00	5.553e+04	4.417e+00	2.143e+00
7	6	4.119e-01	2.828e-02	3.311e+00	1.056e+04	2.947e+00	1.576e+00
8	0	1.538e-01	4.727e-02	1.216e+00	4.521e+01	7.503e+00	3.996e+00
8	1	1.053e-01	4.308e-02	1.484e+00	1.209e+01	2.517e+00	1.239e+00
8	2	4.478e-02	1.098e-01	1.271e+00	1.134e+02	2.555e+00	7.678e-01
8	3	7.220e-03	1.837e-01	7.901e-01	3.738e+03	3.729e+00	5.160e-01

Table III Same as Table I. For the isolated FeIX and FeVIII ions.

n	l	ω_{nl}	A_o	p	B_o	α	ω_{max}
1	0	2.669e+02	1.725e+03	2.659e+00	1.247e+06	2.530e+02	8.000e+02
2	0	3.746e+01	2.427e+00	1.935e+00	3.131e+04	9.087e+01	1.123e+02
2	1	3.350e+01	8.025e+01	2.765e+00	2.134e+06	6.302e+01	1.004e+02
3	0	9.608e+00	2.513e-01	1.735e+00	1.442e+03	3.407e+01	3.348e+01
3	1	8.384e+00	4.166e-01	1.874e+00	2.170e+05	4.692e+01	3.348e+01
3	2	6.366e+00	4.027e+00	2.802e+00	4.679e+05	1.836e+01	1.906e+01
4	0	3.661e+00	5.585e-02	1.620e+00	5.813e+01	1.114e+01	9.601e+00
4	1	3.239e+00	4.567e-02	1.467e+00	6.751e+03	1.985e+01	9.601e+00
4	2	2.581e+00	1.409e-01	2.006e+00	9.788e+04	1.672e+01	9.600e+00
4	3	2.056e+00	2.473e-01	3.213e+00	7.319e+03	5.821e+00	5.690e+00
5	0	1.773e+00	2.591e-02	1.622e+00	1.516e+01	8.211e+00	7.088e+00
5	1	1.596e+00	1.988e-02	1.404e+00	1.546e+03	1.542e+01	6.994e+00
5	2	1.320e+00	4.235e-02	1.743e+00	3.512e+04	1.508e+01	6.994e+00
5	3	1.097e+00	6.480e-02	2.551e+00	5.858e+03	5.973e+00	3.858e+00
5	4	9.675e-01	2.032e-02	3.502e+00	2.289e+02	2.822e+00	2.472e+00
6	0	9.731e-01	1.499e-02	1.624e+00	8.461e+00	8.078e+00	6.994e+00
6	1	8.828e-01	1.254e-02	1.459e+00	7.375e+02	1.457e+01	6.994e+00
6	2	7.399e-01	2.091e-02	1.688e+00	2.295e+04	1.579e+01	6.993e+00
6	3	6.221e-01	2.927e-02	2.208e+00	2.486e+03	5.333e+00	2.744e+00
6	4	5.558e-01	1.221e-02	2.917e+00	1.946e+02	2.767e+00	1.762e+00
6	5	5.193e-01	3.251e-03	3.690e+00	1.942e+01	1.819e+00	1.395e+00
7	0	5.532e-01	9.061e-03	1.556e+00	2.284e+00	4.992e+00	3.966e+00
7	1	5.009e-01	8.573e-03	1.419e+00	1.100e+02	8.673e+00	3.996e+00
7	2	4.170e-01	1.355e-02	1.473e+00	3.170e+03	9.743e+00	3.565e+00
7	3	3.468e-01	1.725e-02	1.892e+00	1.301e+03	4.887e+00	2.007e+00
7	4	3.007e-01	8.889e-03	2.332e+00	2.686e+02	2.994e+00	1.351e+00
7	5	2.865e-01	3.352e-03	2.950e+00	3.618e+01	1.983e+00	1.054e+00
7	6	2.782e-01	8.629e-04	3.512e+00	3.795e+00	1.427e+00	8.371e-01
8	0	3.045e-01	6.100e-03	1.542e+00	1.617e+00	5.072e+00	3.996e+00
8	1	2.715e-01	5.630e-03	1.480e+00	5.507e+01	8.155e+00	3.996e+00
8	2	2.180e-01	9.323e-03	1.324e+00	1.034e+03	7.805e+00	2.474e+00

Table III Continued

n	l	ω_{nl}	A_o	p	B_o	α	ω_{max}
8	3	1.727e-01	1.327e-02	1.620e+00	6.799e+02	4.368e+00	1.450e+00
8	4	1.476e-01	7.649e-03	1.947e+00	1.273e+02	2.657e+00	9.311e-01
8	5	1.339e-01	3.776e-03	2.282e+00	5.597e+01	2.116e+00	7.763e-01
8	6	1.280e-01	1.497e-03	2.632e+00	1.210e+01	1.638e+00	6.280e-01
8	7	1.268e-01	2.552e-04	3.320e+00	6.665e-01	1.201e+00	5.555e-01
9	0	1.449e-01	9.263e-03	9.851e-01	2.552e+01	1.020e+01	3.996e+00
9	1	1.227e-01	8.839e-03	9.978e-01	9.934e+02	1.403e+01	3.996e+00
9	2	8.661e-02	8.261e-03	1.085e+00	9.236e+03	1.263e+01	3.105e+00
9	3	5.570e-02	2.755e-02	8.990e-01	4.238e+05	1.204e+01	1.933e+00
9	4	3.857e-02	8.242e-04	2.243e+00	2.462e+01	2.974e+00	1.269e+00
9	5	2.913e-02	3.974e-03	1.666e+00	1.118e+03	3.389e+00	8.262e-01
9	6	2.480e-02	1.696e-04	2.728e+00	1.239e-01	1.142e+00	4.598e-01
9	7	2.375e-02	8.194e-05	2.949e+00	3.850e-02	9.892e-01	3.863e-01
9	8	2.365e-02	1.906e-02	1.000e+00	1.555e+05	3.294e+00	3.137e-01
10	0	3.621e-02	7.149e-03	1.426e+00	4.446e-02	7.714e-01	5.307e-01
10	1	1.930e-02	1.373e-02	1.267e+00	1.386e-01	8.938e-01	3.503e-01

Table IV Same as Table III. but in a plasma of kT = 10 eV and .1 g/cc matter density, which corresponds to a Debye-Huckel screening length of 1.6 a_0.

n	l	ω_{nl}	A_o	p	B_o	α	ω_{max}
1	0	2.661e+02	2.877e+03	2.752e+00	9.866e+05	2.168e+02	7.979e+02
2	0	3.692e+01	2.206e+00	1.903e+00	3.415e+04	9.285e+01	1.107e+02
2	1	3.290e+01	8.864e+01	2.787e+00	1.988e+06	6.059e+01	9.862e+01
3	0	8.782e+00	1.860e-01	1.647e+00	1.682e+03	3.698e+01	3.287e+01
3	1	7.540e+00	3.579e-01	1.825e+00	2.372e+05	4.818e+01	3.287e+01
3	2	5.492e+00	4.192e+00	2.818e+00	3.407e+05	1.635e+01	1.717e+01
4	0	2.636e+00	3.635e-02	1.469e+00	6.237e+01	1.213e+01	8.775e+00
4	1	2.214e+00	3.140e-02	1.323e+00	7.704e+03	2.108e+01	8.775e+00
4	2	1.568e+00	7.108e-02	1.736e+00	1.434e+05	1.903e+01	8.774e+00
4	3	1.042e+00	7.889e-02	2.624e+00	4.903e+03	5.565e+00	3.767e+00
5	0	8.922e-01	1.182e-02	1.269e+00	3.150e+01	1.229e+01	6.994e+00
5	1	7.208e-01	1.151e-02	1.167e+00	3.257e+03	1.996e+01	6.993e+00
5	2	4.562e-01	1.627e-02	1.327e+00	1.356e+05	2.199e+01	6.993e+00
5	3	2.430e-01	2.545e-02	1.758e+00	7.198e+03	6.340e+00	2.350e+00
5	4	7.425e-02	6.974e-03	1.923e+00	8.090e+02	3.764e+00	1.298e+00
6	0	1.795e-01	4.586e-03	1.110e+00	6.990e+00	8.601e+00	3.996e+00
6	1	1.000e-01	3.530e-03	1.075e+00	2.549e+02	1.273e+01	3.996e+00

References

1. A. Burgess and M. J. Seaton, Mon. Not. R.A.S. 120, 121 (1960).

2. A. Burgess, Mem. R.A.S. 69, 1 (1964).

3. P. M. Fazio and G. E. Copeland, Phys. Rev. A31, 187 (1985).

4. B. F. Rozsnyai, Phys. Rev. 145, 1137 (1972).

5. R. F. Reilman and S. T. Manson, Astr. J. Suppl. 40, 815 (1979).

Figure Captions

Fig. 1 Photoionization cross sections versus incident photon energy for the occupied 1s, 2s and 2p levels of an isolated FeXVII ion, and for the upper nl levels of an isolated FeXVI ion. The principal and angular momentum quantum numbers are indicated by N and L, respectively. The threshold energy is given by ET, all quantities are given in atomic units. All cross sections are normalized to one electron occupancy.

Fig. 2 Computed values of the integral F_{nl} versus temperature for the isolated FeXVII and FeXVI ions, as obtained by using the cross sections of Fig. 1. The dotted curves represent the analytic approximation given by Eq. (10). The quantities FIF and BT stand for the symbols Fi and β of Eq. (10), respectively.

Fig. 3 Photorecombination rates (in units of $cm^3\ sec^{-1}$) versus temperature for an isolated FeXVII ion. The dotted curves labeled by K-s are obtained from Kramer's approximate formulas given by Eqs. (7a) and (7b).

Fig. 4 Same as Fig. 1 but in a plasma environment as given by Table II.

Fig. 5 Same as Fig. 2 but in a plasma environment as given by Table II.

Fig. 6 Same as Fig. 3 but in a plasma environment as given by Table II.

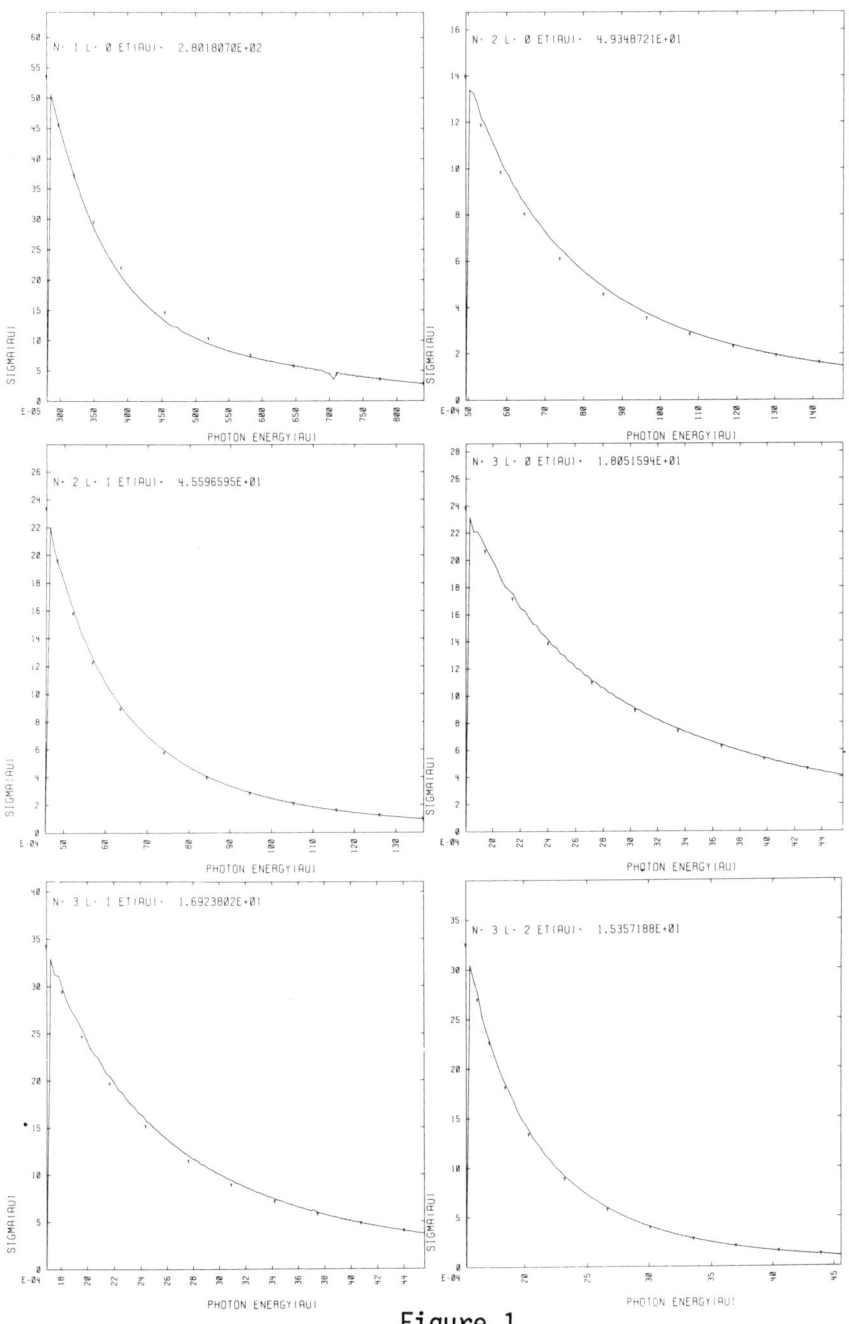

Figure 1

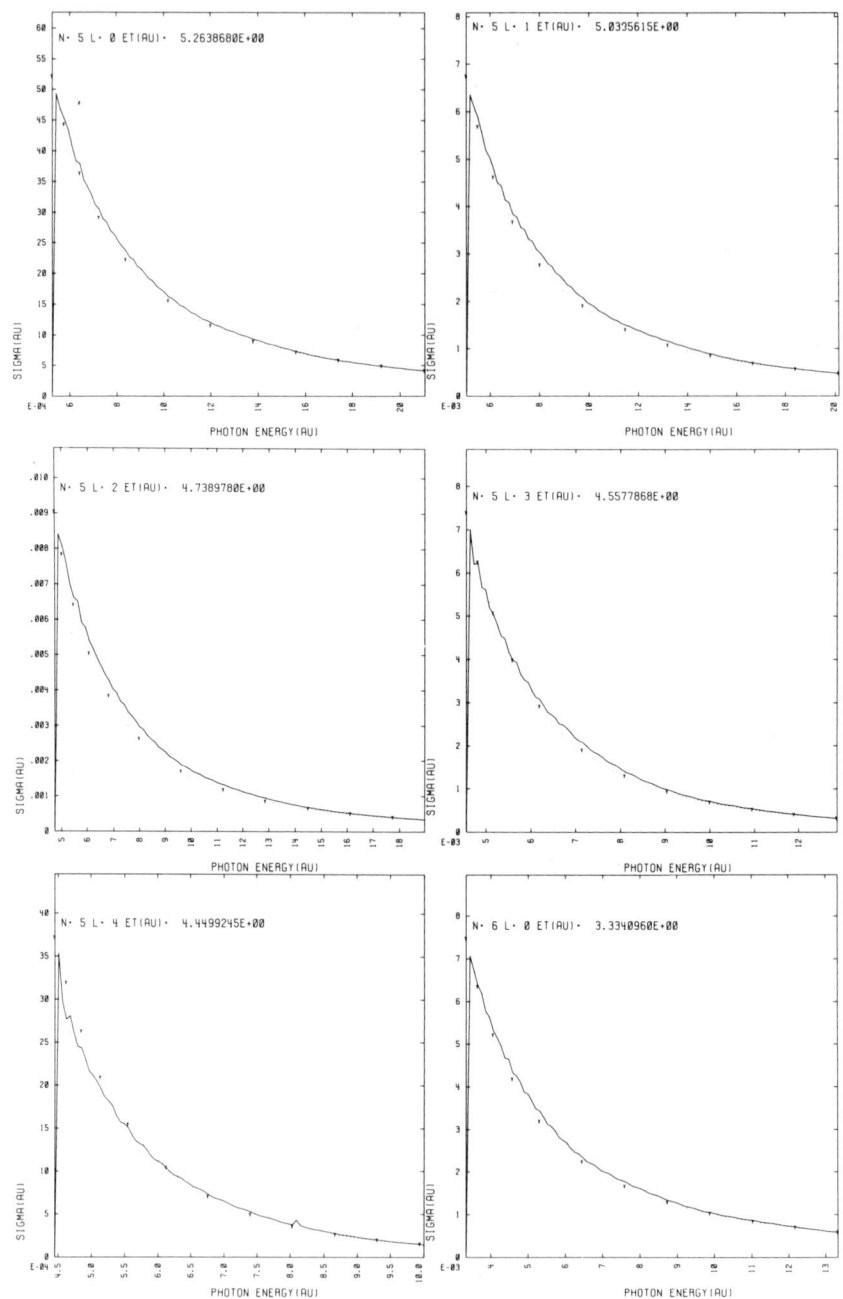

Figure 1 (Continued)

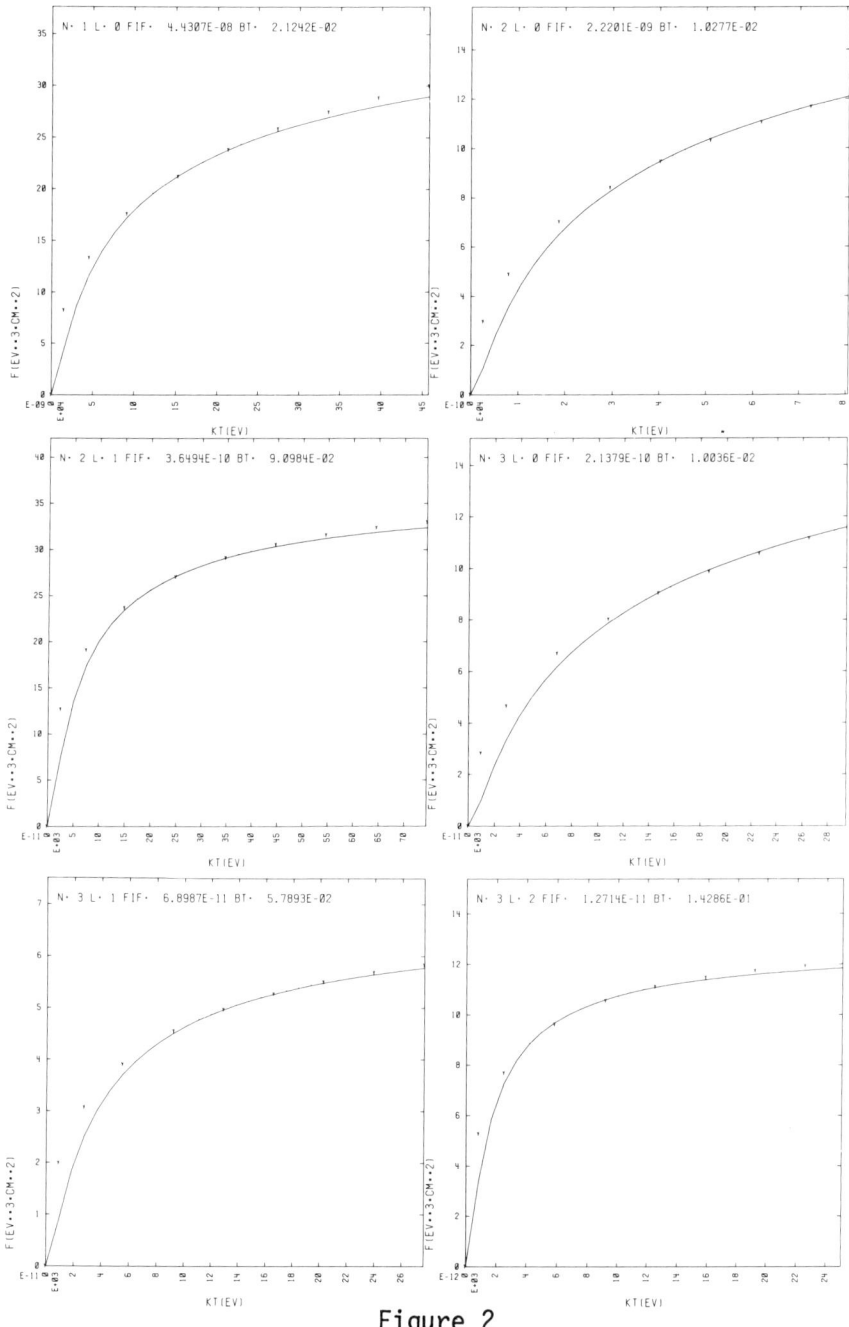

Figure 2

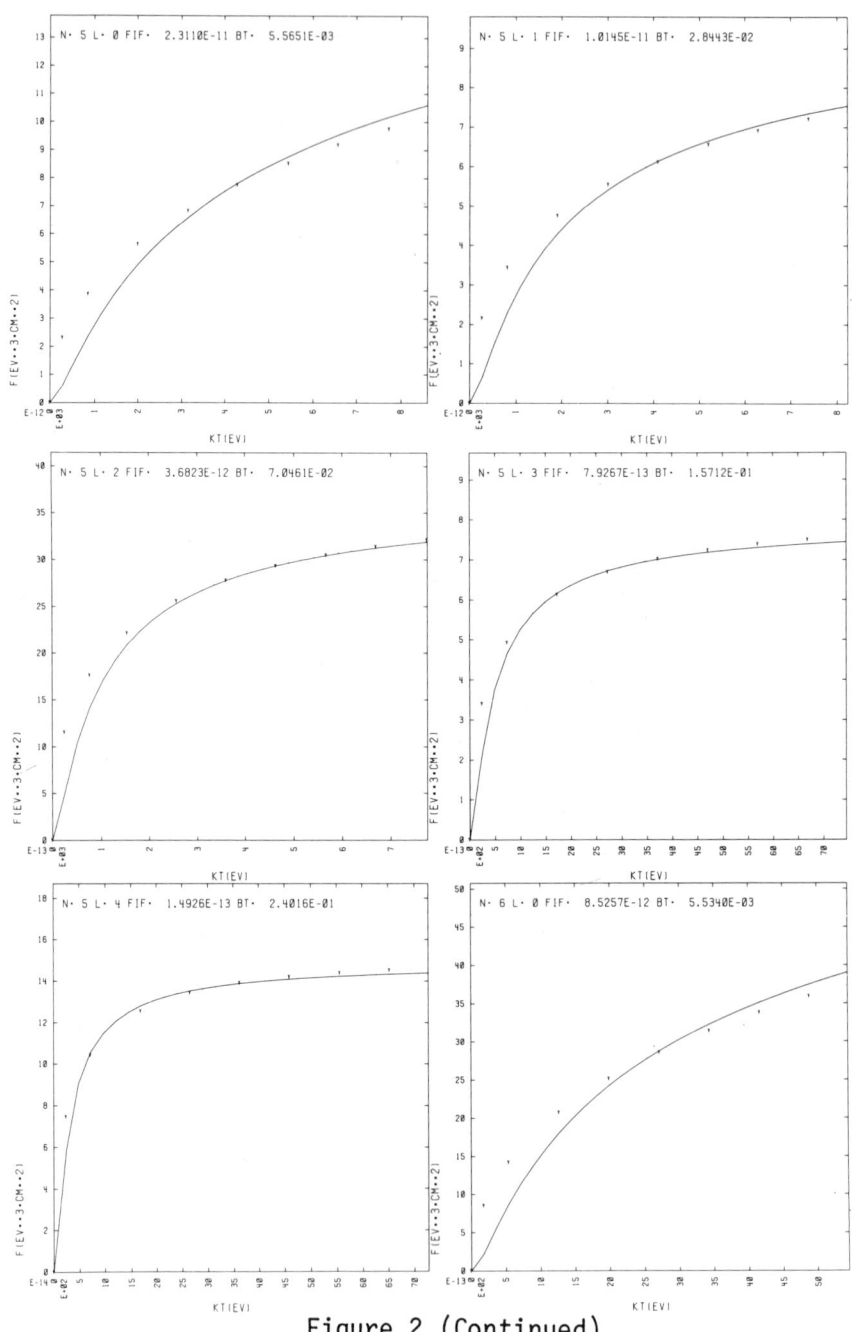

Figure 2 (Continued)

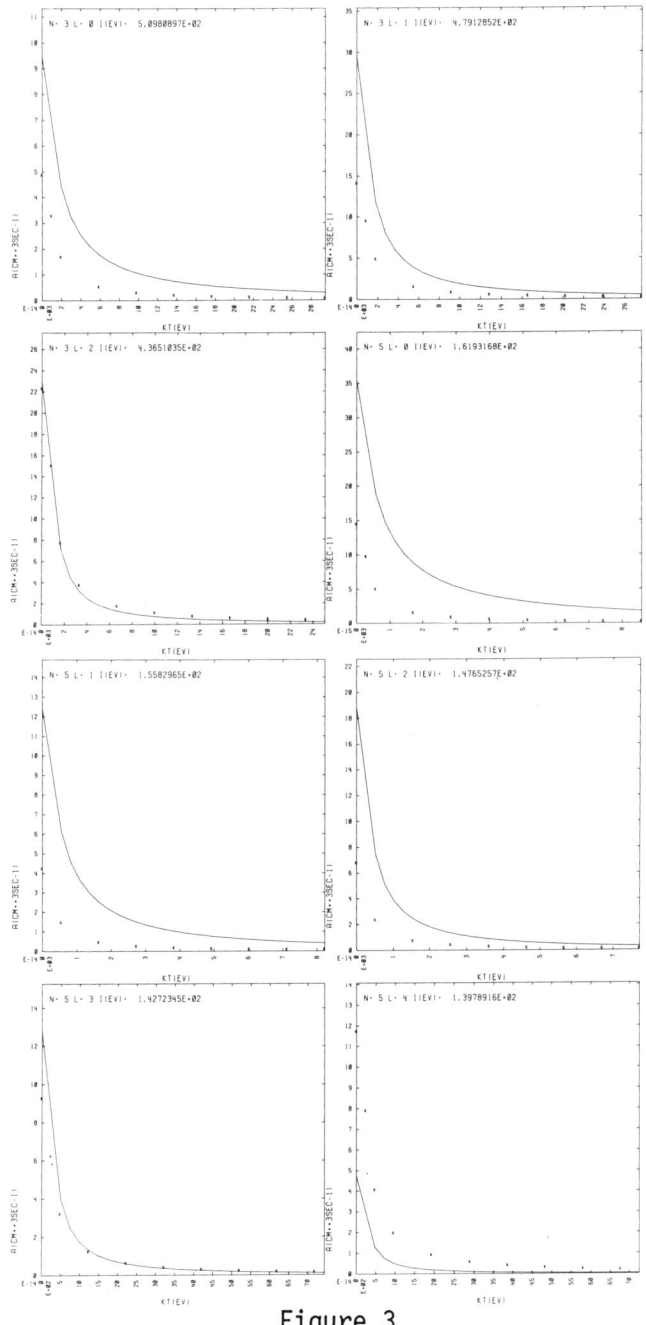

Figure 3

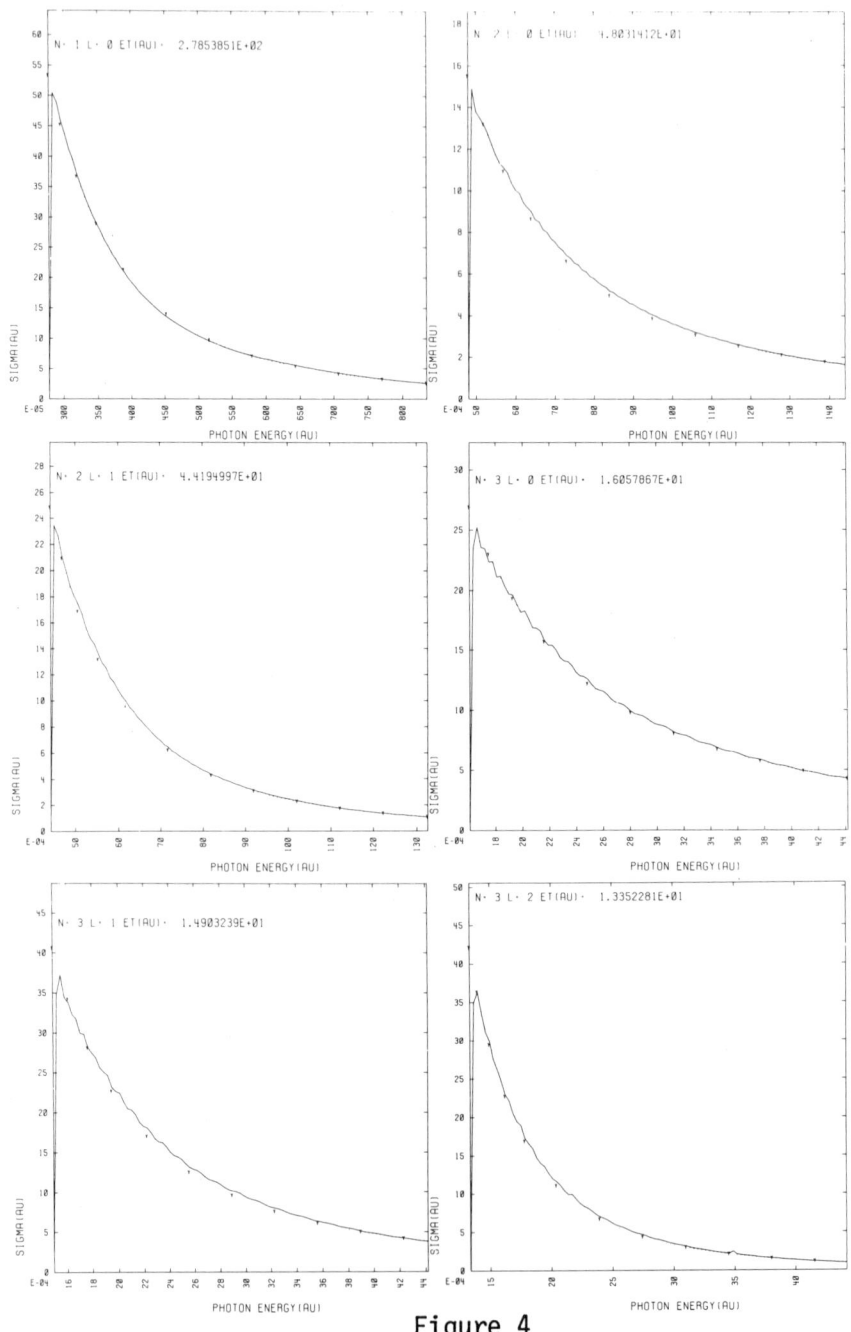

Figure 4

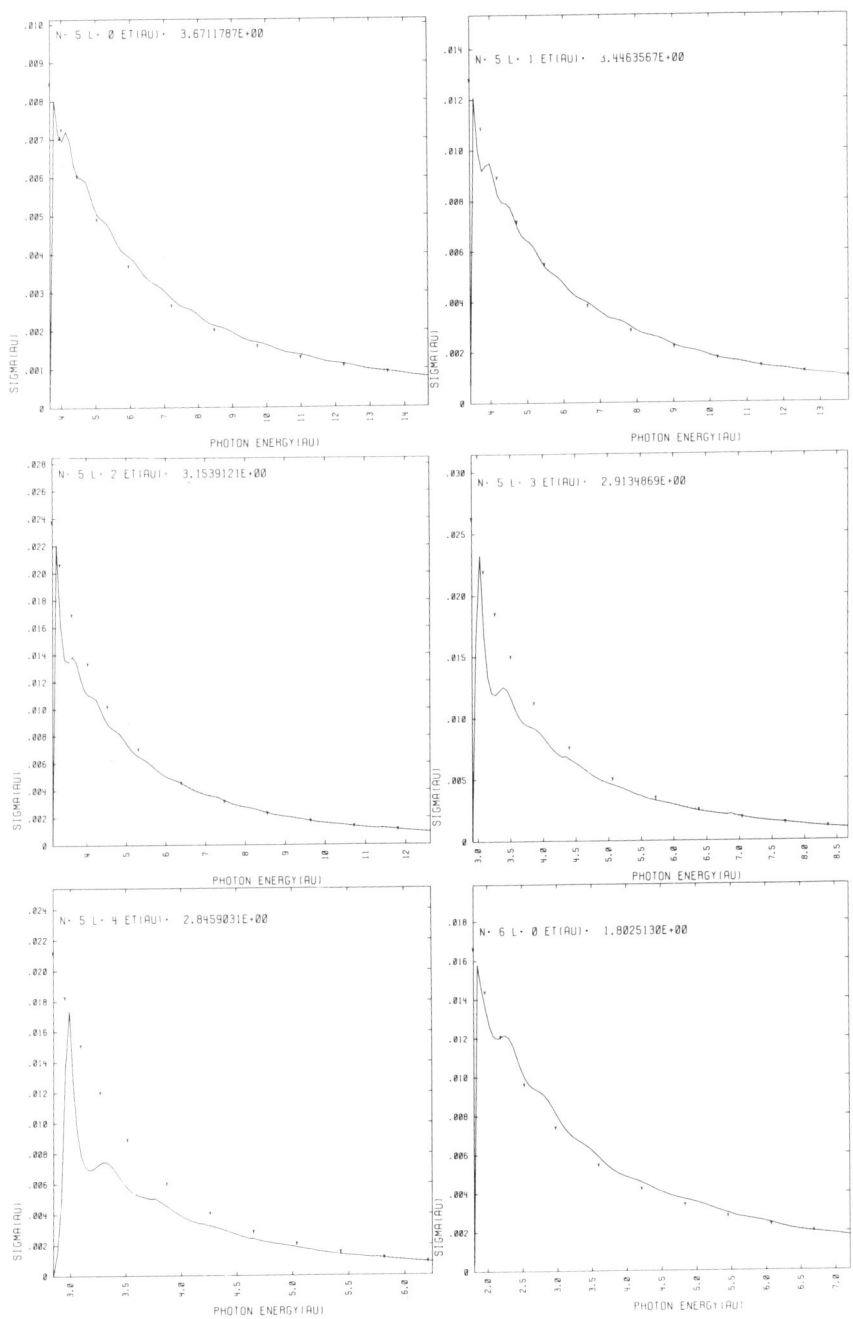

Figure 4(Continued)

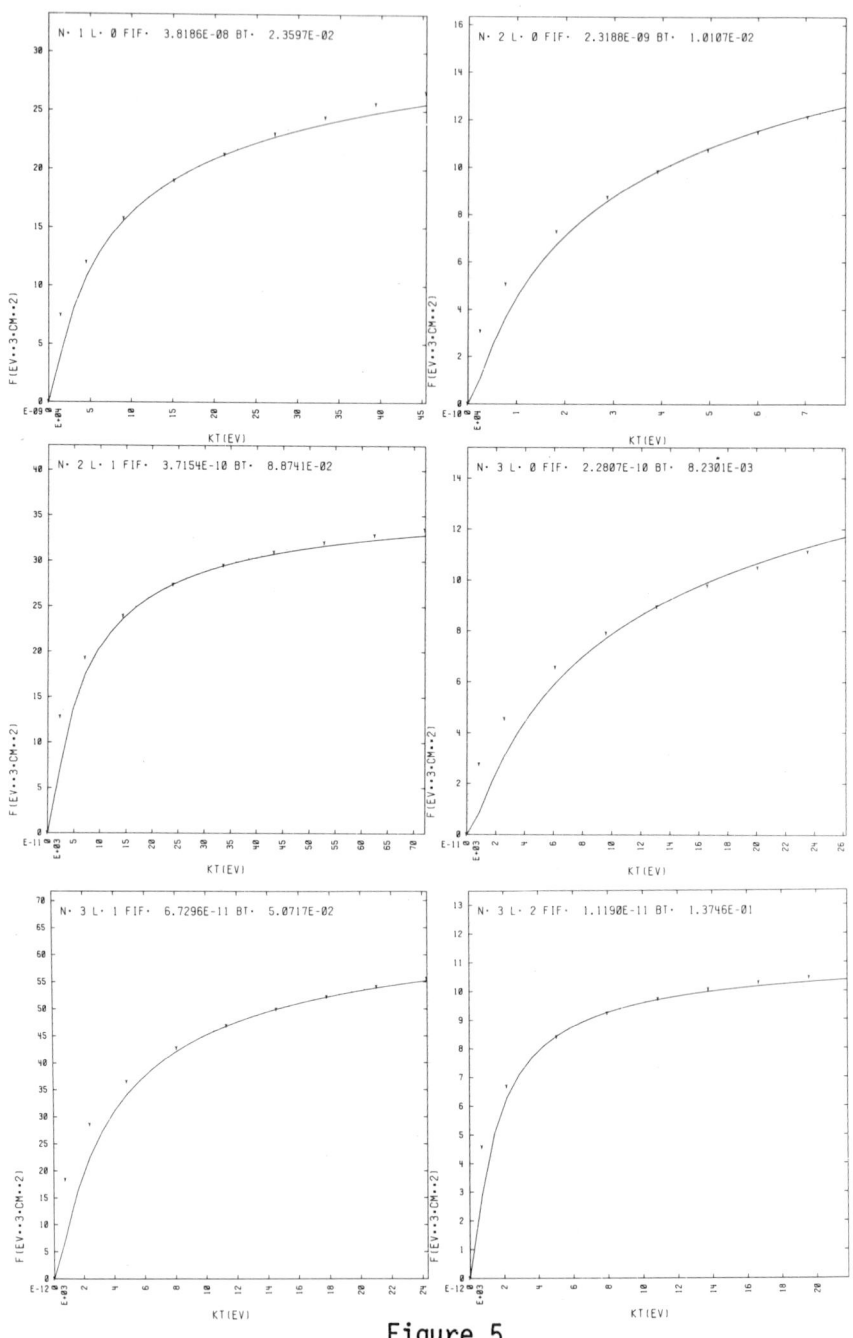

Figure 5

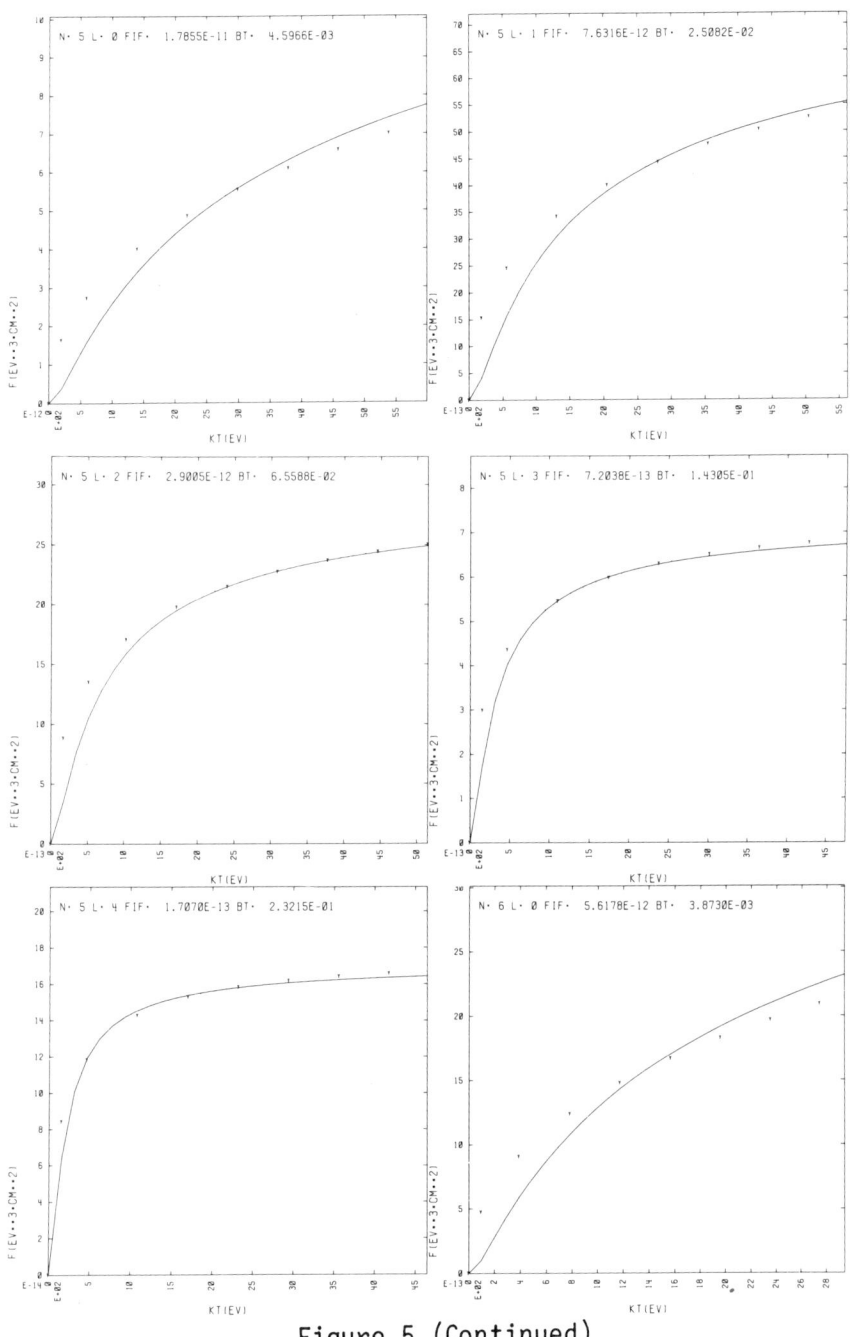

Figure 5 (Continued)

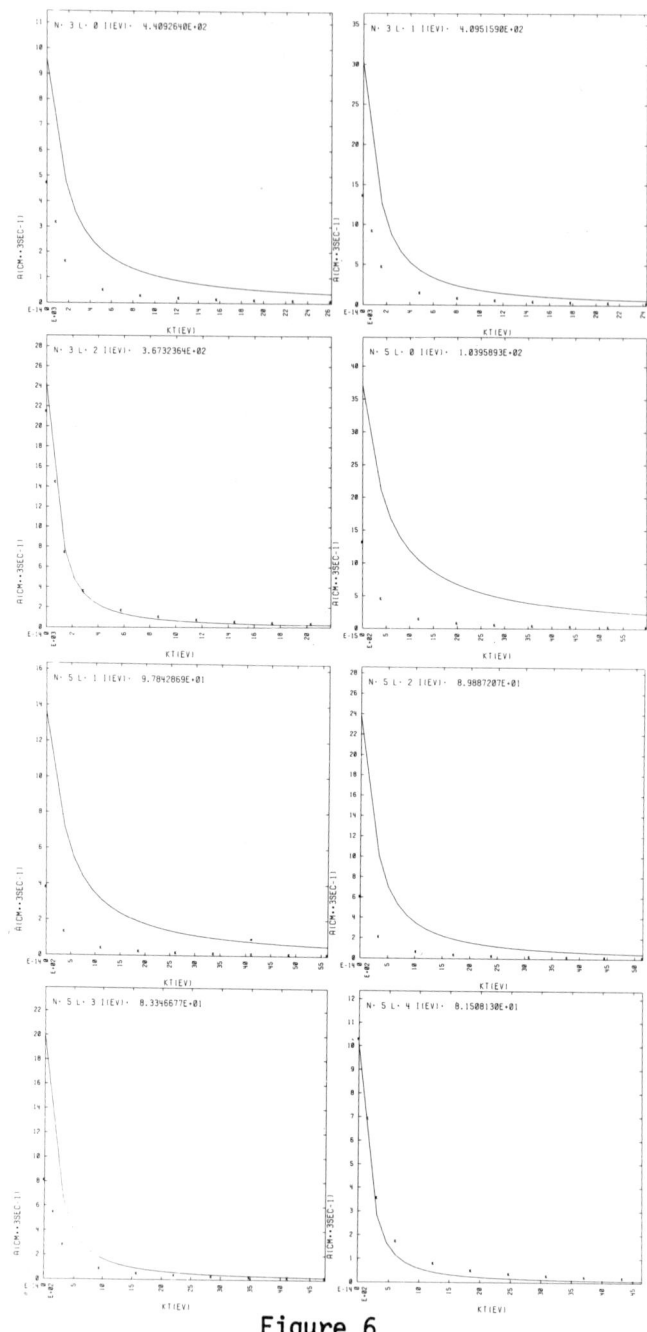

Figure 6

III. RADIATIVE TRANSFER AND OPACITY

Radiation Transport Phenomena Affecting X-Ray Laser Design

J. P. Apruzese

Naval Research Laboratory
Plasma Physics Division
Plasma Radiation Branch
Washington, D.C. 20375-5000

ABSTRACT

Recently, there have been convincing demonstrations of gain in the soft x-ray region at several laboratories. It may be anticipated that this progress will stimulate further experimental and theoretical work. On the theoretical side, several subfields of physics are relevant to x-ray laser design - such as hydrodynamics, plasma physics, atomic physics, etc. The purpose of this paper is to review one such area - the effects of various radiation transport phenomena on the viability and specific configurations of x-ray laser designs. It is shown that some effects (such as opacity broadening and radiative cooling) are positive in the sense that they increase the likelihood of achievement of x-ray lasing, while other effects are neutral or deleterious to the achievement of population inversions. Quantitative results are presented along with qualitative, intuitive descriptions of the phenomena. Simple parameterizations are also reviewed which, for uniform plasmas, are useful for quantitative calculations.

I. Introduction

The recent dramatic successes in achieving x-ray lasing in the laboratory by Jacoby, Pert and co-workers at Hull,[1] Matthews, Rosen, Hagelstein, and co-workers at Livermore,[2,3] Suckewer, et al., at Princeton,[4] and, most recently Seely and co-workers[5] at the Rochester laser facility have greatly stimulated further theoretical and experimental work in this field. The purpose of this paper is to review and focus upon one important aspect of the design of x-ray lasers: the effects of radiation transport upon the conditions required and the conditions obtainable for demonstrable gain in laboratory experiments. First, the general effects on the plasma state caused by the presence of significant optical depth in spectral lines are reviewed. As will be made evident, only lines, not continua, will exhibit opacity effects in plasmas likely to be of interest for x-ray lasers. Next, those effects which are deleterious, in some sense decreasing the chances of achieving gain, will be considered. These effects include the undesirable enhancement of the lower lasing level by photon trapping, limiting of the photon pump power by self-absorption in photopumped schemes, and the enhancement in the levels of plasma ionization attendant to photon trapping. This latter problem, if unsuspected by the experimenter, can result in a plasma too highly ionized for a given scheme to be successful. Finally, we will explore some beneficial effects arising from radiation transport phenomena. These effects are first, opacity broadening of pumping lines which greatly reduces the stringency of wavelength matching requirements for photon pumped schemes; second, radiative cooling, which has already been demonstrated to be effective for recombination x-ray lasers, and, third, the increased plasma ionization brought about by photon trapping, mentioned as deleterious above, but which can reduce the power requirements for the driver to reach a given ionization stage.

Of course, the second of these positive effects, radiative cooling, occurs most profusely in a plasma optically thin at all energies. However, even in x-ray laser geometries some opacity is unavoidably present and a technique for increasing the cooling in the presence of this opacity will be described below.

II. Basic Physics and Parameterizations

First, we consider the magnitudes of various photon absorption mechanisms which are present in plasmas which might be considered suitable for x-ray lasing. These opacity mechanisms are bound-bound absorption (lines), and the free-bound (ionization) and free-free (bremsstrahlung) continua. Consider an ion density of 10^{20}, which at temperatures of a few hundred eV or so would produce a roughly critical (at $1.06\mu m$) electron density. A typical strong resonance line has an oscillator strength of ~0.5, and at an energy of ~2 keV, such a line has an absorption cross-section, at line center, of ~7×10^{-17} cm^2, requiring, at the above ion density, only $1.5 \mu m$ of plasma for optical depth unity. Of course not all the ions will be in the same stage, and the ion density may be somewhat below 10^{20} cm^{-3}, but a plasma with a minimum width of $100 \mu m$ - typical of current x-ray laser experiments - will normally contain at least several optical depths in the strongest resonance lines. The bound free cross sections are typically three orders of magnitude below those of the strongest lines, and the free-free cross sections (in the x-ray region) are three or more orders of magnitude smaller than the bound-free. In summary, radiation transport in x-ray lasing plasmas is almost entirely due to spectral lines, to which the remainder of this paper is confined.

This is fortunate in some respects since some useful analytical parameterizations have been developed for line transport in laboratory plasmas.[6,7] These parameterizations illuminate the effects produced by line absorption as well as provide, in some cases, practical computational tools. At first let us establish notation. A spectral line arises from the decay of an upper level of statistical weight g_u at a rate A_{ul} sec^{-1} to a lower level of statistical weight g_1. The population densities for ions in the upper and lower levels are N_u and N_1 cm^{-3}, respectively. The upper level is populated collisionally by electrons which in principle may arise from any other level or ion. The upper level may also be depopulataed collisionally to other levels or ionic stages. The collisional population and depopulation rates of the upper level of the spectral line are referred to as C_u and D_u, in cm^{-3}sec^{-1} and sec^{-1}, respectively. The radiative population and depopulation may be treated by counting as depopulating only those downward transitions which result in the photon escaping the plasma. When the photon is reabsorbed in the line, this standard "bookkeeping" system is simply to pretend that the transition never occured. In keeping with the extensive astrophysical literature the line quenching probability is defined as the rate of collisional to total de-excitation of the level, i.e.

$$P_Q = \frac{D_u}{A_{ul} + D_u} \qquad (1)$$

The line photon escape probability along a single, specific path from a specific location in the plasma is defined as

$$P_e = \int \phi(\nu) \exp(-\tau_0 \phi(\nu)/\phi(o)) \, d\nu \qquad (2)$$

where the line photon emission-absorption profile (whether pure Doppler, Voigt, or Stark), is normalized, i.e.,

$$\int \phi(\nu) \, d\nu = 1$$

and the line center optical depth along the specific path considered is τ_o (dimensionless). The single path escape probability has proven useful for detailed multicell computations. Figure 1 shows the exact probability for a Voigt profile as a function of line center optical depth for two line broadening parameters. The dotted lines are the results obtained from very simple analytic approximations[8] to P_e which are in use in our detailed radiation-hydrodynamics codes at NRL. A photon cell-to-cell coupling technique which depends upon differencing of the escape probabilities along various paths is employed. This technique has been described in detail elsewhere.[9,10]

At slight risk of some confusion, we now modify our notation for P_e. From now on, P_e is defined as the quantity in Eq. (2) averaged over all possible paths and all possible points of photon origin within the plasma. Only by taking such a global plasma average can this concept be utilized for simple quantitative calculations for uniform laboratory plasmas.

Let us first consider the equilibrium of the upper level -- employing the above concepts, bookkeeping conventions, and notation.

$$C_u = N_u(D_u + A_{ul} P_e) \quad (4)$$

Noting that an optically thin plasma is characterized by $P_e = 1$, a few simple algebraic manipulations lead to

$$\frac{N_u(\text{thick})}{N_u(\text{thin})} = \frac{D_u + A_{ul}}{D_u + A_{ul} P_e} = \frac{1}{P_Q + P_e(1-P_Q)} \geq 1 \quad (5)$$

Thus line photon reabsorption increases the population of the upper level in a fashion which quantitatively depends on P_Q and P_e - both the size and the density of the plasma, as well as the atomic rates. At this point, still another type of escape probability must be defined and considered. In an optically thick plasma, most photons do not escape the plasma in the first flight following collisional creation of the level. However, after few or many radiative reabsorptions and re-emissions, the "originally collisionally created" photon may ultimately leave the plasma, usually when it is emitted on the wings of the line profile where optical depth is small. The advantage of the "ultimate escape" accounting system is that the photon escape is related directly to the collisional creation rate. The ultimate escape probability P_u, is thus equivalent to

$$P_u = \frac{N_u A_{u1} P_e}{C_u} = \frac{P_e(1-P_Q)}{P_Q + (1-P_Q)P_e} \qquad (6)$$

Therefore, the attenuation rate depends upon the interplay of optical depth (through P_e) and density and atomic rates (through P_Q). This is true because photon escape on subsequent flights after reabsorption depends to a significant degree on the collisional destruction probability per reabsorption, P_Q.

In summary, the presence of optical depth in spectral lines leads to two significant effects in x-ray lasing plasmas. These are: the enhancement of the upper level population and the reduction in photon emission - both of which depend not only on the optical depth (i.e., P_e), but also explicitly on the atomic rates and density - as embodied in the collisional destruction probability per reabsorption P_Q.

III. Potentially Deleterious Effects on X-Ray Lasers

Why does this specifically affect x-ray laser design? Every x-ray lasing line is of course characterized by an upper level (2) and lower level (1). In the vast majority of schemes, and all of those successful to date, neither of these levels is the ground state (G), which lies below both of these levels. To achieve an inversion between levels 2 and 1 the population rate for level 2 must greatly exceed that for level 1, or the depopulation rate for level 1 must greatly exceed that for level 2, or, preferably both effects will be present and act synergistically. In one generic scheme (typified by neonlike,[2,3,11-19] hydrogenlike,[1,4,5] carbonlike,[20] or nickel-like[21] ions) level 2 is pumped by electrons – either through collisional excitation from the ground state or recombination from the adjacent, more highly charged ionic species. This type of scheme is the one which has conclusively proved to be successful at the present time in the experiments of Refs. 1-5. In another proposed technique, two ionic species, usually from different elements, are employed. The requirement for this scheme is that a transition from pumping ion A have a wavelength which closely matches that of transition 2→G in the pumped (lasing) ion B. Just how close the match must be is considered below. No x-ray lasing has yet been demonstrated with such a matched-line scheme, although Trebes and Krishnan[22] have taken a very important step in this direction with their demonstration of flourescence in the C II-Al III system. Bhagavatula[23,24] has also reported inversions using a carbon-magnesium scheme.

In both types of scheme the transition 1→G in the lasing ion is usually strongly permitted in order that the lower lasing level (1) empty quickly to sustain the 2→1 inversion. This also (unfortunately) results in the 1→G transition being susceptible to trapping effects and the subsequent enhancement of the population of level 1 as given by Eq. (5). At NRL we have considered in detail[7,25] the system which is perhaps the most promising and widely known of the matched line schemes.[19,26-29] In this approach the heliumlike Na X transition $1s^2\ ^1S_0 - 1s2p\ ^1P_1$ at 11.0027 Å pumps the heliumlike Ne IX transition $1s^2\ ^1S_0 - 1s4p\ ^1P_1$ at 11.0003 Å, resulting in lasing in the Ne IX 4→2, 3→2, and 4→3 transitions near 58 Å, 82 Å, and 230 Å, respectively. Detailed, multicell, multifrequency calculations of the gain in the neon (with an assumed fixed sodium pump power) have been compared[7] to simpler calculations employing the average escape probabilities and quenching parameters as embodied in Eq. (5). We find that the analytic model does a very good job of describing the average level populations (hence:gain) in the pumped plasma as long as the density and temperature are uniform in this plasma - as indeed was accomplished in Livermore's exploding foil experiments.[2,3] Figure 2 displays the gain in the Ne IX 3d ^1D - 4f ^1F transition (solid line) and the 2p ^1P - 3d ^1D (dotted line) transition as a function of optical depth in the $1s^2\ ^1S_0 - 1s2p\ ^1P_1$ resonance line. The optical depth in the $1s^2\ ^1S_0 - 1s3p\ ^1P_1$ line is about one-fifth of this value. Note that at lower electron density higher optical depths in the trapped line can be tolerated before the gain is quashed. This is due to the higher fractional inversion which is characteristic of a lower density plasma where less collisional mixing of the lasing levels occurs. If the inherent inversion fraction is greater, more reduction of this fraction through trapping effects is obviously tolerable. The 2-3 transition is not as highly inverted as the 3-4, and is therefore more susceptible to trapping problems. Our assumed (one-sided) sodium line pump flux is that of a 170 eV blackbody, which we know from line yield measurements is energetically attainable in the NRL Gamble-II Z-pinch, if a suitable design for incorporating sodium can be found. On large machines, such as Physics International Company's DOUBLE EAGLE or Maxwell Laboratories' BLACKJACK 5, an order of magnitude more pump flux, with subsequent enhancement of the resulting gain, is reasonably to be expected. Hagelstein has discussed these trapping effects in his thesis,[15] and concludes that optical depths up to 5 in the trapped 1→G transition may be tolerated. Our work is generally consistent with his conclusion, although, as stated above, density effects do modify the precise cutoff point.

One way that the effects of this deleterious trapping may be reduced is through enhancement of the trapped line photon's escape probability through the frequency Doppler shift brought about by plasma mass motions. For an element of atomic mass M (amu) in a plasma of ion temperature T(eV), the ion thermal velocity is $1.4 \times 10^6 \sqrt{T/M}$ cm sec^{-1}. In plasmas where the expansion or implosion velocity substantially exceeds this value the enhancement of escape probability is quite significant. In effect, when the line photon enters a region of plasma whose local velocity has shifted the apparent frequency of the line away from the core absorption region, the photon is "free" to escape the entire plasma. Such effects were treated by Sobolev in 1957 in a classic paper,[31] and much important work in the plasma context has been accomplished by Irons[32-34] and also by Tallents[35] and Malvezzi and co-workers.[36] These motional Doppler effects contributed to the success of the experiments of Jacoby, Pert, Shorrock, and Tallents.

At NRL, Young, Stephanakis and colleagues in the Plasma Technology Branch are attempting to create gas-puff, neonlike argon plasmas using the pulsed-power machine Gamble-II. The first goal is to produce a uniform, repeatable, linear plasma, but ultimately it is hoped to achieve lasing in the 3p-3s transitions near 30 eV analagous to the successful Livermore experiments in which 3p-3s lasing was accomplished in neonlike selenium and yttrium.[2,3] Perhaps the major problem with x-ray lasing in Z-pinches is the difficulty of producing transverse plasma dimensions as small as 100μm. Rather, dimensions of ~1 mm are more realistic. On the positive side, pulsed-power generators can currently provide much more energy to plasmas than high power lasers and may be able to achieve gain lengths of 4 cm or longer. There are two especially favorable aspects to the use of neonlike argon in a relatively modest machine such as Gamble-II. The first of these is the high ratio of implosion velocity (10^7 cm sec^{-1}) to thermal velocity (2.0×10^6 cm sec^{-1} at T=80eV). This greatly alleviates the trapping problem, as shown in Fig. 3. There the gain in the J=0 to 1 transition at 434 Å is shown versus temperature for different velocities, at an ion density of 5×10^{17} cm^{-3}. Note the threefold-or-more reduction in gain at zero velocity. The other favorable aspect has been discussed, quantified and emphasized by Elton.[30] It is simply easier to obtain lasing at longer wavelengths. The gain coefficient for a Doppler-broadened line α(cm^{-1}) is given by

$$\alpha = 10^{-16} f_{osc} \lambda(\text{Å}) (M/T)^{1/2} (N_2 - \frac{g_1}{g_2} N_1). \quad (7)$$

For neonlike argon the low temperature and long wavelength contribute substantially to the possibility of achieving measurable gain as seen in Eq. (7). Thus this ion is an excellent candidate for soft x-ray lasing in a low-to-moderate power Z-pinch.

The next "deleterious" radiation transport effect to be discussed is the limiting of photon pump power in photon-pumped schemes. In any such scheme, the gain scales nearly linearly with the input flux of the pumping line radiation. To increase the line power arising from the plasma one would generally like to make the pumping plasma denser and/or hotter. The collisional excitation rate for the pumping transition increases substantially with both density and temperature. Of course, increasing the temperature to the point where the pumping ion largely ceases to exist is obviously self-defeating. However, increasing the density usually brings an increase in optical depth and resulting decrease in P_e, along with an increase in the photon destruction probability per reabsorption P_Q. Just how does the interplay of those various factors come out when one attempts to increase pumping power through compression of the pumping plasma? Figure 4 supplies at least a partial answer to this question for the Na X $1s^2$ ^1S_o - 1s 2p $^1P_1$ pumping line. The pumping flux at the frequency position of the pumped Ne IX $1s^2$ ^1S_o - 1s 4p $^1P_1$ line was calculated as a function of density with a multifrequency, multicell, collisional-radiative equilibrium code. Details are given in Ref. 7. Figure 4 shows that the pumping flux increase with density varies from an N^2 dependence at a few X 10^{18} ions cm^{-3}, to weaker than linear at $N_I = 10^{21}$ cm^{-3}. That this is consistent with the (P_e, P_Q, P_u) parameterization is demonstrated in Fig. 5, where the exact computed multifrequency results (integrated over line profile) compare favorably to the "ultimate escape probability" formulation of Eq. (6). The gains in the pumped neon plasma obtainable from various pumping fluxes are substantial and are discussed in Refs. 7 and 25. We conclude that the "ultimate escape probability" concept is quantitatively supportable as a computational tool for uniform plasmas.

IV. Beneficial Effects of Radiation Transport

Radiation transport phenomena may also be harnessed by the experimenter to enhance the chances of achieving x-ray lasing with certain schemes. Foremost among the positive effects is that of line opacity broadening, discussed first in this connection by Norton and Peacock[27] and quantified with multifrequency numerical calculations at the Naval Research Laboratory. These latter calculations are presented and discussed in detail in Refs. 7 and 25. Line photons emitted on the wings escape the plasma more easily due to the lower monochromatic optical depth. Hence, sampling the photons which have escaped an optically thick plasma, one finds a disproportionate number far from line center, i.e., the emitted profile is "opacity broadened". An example of a detailed calculation of this phenomenon is found in Fig. 6, where the emitted line profile for Si XIII $1s^2 - 1s2p\ ^1P$ at 6.650 Å is broadened by an optically thick silicon plasma to overlap the Al XII $1s^2-1s3p\ ^1P$ line at 6.635 Å. Thus the Si/Al scheme - which at first glance would appear to be impossible due to the 15 mÅ line mismatch - is quite viable if an optically thick (τ_o of several hundred) silicon pump plasma can be produced. Of course, too large a pump plasma density and/or size will produce a continuum emission approximating the intensity of the lines. This would destroy the inversion in the pumped plasma by enhancing all levels. In general, our calculations at NRL have shown that the following line mismatch may be taken as a safe upper limit for the viability of a matched-line x-ray laser scheme

$$\frac{\lambda_{pump} - \lambda_{pumped}}{\lambda_{pump}} \leq 6.9 \times 10^{-4} \times \left(\frac{T_{pump}(eV)}{M_{pump}(amu)}\right)^{1/2} \qquad (8)$$

The square-root quantity of Eq. (8) is, of course, proportional to the pump-line Doppler width.

Another design consideration for x-ray lasers which is substantially influenced by radiation transport phenomena is the necessity of achieving the proper ionization stage(s). In single-component schemes such as the neonlike ion, the presence of optical depth increases excited state populations according to Eq. (5). In many instances, the ionization rates from excited states are two-to-three orders of magnitude above that of the ground state. Thus, by enhancing excited state populations, line photon trapping may substantially increase the overall ionization state of the plasma. This is seen clearly in the results shown in Fig. 7. The neonlike argon state persists at much higher temperatures for an optically thin plasma, or one in which velocity gradients produce minimal trapping. An even more pronounced effect occurs in two-component, photon-pumped schemes, such as Na/Ne. The 1s4p levels in heliumlike Ne IX may be pumped so intensively by the impinging Na X $1s^2$-$1s2p$ ^1P line that rapid ionization from the 4p Ne IX level produces domination of hydrogenlike Ne X at abnormally low plasma temperatures. An example of this effect has been provided by our previous analysis[37] of an imploded microballoon experiment carried out some years ago[38] by Yaakobi and co-workers at the University of Rochester. In their experiment, sodium impurities were present (unintentionally) in the glass microballoon and produced a strong signature of the sodium pump line (Fig. 8) which our theoretical spectrum[37] (also in Fig. 8) showed has an equivalent blackbody brightness of 227eV. Of course, the neon in their experiment was compressed to much too high a density to promote lasing. However, if one could bathe a lower density neon plasma in this equivalent pump intensity, the effects on its ionization state would be considerable, as demonstrated in Fig. 9. Figure 9 (from Ref. 25) shows that, under the influence of the pumping radiation, the neonlike stage peaks at only 50-65 eV, less than half the temperature at which it would normally maximize. One may consider this effect beneficial in two senses. First, the experimenter need supply much less energy to excite the neon to the appropriate stage. Second, the relatively cold neon required results in higher gain due to the narrow line Doppler width and minimal collisional excitation rates to the lower lasing levels.

The final beneficial effect of radiation transport discussed here is that of radiative cooling. Such radiative cooling often dominates that due to hydrodynamic expansion, and played a decisive role in the success of at least two recombination-pumped laser experiments,[4,5] and probably was very important also in the Livermore experiments.[2,3,39] Rapid radiative cooling allows recombination pumping to occur at high densities suitable for large gain. Figure 10 illustrates one technique by which transport effects may be used to enhance radiative cooling. In this figure the spectral line intensities calculated to arise from a sodium/neon Z-pinch plasma are plotted in J/ns as a function of the relative abundances of sodium and neon. Note that, when a pure neon plasma is replaced by one with the same total (sodium plus neon) ion density, with 75% neon and 25% sodium, the total emitted power increases dramatically. The neon lines, already quasi-saturated by opacity, decrease very little, while the relatively thin sodium lines are very strong. Thus, to enhance cooling, lots of thin lines from two or more elements are more effective than a relatively few very optically thick lines from one element. One can envision these effects utilized to tailor radiatively-cooled, recombination lasing experiments for optimum gain. A complete discussion of these opacity/cooling effects is given in Ref. 40.

V. Summary

It has been demonstrated that the effects of line radiation transport may critically influence the success or failure of specific x-ray laser schemes and/or experiments. Deleterious effects, reducing the gain, include enhancement of the populations of lower lasing levels. Opacity broadening, increased ionization, and radiative cooling, are some of the effects which may be deemed to be beneficial. Useful parameterizations of line transport effects, a limit for the closeness of useful line matching in opacity-broadened, photon-pumped schemes, as well as detailed calculations of the above effects have been presented and reviewed. Finally, it is suggested that opacity effects may be harnessed to enhance radiation cooling by employing mixtures of two or more elements of appropriate atomic number in recombination-lasing experiments.

ACKNOWLEDGMENTS

Much of the work discussed in this review article has been presented elsewhere and the reader is urged to consult the original references. The contributions of my co-workers at NRL, Jack Davis, Ken Whitney, Dwight Duston, Paul Kepple, Milan Blaha, John Rogerson, and Chris Agritellis, have been invaluable. Support for much of the work discussed here has been provided by the Defense Nuclear Agency, the Office of Naval Research, and the Strategic Defense Initiative Organization/Innovative Science and Technology Office.

REFERENCES

1. D. Jacoby, G. J. Pert, L. D. Shorrock, and G. L. Tallents, J. Phys. B 15, 3557 (1982).

2. M. D. Rosen, et al., Phys. Rev. Lett. 54, 106 (1985).

3. D. L. Matthews, et al., Phys. Rev. Lett. 54, 110 (1985).

4. S. Suckewer, et al., Phys. Rev. Lett. 55, 1753 (1985).

5. J. F. Seely, et al., Optics Commun. 54, 289 (1985).

6. J. P. Apruzese, J. Davis, D. Duston, and R. W. Clark, Phys. Rev. A 29, 246 (1984).

7. J. P. Apruzese and J. Davis, Phys. Rev. A 31, 2976 (1985).

8. J. P. Apruzese, J. Quant. Spectrosc. Radiat. Transfer 34, 447 (1985).

9. J. P. Apruzese, J. Davis, D. Duston, and K. G. Whitney, J. Quant. Spectrosc. Radiat. Transfer 23, 479 (1980).

10. J. P. Apruzese, J. Quant. Spectrosc. Radiat. Transfer 25, 419 (1981).

11. A. N. Zherikhin, K. N. Koshelev, and V. S. Letokhov, Sov. J. Quantum Electron, 6, 82 (1976).

12. A. V. Vinogradov, I. I. Sobel'man and E. A. Yukov, Sov. J. Quantum Electron, 7, 32 (1977).

13. L. A. Vainshtein, A. V. Vinogradov, V. I. Safronova, and I. Yu. Skobelev, Sov. J. Quantum Electron, 8, 239 (1978).

14. A. V. Vinogradov and V. N. Shylaptsev, Sov. J. Quantum Electron 13, 1511 (1983).

15. P. L. Hagelstein, Lawrence Livermore National Laboratory Report No. UCRL-53100, 1981.

16. U. Feldman, A. K. Bhatia, and S. Suckewer, J. Appl. Phys. 54, 2188 (1983).

17. J. P. Apruzese and J. Davis, Phys. Rev. A 28, 3686 (1983).

18. U. Feldman, J. F. Seely, and A. K. Bhatia, J. Appl. Phys. 56, 2475 (1984).

19. R. H. Dixon and R. C. Elton, J. Opt. Soc. Am. B 1, 232 (1984).

20. K. G. Whitney, J. Davis, and J. P. Apruzese, "Some Effects of Radiation Trapping on Stimulated VUV Emission in Ar XIII," in Cooperative Effects in Matter and Radiation, C. M. Bowden, D. W. Howgate, and H. R. Robl, editors, Plenum, New York (1977).

21. S. Maxon, P. Hagelstein, K. Reed, and J. Scofield, J. Appl. Phys. 57, 971 (1985).

22. J. Trebes and M. Krishnan, Phys. Rev. Lett. 50, 679 (1983).

23. V. A. Bhagavatula, Appl. Phys. Lett. 33, 726 (1978).

24. V. A. Bhagavatula, IEEE J. Quantum Electron. 16, 603 (1980).

25. J. P. Apruzese, J. Davis, and K. G. Whitney, J. Appl. Phys. 53, 4020 (1982).

26. A. V. Vinogradov, I. I. Sobelman, and E. A. Yukov, Sov. J. Quantum Electron. 5, 59 (1975).

27. B. A. Norton and N. J. Peacock, J. Phys. B 8, 989 (1975).

28. P. L. Hagelstein, Plasma Phys. 25, 1345 (1983).

29. D. L. Matthews, et al., IEEE J. Quantum Electron. 19, 1786 (1983).

30. R. C. Elton, Opt. Eng. 21, 307 (1982).

31. V. V. Sobolev, Sov. Astron. 1, 678 (1957).

32. F. E. Irons, J. Phys. B 8, 3044 (1975).

33. F. E. Irons, J. Phys. B 9, 2737 (1976).

34. F. E. Irons, Aust. J. Phys. 33, 25 (1980).

35. G. J. Tallents, J. Phys. B 13, 3057 (1980).

36. A. M. Malvezzi, et al., J. Phys. B 12, 1437 (1979).

37. J. P. Apruzese, P. C. Kepple, K. G. Whitney, J. Davis, and D. Duston, Phys. Rev. A 24, 1001 (1981).

38. B. Yaakobi, et al., Phys. Rev. Lett. 39, 1526 (1977).

39. J. P. Apruzese, J. Davis, M. Blaha, P. C. Kepple, and V. L. Jacobs, Phys. Rev. Lett. 55, 1877 (1985).

40. J. P. Apruzese and J. Davis, J. Appl. Phys. 57, 4349 (1985).

FIGURE CAPTIONS

Fig. 1. Escape probability along a single path is plotted against line center optical depth of the path, averaged over a Voigt profile. Both the exact probability, and analytic approximations (from Ref. 8) are given, for two different broadening parameters (a).

Fig. 2. Gain in the Ne IX $3d^1D-4f\ ^1F$ (solid line) and $2p\ ^1P-3d\ ^1D$ (dotted line) is plotted against optical depth in the $1s^2-1s2p\ ^1P$ line. The planar neon plasma is assumed to have a temperature of 65 eV and is pumped on one side by a matched-line sodium flux of 1.3×10^3 ergs/cm^2 sec Hz.

Fig. 3. Gain coefficient in the Ar IX transition $2p^5\ 3p\ (3/2,3/2)_0 - 2p^5\ 3s\ (1/2,1/2)_1$ at 434 Å is shown as a function of temperature, for various velocities, at an argon ion density of 5×10^{17} cm^{-3}, and cylindrical plasma diameter 0.9 mm. Also shown is a case where the argon ions are mixed with a 1.0×10^{19} cm^{-3} concentration of neon ions.

Fig. 4. Output flux in the Na X $1s^2-1s2p\ ^1P$ line at the frequency position of the pumped Ne IX $1s^2-1s4p\ ^1P$ line is shown as a function of ion density for a cylindrical plasma of temperature 400eV and diameter 1.4 mm. Also shown is the internal energy per unit volume, as well as the equivalent blackbody brightness temperature of the emitted flux.

Fig. 5. The spatially averaged, solid-angle-averaged single flight photon escape probability P_e, the collisional photon-quenching probability per scattering P_Q, and the approximate analytic ultimate escape probability are plotted as a function of ion density for a 400 eV cylindrical sodium plasma of diameter 1.4 mm. Also shown is the exact numerically computed ultimate photon escape probability P_u.

Fig. 6. Line profile of the silicon pump line $1s^2-1s2p$ ^1P at 6.650 Å arising from a 1.5 mm, 400 eV silicon plasma of ion density 8 × 10^{19} cm^{-3} is plotted on the same scale as the intrinsic absorption profile of the pumped Al XII $1s^2-1s3p$ ^1P at 6.635 Å, demonstrating opacity broadening of the emitted line overcoming the wavelength mismatch.

Fig. 7. Neonlike argon fraction vs. temperature at an argon ion density of 5 × 10^{17} cm^{-3}, for the various indicated conditions.

Fig. 8. Compressed neon-filled microballoon spectrum obtained at Rochester by Yaakobi and colleagues (Ref. 38) is plotted alongside NRL model calculations (Ref. 37).

Fig. 9. Ionization balance for neon in the presence of a 227 eV pumping sodium line, at a neon ion density of 10^{20} cm^{-3}. Note the rapid disappearance of the neonlike stage for T>100 eV due to the presence of the sodium pump radiation.

Fig. 10. The K-shell spectrum of sodium-neon mixtures is shown. No attempt is made to display the line profiles: intensities are directly proportioned to the line heights. Electron temperature is 400 eV, diameter 5.6 mm, and the sum of the sodium and neon ion densities is 5 × 10^{19} cm^{-3}. Sodium ion fraction increases from top to bottom.

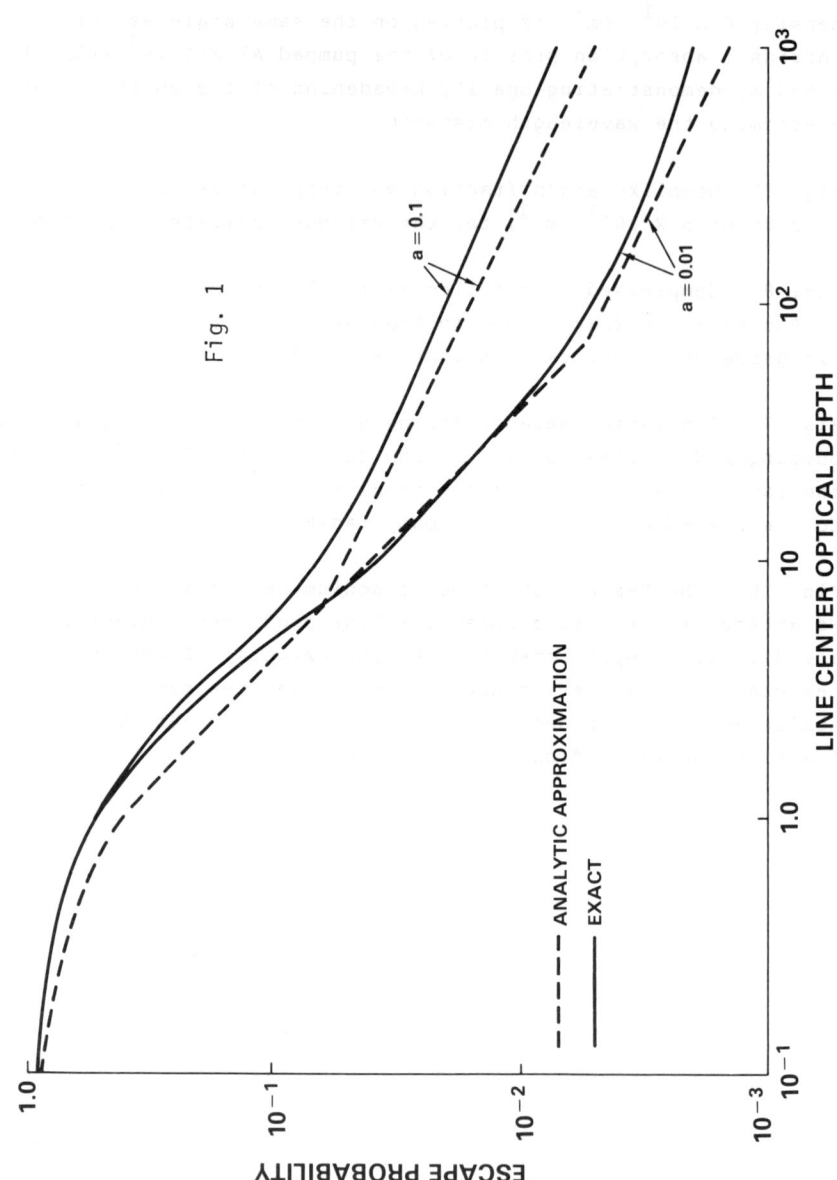

Fig. 1

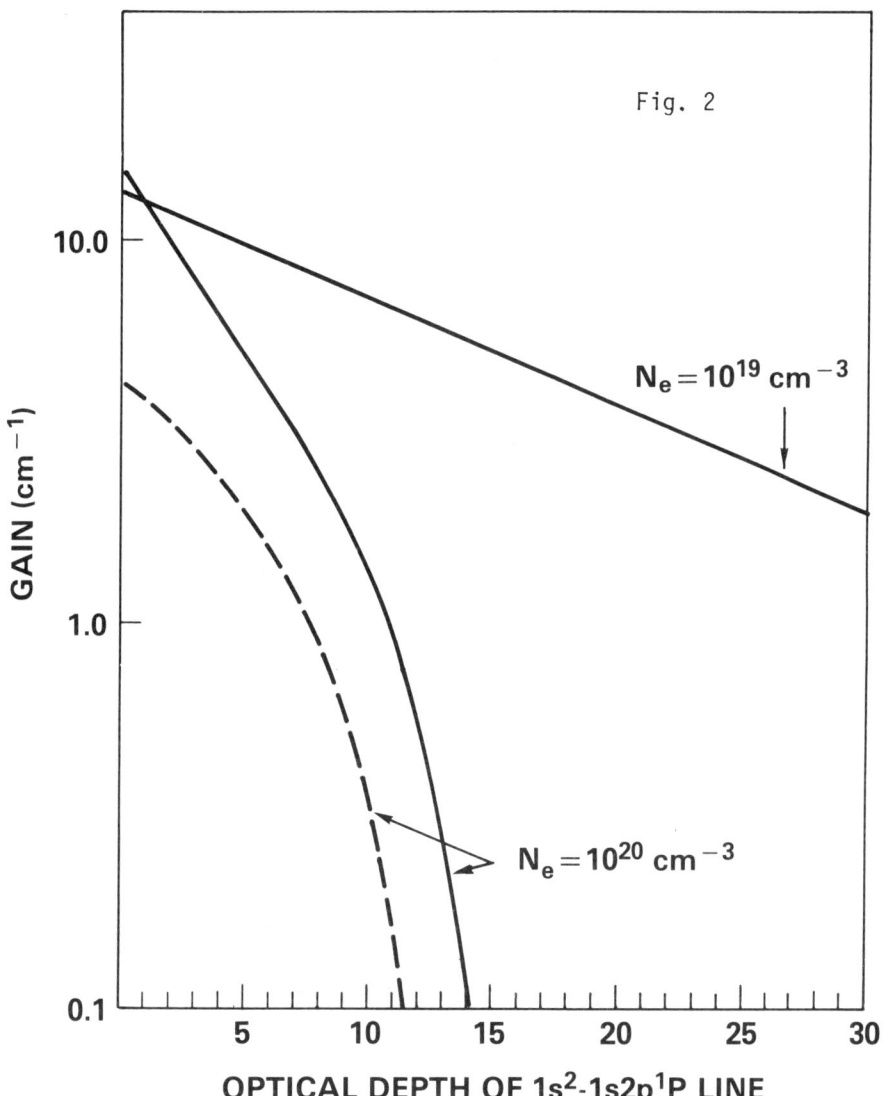

Fig. 2

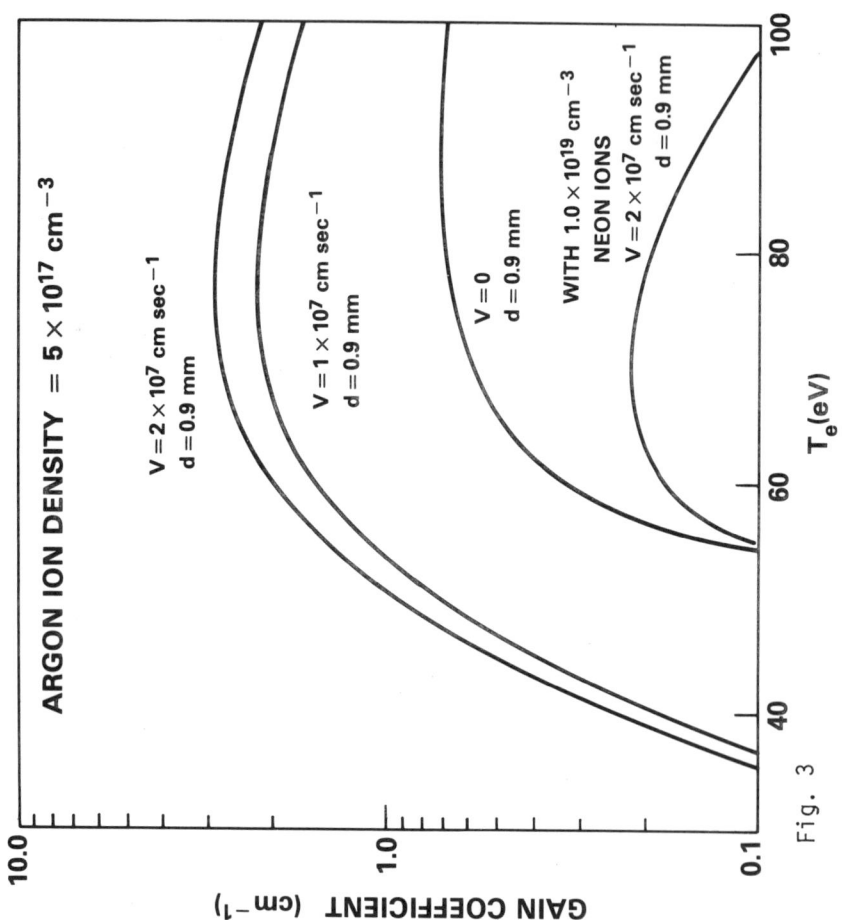

Fig. 3

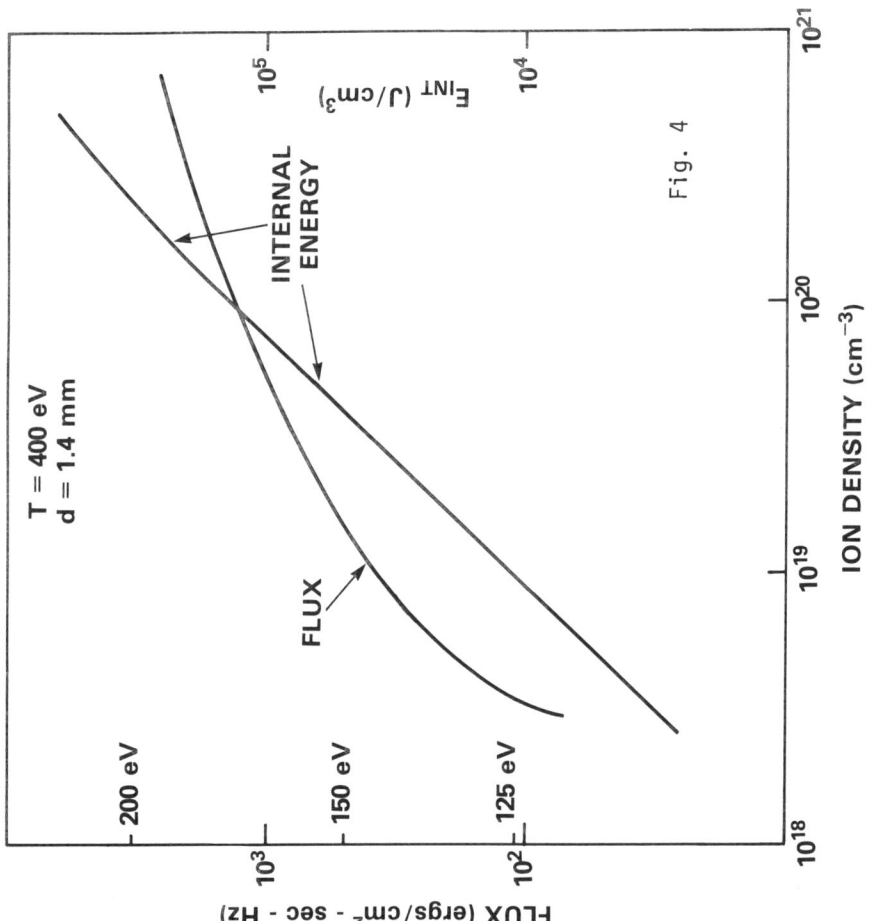

Fig. 4

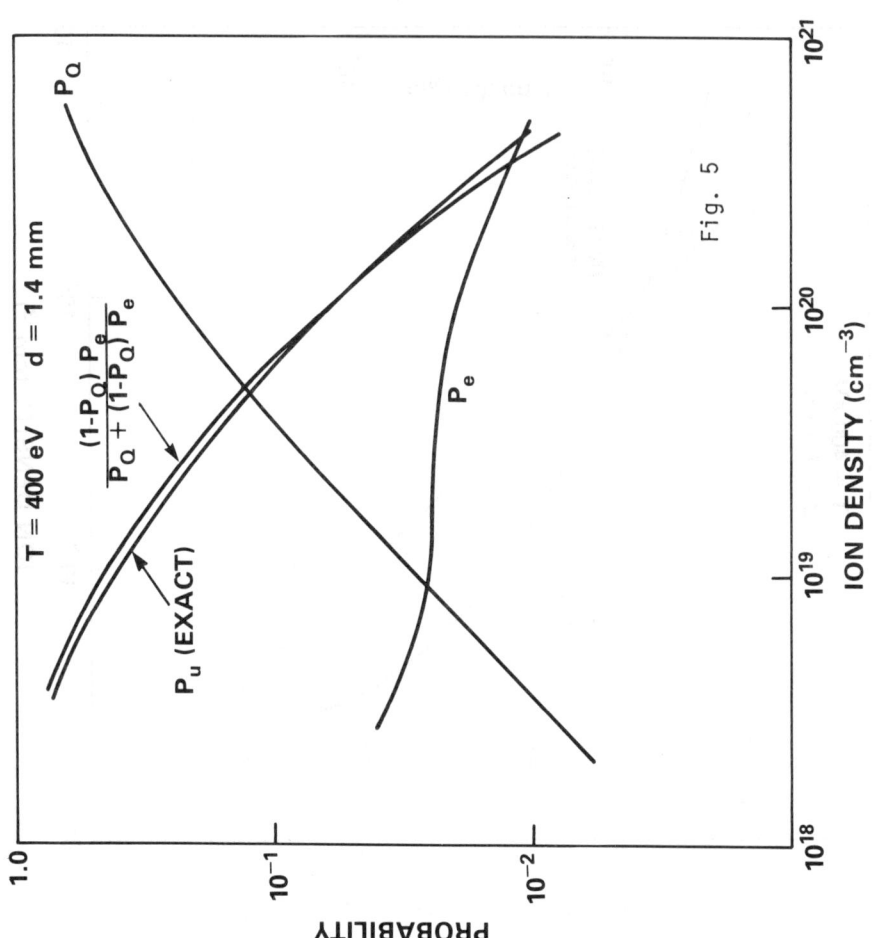

Fig. 5

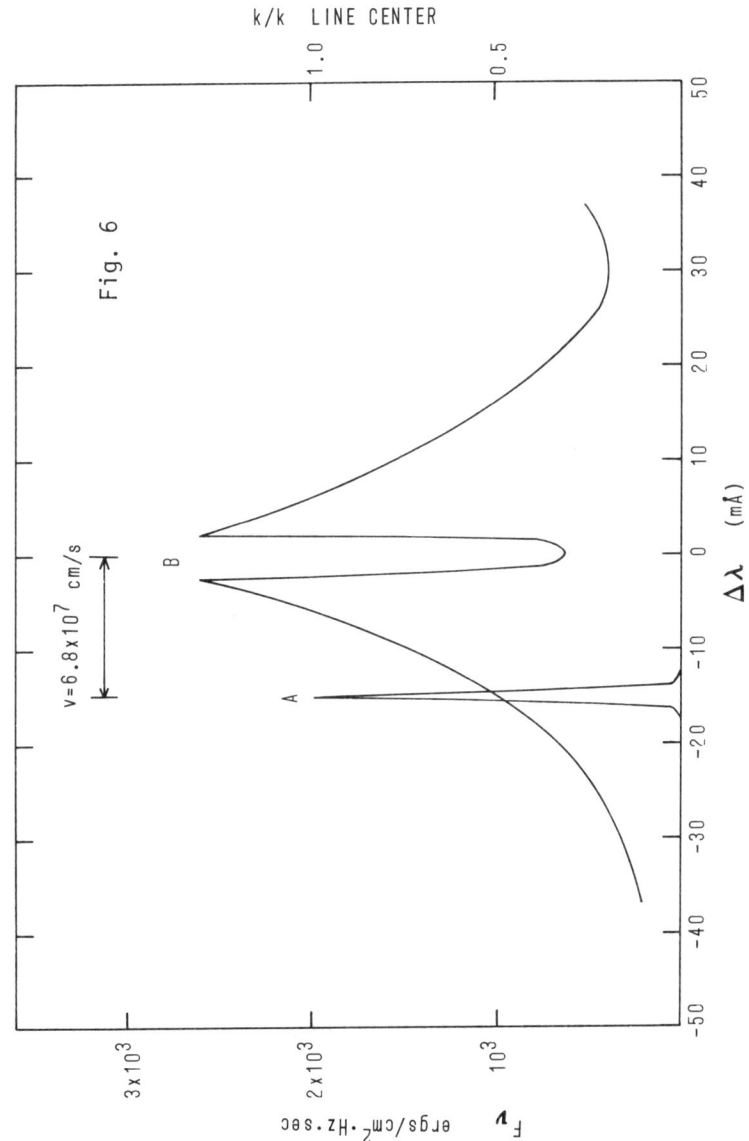

Fig. 6

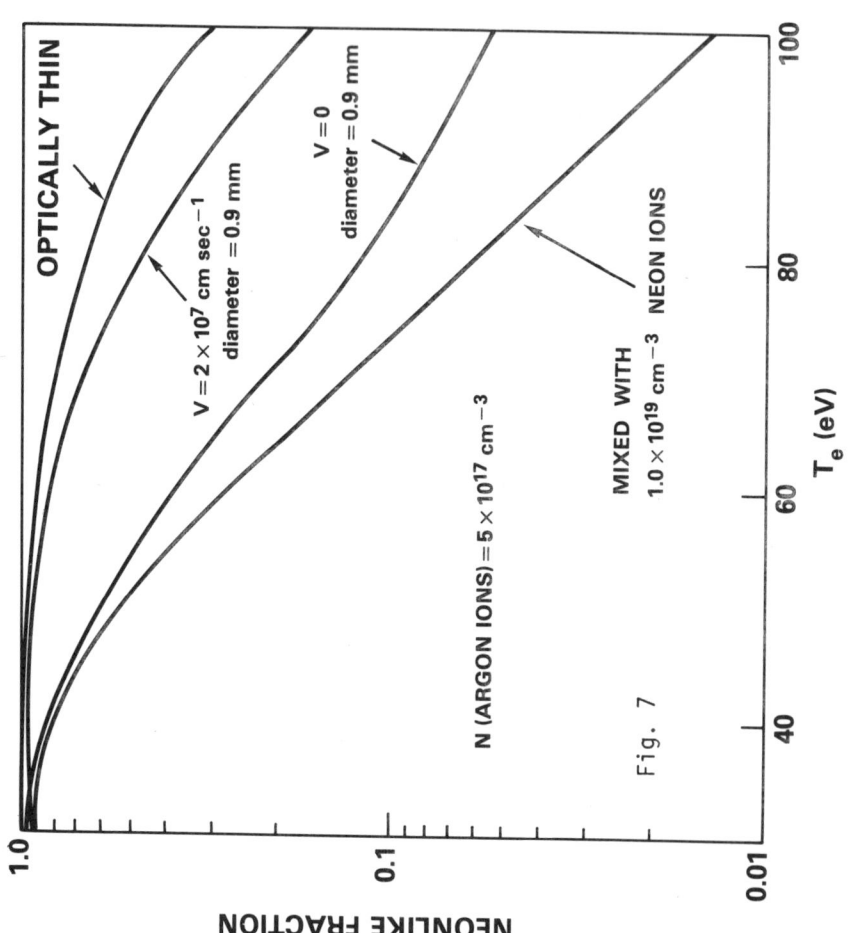

Fig. 7

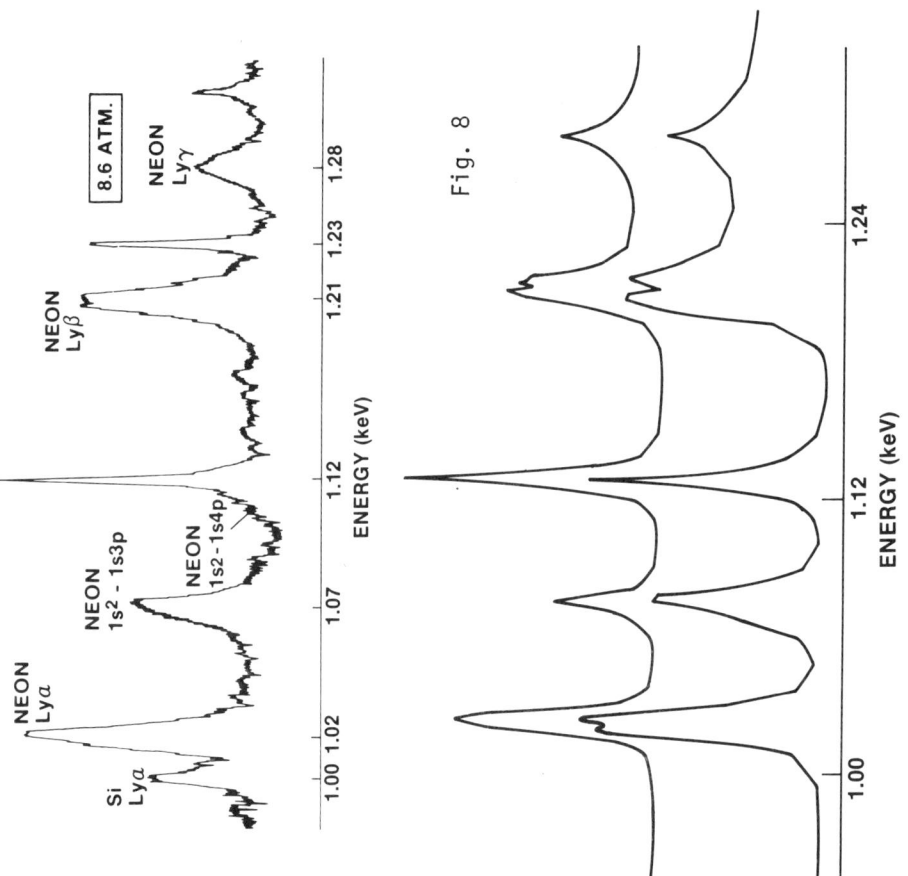

Fig. 8

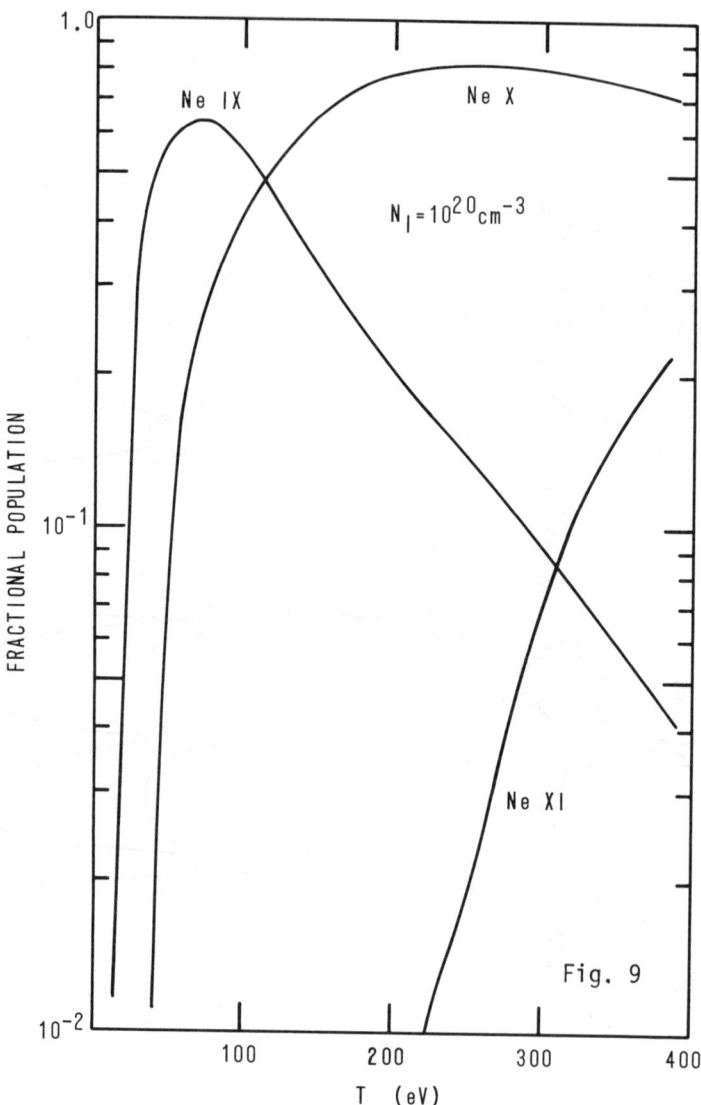

Fig. 9

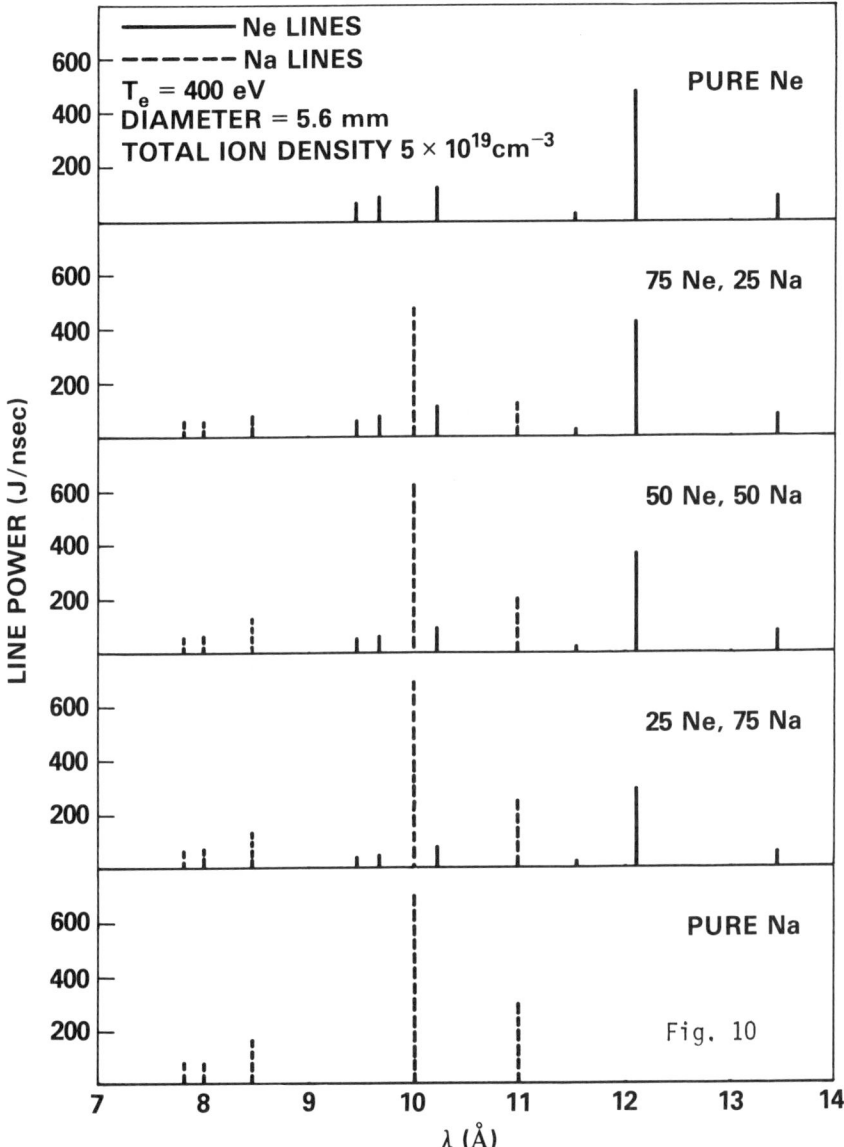

Fig. 10

STUDIES OF THE SOLUTION OF A RADIATION TRANSFER BENCHMARK*

by

L. Yobs, J. M. Salter**, and R. W. Lee

University of California
Lawrence Livermore National Laboratory
P.O. Box 808 / L-23
Livermore, California 94550

**Plasma Physics
Blackett Laboratory
Imperial College
London, SW7-2BA, England

ABSTRACT

The radiative transfer benchmark which was defined previously is discussed in the light of the solution which uses the Equivalent Two-Level Atom scheme to handle the multitransition aspects of the problem and Variable Eddington factors to assist in the solution of the transfer equation. Discussion is presented of the character of the benchmark solution. The solution exhibits slow convergence due to two distinct conditions. Further, a discussion is presented of two simplified solutions which attempt to accelerate the solution. Finally, correction to the previously published benchmark are presented in an appendix.

*This work was performed under the auspices of the U.S. Dpartment of Energy by Lawrence Livermore National Laboratory under Contract No. W-7405-ENG-48.

I. INTRODUCTION

The need for a method to compare different radiation transfer techniques for multi-ion stage, multiple transition problems gave rise to the benchmark which was presented in the last conference on "Radiative Properties of Hot Dense Matter."[1] The solution of this benchmark was published recently and compares facets of the solutions for three different radiative transfer techniques.[2] The three techniques are lambda (Λ) iteration, an accelerated Λ method, and the Variable Eddington (VE) factor technique. All of these methods used the equivalent two-level atom to treat the multilevel aspect of the benchmark. The solution showed some surprising characters. First, the Λ and accelerated Λ solutions converge in over fifty iterations with the VE method taking approximately one half the iterations. This would seem unusual for a problem with peak optical depths ~200. Second, this simple problem seems to have at least two major difficulties which may be typical for straightforward line transport problems in laboratory systems.

Below we will focus two aspects of the radiative transfer solution. First, we will discuss the encountered "stiffness" in the solution of the problem. This arises in at least two ways: (1) due to a large disparity in the rates feeding the auto-ionizing states and (2) due to an arrangement of optically thick and optically thin lines in the Lyman and Balmer Series. Second, we will discuss two simple techniques to speed up the calculations.

II. CONVERGENCE PROPERTIES

    A. AUTO-IONIZING TRANSITIONS

    The convergence rates for the VE method shows that the

lithium-like auto-ionizing level 1s2p² ²P converges most slowly and that this level has a departure coefficient of 10^{-6} compared to the helium-like ground. This is in distinction to the other auto-ionizing level (²D) which has only 6 eV separation from the ²P and has a departure coefficient of ~1. The main problem is that ²P level is connected to the helium-like ground state by an auto-ionization rate that is ~10^{10} sec^{-1} which is 10^3 smaller than the other rates into the auto-ionizing complex. This low coupling together with the relatively large population in the helium-like ground state, where approximately 40% of the total population is found (i.e., 10^4 times the population of the lithium-like auto-ionizing levels), gives rise to a very slow transfer of population between these two states. Figure 1a shows a schematic of the pertinent level populations. The difficulty entailed is due to the disparity of rates and populations. This type of energy level structure should be expected in ion spectra produced in the laboratory. An increase in either the collisionality of the system with higher electron density or optical depth would make these levels converge more rapidly.

B. LYMAN SERIES

The second factor which will slow down the convergence is the fact that the optical depth in the Lyman β (3 to 1) and Balmer α (3 to 2) are of the same order, 26 and 15, respectively, while the Lyman α (2 to 1) transition has an optical depth of ~180, but the system is effectively thin. This leads to a difficult radiative transfer problem to solve and is illustrated in Fig. 1b. The convergence rate is very slow due to this grouping of transitions and, again, the situation would be simplified were the Lyman α transition effectively thick, which would occur at higher optical depths.

These two cases indicate that the choice of a simple benchmark provides complications without recourse to the effects of gradients in temperature and density or to other physical complications like mass motion.

III. MODIFIED METHODS

With the solution of the benchmark confirmed by calculation we have attempted to improve the convergence CPU time and test a highly simplified version of the single-flight escape factor method.

A. DECREASING THE VARIABLE EDDINGTON FACTOR RUN TIME

The problem as stated has 17 continua and 31 lines, where none of the continua are optically thick and we have only a few of the line transitions with optical depths greater than 10. Therefore, we tested a number of modifications of the basic VE factor scheme. First, we tried to limit the number of transitions on which we performed the full VE calculations. A scheme which reproduced the complete results divided the transition into three types. First, for transitions with peak optical depth below, τ_Λ, we solve only the radiative transfer equation once and use this radiation field for all other iterations. Second, for transitions greater than τ_Λ but less than τ_{VE} we perform Λ iterations for all steps. Finally, for transitions with optical depths greater than τ_{VE} we use the complete VE method. A number of values of τ_Λ and τ_{VE} were tested and a safe choice, which produced the same accuracy as the complete VE calculation, was

$$\tau_\Lambda = 0.1$$
$$\tau_{VE} = 5.0.$$

The results of the simulation were the same, in that the number of iterations was unchanged; however, due to the savings in performing the simplified Λ, or no radiative transfer, we had reduction in computation time by <u>60%</u>. This is clearly an advantage to those interested in complex problems.

B. SINGLE-FLIGHT ESCAPE PROBABILITIES

Since the problem described here is a uniform slab, the concept of an escape probability should be as valid as one could expect to find. Therefore, since we can test the single-flight escape concept, we performed the following calculation. We assume that the continua are elow the τ_Λ limit and do not calculate the radiation field in them. For the lines we solve the rate equations for the level populations and then use these to estimate the optical depths in the lines and the source functions. Normally we would continue on to the radiative transfer equation; however, here we use the optical depth and source function to determine the radiation field in the lines by the following analogy to the two-level atom.

In the two-level atom we can write the radiative rates in one transition as

$$R_{UL} = n_U A_{UL} + n_U B_{UL} \bar{J}_{UL} - n_L B_{LU} \bar{J}_{UL}$$

where the A and B are the Einstein spontaneous emission and absorption coefficients, the n_U, and n_L are the upper- and lower-level populations, and J is the intensity averaged over angle and frequency. Using the definition of the two-level atom source function, S, with complete redistribution

$$S = \frac{n_U A_{UL}}{n_L B_{LU} - n_U B_{UL}}$$

We can obtain

$$R_{UL} = n_U A_{UL} P_e$$

Here P_e is the escape factor and is given by

$$P_e \equiv (1 - \frac{\bar{J}}{S}).$$

We normally require the solution of the radiative transfer equation to get J but instead we invoke the relationship between P_e and the optical depth,[3] this gives us a method of finding J. Therefore we iterate by using J defined as

$$J = S(1 - P_e(\tau))$$

where the S, τ are previously determined and the functional dependence of P_e on τ is prescribed. This J is then used in the rate equations and the iterations are continued until the solution converges. The only modification of the scheme is that account is taken of the multilevel structure using the ETLA approach for the source function.

This method converged to an answer for the populations which is approximately a factor of *two* different at the boundary, with factors of 10-30% on the slab center. The results took approximately the same number of iterations as the pure Λ iteration, i.e., 46, but the time taken was much less, i.e., 50% of the Λ iteration time and 35% of the modified VE terms.

These two calculations show that a great improvement in speed can be made over straight forward techniques. Further studies are now underway.

REFERENCES

1. R.W. Lee and J.M. Salter, "Radiative Properties of Hot Dense Matter", page 37 (World Scientific/Singapore) 1985.

2. J.M. Salter, R.W. Lee, L.A. Yobs, submitted JQSRT, 1985.

3. F.E. Irons, JQSRT 22, 1 (1979) -- three papers.

APPENDIX

CORRECTION TO THE BENCHMARK

The published benchmark contains four errors or omissions. Firstly, in the helium-like A-values, the rates from 4, 5, and 6 to level Z should be

1.24+12, 3.90+11, 1.20+11,

respectively. Secondly, the boron-like, 3-body recombination rate is required and is

4.20-36.

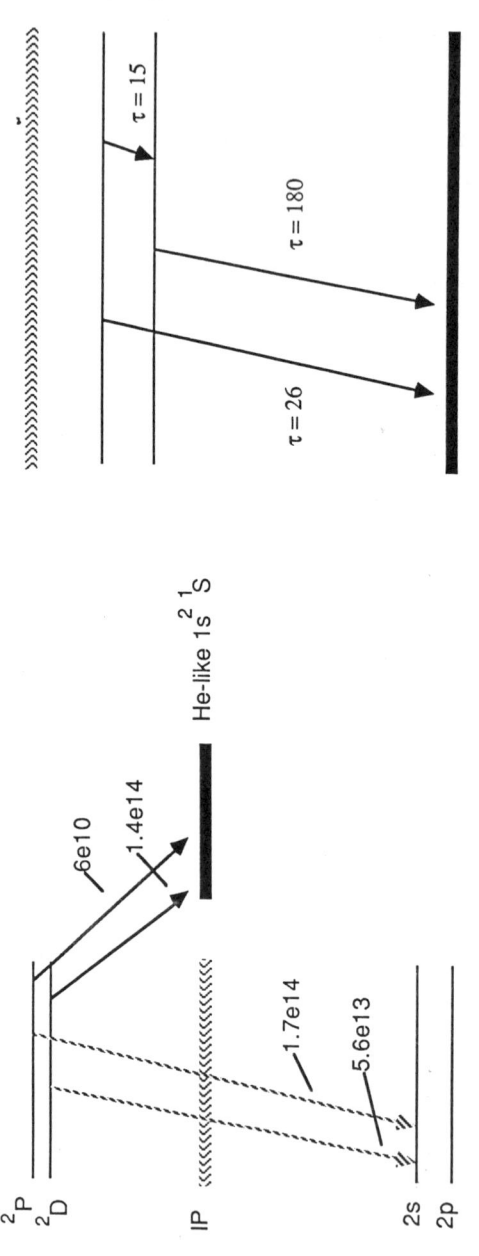

FIGURE 1b The hydrogenic series showing the optical depths of the critical transitions

FIGURE 1a. The Li-like ionization stage including the autoionizing states. (———) = autoionization (·······) = A - values

PHYSICS OF PARTIAL REDISTRIBUTION AND RADIATIVE TRANSFER IN STELLAR ATMOSPHERES

Jeffrey L. Linsky[*]

Joint Institute for Laboratory Astrophysics
National Bureau of Standards and University of Colorado
Boulder, Colorado 80309-0440

Abstract

The term partial redistribution refers to the redistribution in frequency and angle in the observer's frame when the scattering process in the atom's frame contains both coherent and noncoherent components. Astronomers often must solve the radiative transfer equation including partial redistribution when they wish to derive the temperature and density structure of a stellar atmosphere from the analysis of the observed profiles of strong resonance lines. In this review paper I will first discuss the physical processes that lead to coherent and noncoherent scattering in the atom's frame, and the various types of redistribution functions commonly used in astrophysics. Then I will describe several important examples of partial redistribution calculations of the hydrogen Lyman alpha and Mg^+ resonance lines formed in static plane-parallel models for the solar atmosphere and stellar atmospheres. Finally, I will show computed profiles for geometrically extended and expanding atmospheres when the radiative transfer equation, including partial redistribution, must be solved in the comoving frame of the fluid.

---

[*]Staff Member, Quantum Physics Division, National Bureau of Standards, Boulder, Colorado.

"Most of what we know about stars and systems of stars is derived from an analysis of their radiation and this knowledge will be secure only as long as the analytical technique is physically reliable," D. Mihalas (Ref. 1).

"Non-LTE theory is not conceptually that complex," R. N. Thomas (Ref. 2).

I would like to begin with a question. Of what relevance is the topic of partial redistribution, which has been developed to study low temperature, low density astrophysical plasmas, at a conference on high temperature, high density plasmas? It seems to me that the relevance is not in applications but in the basic physical concepts of astrophysical partial redistribution (hereafter PRD) that may be applicable in a very different region of parameter space. Since PRD is a reasonably well-developed theory that has been tested against real data many times, the physical insight and mathematical developments may turn out to be helpful in understanding high temperature and density plasmas. It is in this spirit that I would like to provide an overview of the physics of PRD and discuss a few of its applications to stellar atmospheres.

I. THE ASTROPHYSICAL SETTING

Astrophysical plasmas come with a vast range of physical properties but have one characteristic in common -- inaccessibility. Almost by definition, astrophysical plasmas are beyond the reach of direct (that is <u>in situ</u>) measurement, and the only empirical tool available to the astronomer for the characterization of these plasmas is their emergent radiation. Thus the development of modern spectroscopic diagnostic techniques and methods for the solution of the radiative transfer equation with a sound basis on the essential physics is of vital concern to astrophysics.

Since the main topic of this meeting is high temperature, high density laboratory plasmas, I should perhaps begin with a crude roadmap of the type of plasmas astronomers routinely deal with. In Table 1, I list typical parameters of such plasmas including total densities, ratios of electron to total hydrogen (protons, neutrals, and molecules) densities, temperatures, and typical scale lengths. In this list are the three phases of the interstellar medium (gas between the stars) and seven common regimes that occur in the atmospheres of stars. In all cases the relevant scales vastly exceed those possible in the laboratory. Also astrophysical plasmas are commonly in violent motion with both turbulent and systematic (e.g. outflows) velocities and photons often can readily leak out into cold space.

Astronomers, like laboratory physicists attempt to characterize their plasmas, but for the astronomers there are many more properties to determine including:

(1) <u>Thermodynamic properties</u>: electron temperature, density of important species (including electrons), pressure, and the population of all important excitation and ionization states. The latter are generally far from their thermodynamic equilibrium values. A major difficulty is that the thermodynamic properties generally vary along the line of sight.

(2) Properties of the <u>radiation fields</u>, especially in important spectral lines and photoionization continua. The ability of photons to leak out into space and the long range nature of the scattering process (see below) conspire to produce radiation fields that commonly depart greatly from their thermodynamic equilibrium values.

(3) <u>Velocity fields</u>. These include both systematic motions like outflows (commonly called stellar winds) and turbulent motions on scales both large and small compared to photon mean free paths and thermalization lengths (see below).

(4) <u>Geometry</u>. Astrophysical plasmas are often approximated as a plane-parallel stratified medium or a spherically symmetric medium.

(5) <u>Degree of inhomogeneity</u>. In general, astrophysical plasmas do not exhibit a smooth variation in physical parameters with positions, but rather include unresolved inhomogeneities which may be due to propagating shock waves or embedded thin magnetic flux tubes that confine a plasma, which is commonly hotter and more dense than the surroundings. Radiation of course can leak out of these hotter regions and photoionize and photoexcite the surrounding plasma. A rather insidious form of inhomogeneity can occur when there are very steep temperature gradients, for example at the interface (transition region) between the 10^6 K solar corona and 10^4 K chromosphere. In this case some thermal electrons from the corona can make it across the interface to become a high energy tail to the electron energy velocity distribution and enhance the local collisional excitation and ionization rates which are usually produced only by electrons with several times the mean thermal energy.

(6) <u>Energy balance</u>. A major goal in diagnosing astrophysical plasmas is to assess the energy balance; that is, to determine the energy losses through radiation, thermal conduction, and expansion against the surrounding gas pressure and the stellar gravitational field. The total energy loss then measures the local heating rate and provides information on the heating mechanism which may be by shocks or the annihilation of magnetic fields.

(7) <u>Momentum balance</u>. This means estimating the forces needed to provide the systematic flows and the mechanisms responsible for these forces.

II. BASIC CONCEPTS, TERMINOLOGY, AND APPROXIMATIONS

The methods that astronomers have generally used to model stellar atmospheres have been guided more by phenomenological than physical considerations. That is, the goal has been to include all relevant line and continuum opacity sources and energy levels in order to model the distribution of thermodynamic and velocity distributions along the line of sight for an appropriate geometry so as to "understand" the energy and momentum balance. In the process the physics has often been oversimplified to a degree that the thermodynamic variables are often not accurately measured and the energy and momentum balance are not adequately "understood." When such blunders are recognized, and often many years after they are first recognized, an attempt is made to include a more accurate physical description of the radiative transfer process.

In this section I will give a superficial description of the different levels of approximation that astronomers have used to describe the interaction of radiation and matter in a stellar atmosphere. For more detailed reviews of this topic and to provide an entrance into the literature, I call your attention to the recent reviews by Hubeny,[3] Oxenius,[4] and Linsky[5] and to the very interesting workshop proceedings <u>Progress in Stellar Spectral Line Formation Theory</u> edited by Beckman and Crivellari.

A. The standard nonLTE approximation

The radiative transfer equation for the monochromatic specific intensity $I(\nu,\vec{n})$ (ignoring polarization) is generally written

$$\vec{n} \cdot \nabla I(\nu,\vec{n}) = -k(\nu,\vec{n}) I(\nu,\vec{n}) + j(\nu,\vec{n}) \quad , \tag{1}$$

where $k(\nu,\vec{n})$ is the macroscopic absorption coefficient and $j(\nu,\vec{n})$ is the macroscopic emission coefficient. It is important to recognize that this equation is a phenomenological (heuristic) semiclassical description of what is inherently a quantum statistical mechanical phenomenon.[6,7] The ratio of emission to absorption is generally defined as the source function,

$$S(\nu,\vec{n}) \equiv \frac{j(\nu,\vec{n})}{k(\nu,\vec{n})} \qquad (2)$$

and the emergent specific intensity is given by

$$I(\nu,\vec{n}) = \int S(\nu,\vec{n})\, e^{-\tau_\nu}\, d\tau_\nu \quad , \qquad (3)$$

where τ_ν is the monochromatic optical depth along the line of sight.

The fundamental question in stellar atmospheres theory is then how to evaluate the source function or, equivalently, the emission coefficient. The first approach tried was to assume that the plasma behaves locally as if it were in thermodynamic equilibrium -- an approximation called local thermodynamic equilibrium or LTE. In this case, $S(\nu,\vec{n})$ is replaced by the Planck function (i.e., Kirchhoff's Law), and the excitation and ionization equilibrium of all species depend only on the local temperature and density. The assumption of LTE, of course, enormously simplifies the computational problem so that one can include 10^7 spectral lines or solve for the emergent radiation from a three-dimensional time-dependent atmosphere including propagating shock waves or other embedded inhomogeneities.

More than 50 years ago it was recognized that when densities are low, radiative processes must dominate over collisions in determining the population of excited states or the degree of ionization. For example, for a two-level atom consisting of a lower (L) and upper (U) state, spontaneous de-excitation will dominate over collisional excitation ($A_{UL} \gg C_{UL}$) when the electron densities are sufficiently low. The plasma will then not be in LTE and is called a nonLTE plasma. This condition was first recognized for gaseous nebulae and only later for stellar atmospheres.

The new physics introduced by the nonLTE approximation is that the emission coefficient includes a scattering term. For a simple two-level atom the source function now consists of two terms (see Ref. 8, pp. 433-438 for a derivation from the statistical equilibrium equation)

$$S(\nu,\vec{n}) = \frac{(1-\varepsilon)}{\phi(\nu)} \iint R(\nu',\vec{n}';\nu,\vec{n}) \, I(\nu',\vec{n}') \, d\nu' \, d\vec{n}' + \varepsilon B_\nu(T) \quad , \quad (4)$$

where $\varepsilon \approx C_{UL}/(A_{UL}+C_{UL})$, $B_\nu(T)$ is the Planck function, $\phi(\nu)$ is the normalized absorption profile, and the redistribution function R is the probability that a photon with initial frequency ν' and direction vector $\vec{n}'$ will be scattered (in the observer's frame) into frequency ν and direction vector $\vec{n}$. There are many assumptions that go into the derivation of this ostensibly simple equation that are discussed in detail by Mihalas[8] and Hubeny,[3,6] but for the present discussion we can think of it in probabilistic terms. The quantity ε is the probability that a photon is destroyed (by a collisional de-excitation), and $(1-\varepsilon)$ is therefore the probability of scattering. Thus the first term describes creation of $(\nu,\vec{n})$ photons by scattering from all other $(\nu',\vec{n}')$ in a line, and the second term describes the creation of $(\nu,\vec{n})$ photons by thermal processes (i.e. detailed balance). For many important lines in astrophysical plasmas such as the resonance lines of H, Ca^+, and Mg^+, ε is of order 10^{-4} to 10^{-6} so that scattering dominates and has a long range character.

NonLTE problems in stellar atmospheres are often solved in one of two approximations. The first, called complete redistribution (CRD), assumes that the frequency distribution of photons after and before scattering is totally uncorrelated in the observer's frame. In this case the normalized emission profile $\psi(\nu)$ is identical to the normalized absorption profile $\phi(\nu')$ so that

$$R(\nu',\vec{n}';\nu,\vec{n}) = \phi(\nu')\phi(\nu) \tag{5}$$

and

$$S(\nu) = (1-\varepsilon)\bar{J}(\nu) + \varepsilon B_\nu(T) \quad, \tag{6}$$

where

$$\bar{J}(\nu) = \iint \phi(\nu') I(\nu',\vec{n}') \, d\nu' \, d\vec{n}' \quad. \tag{7}$$

This approximation allows for considerable computational simplification because one does not have to compute the emission profile or equivalently the distribution of atoms among upper level frequency substates. This simplification facilitates computations for multilevel, multiline atoms and time-dependent spatially inhomogeneous atmospheres.

The CRD approximation is valid throughout the line profile only when elastic collisions cause the excited atom to lose all memory of the initial frequency of the photon before scattering. This approximation is also valid in the Doppler line core (within ±3 Doppler half-widths from line center) because thermal motions of the scattering atom effectively redistribute scattered photons in frequency within this range in the observer's frame. For example, Thomas[9] calculated the ratio $\psi(\nu)/\phi(\nu)$ for the case of coherent scattering in the atom's rest frame and Doppler redistribution as seen by the observer. He showed that whereas $\phi(\nu)$ decreases four orders of magnitude from line center to three Doppler widths away, the $\psi(\nu)/\phi(\nu)$ ratio remains within a factor of 4 of unity over this whole range. Therefore, in the Doppler core $\psi(\nu)/\phi(\nu) = 1$ is a decent approximation and the source function is frequency-independent (as first shown by Kenty, Ref. 10) with the simple form of Eq. (7).

It is important to recognize that Thomas was primarily concerned about the flux in the core of the line and whether this flux can be used to infer the properties of a stellar atmosphere, in particular the existence and location of a chromosphere. From this point of view the CRD assumption is often justified as most of the flux in a stellar profile for a dwarf star like the Sun is often located in the Doppler core. Unfortunately, this justification of the use of CRD for a limited range of purposes was often misused as justification for the CRD assumption when analyzing the whole line profile including the inner line wings, where the probability of Doppler motions producing redistribution to the line core is extremely small!

This fundamental point was beginning to be appreciated during IAU Colloquium No. 19 on Stellar Chromospheres held at NASA Goddard Space Flight Center on February 21-24, 1972. At this Colloquium, Jefferies and Morrison[11] reviewed calculations of line profiles formed in plane-parallel atmospheres characterized by a sharp rise in temperature with height (i.e., chromospheres), showing that for strong resonance lines the chromospheric rise in temperature can produce emission in the line cores. Avrett[12] then described computed Ca II H and K line profiles for a solar model and a model scaled to simulate a lower gravity star. Both sets of calculations assumed complete frequency redistribution (CRD) over the whole line profile, which implies a frequency-independent source function. Essentially all previous calculations for nonisothermal model atmospheres[13-17] had employed the CRD assumption. Avrett[12] showed that it is possible to choose a solar chromospheric model that leads to computed Ca II H and K profiles that are close to the observed profiles at the center of the solar disk. He called attention, however, to the inner wings of the line profile, where the scattering process should be nearly coherent in the observer's frame and the computed line profile properly taking coherence into account should be significantly darker than in the CRD case.[17] He proposed that a good test of the importance of coherent scattering in the line wings would be to match solar plage profiles where the drop in intensity from the K_2 peak to the K_1 minimum is particularly sharp. Subsequently, however, Shine and Linsky[18] had no difficulty fitting solar plage profiles assuming CRD. Avrett also noted that the computed K line profile for a solar model scaled to the gravity of a giant star showed a decrease in intensity from K_2 to K_1 that is much shallower than is generally observed. He suggested that the cause for this discrepancy is likely the failure to include coherent scattering in the line wings in the CRD calculations. Subsequent work has shown that his suggestion was correct.

The second approximation sometimes used in computing line profiles formed in stellar atmospheres is coherent scattering (CS). In this approximation photons are assumed to be scattered coherently as seen by the observer, even though thermal motions are known to redistribute photons within the Doppler core. In this approximation

$$R(\nu',\vec{n}';\nu,\vec{n}) = \delta(\nu'-\nu) \tag{8}$$

and

$$S(\nu) = \frac{(1-\varepsilon)}{\phi(\nu)} J(\nu) + \varepsilon B_\nu(T) \quad , \tag{9}$$

where

$$J(\nu) = \int I(\nu,\vec{n}) \, d\vec{n} \quad . \tag{10}$$

Again we arrive at a computationally simple approach which is nearly valid in one portion of the line profile (the wings) but not another (the core).

B. Partial coherent scattering

The next level of approximation was introduced by Jefferies and White,[19] who re-examined the validity of the CRD assumption for the case of radiative and collisional damping in the atomic frame and Doppler redistribution. They concluded (see Fig. 1) that photons within about three Doppler half-widths of line center are completely redistributed in frequency, but that Doppler motions only slightly change the initial frequency of photons further from line center and rarely reshuffle these photons to the line core. In other words, scattering in the wings is nearly coherent over a very wide range of damping parameters. They therefore proposed that to a good approximation the redistribution function may be written as

$$R(x',x) = a(x')\delta(x'-x) + [1 - a(x')]\phi(x')\phi(x) \quad , \qquad (11)$$

where x and x' are frequency shifts measured in Doppler half-width units, e.g. $x = (\nu-\nu_0)/\Delta$, ϕ is the normalized absorption line profile, and $a(x')$ is a function that is essentially zero for $x' < 3$ and nearly unity for $x' \geq 3$. This approximation is often called "partially coherent scattering" (PCS). Subsequently, Kneer[20] pointed out that this approximation is neither normalized nor symmetric, and he proposed a revised form of equation (2) that meets both requirements. Then Basri[21] argued that neither this equation nor Kneer's revision takes into account Doppler diffusion (drifting) in the optically thick line wings, which can be important in very thick, low density atmospheres, and that accurate radiative transfer solutions require use of the exact $R(x',x)$. Ayres[22] has proposed a revision of equation (11) that includes the effects of Doppler diffusion.

The PCS approximation is an interesting one because it includes both coherency in the wings and noncoherency in the core in a computationally simple way. A further refinement by Hubeny[23] that permits the division between the core and wings to depend on depth approximately takes into account Doppler drifting. This modification to PCS now makes it a good approximation to the exact redistribution functions described below.

Jefferies and White also called attention to the depression of the line profiles in the inner wing that occurs as the scattering in the wing is taken to be more nearly coherent (a_{max}, the fractional coherency, approaches 1.0). This is shown for an isothermal atmosphere in Fig. 2, where for $a_{max} = 1.0$ the computed profile merges with the profile for coherent scattering in the observer's frame at x ($\equiv v_2$ in their notation) of about 4.6. They concluded that the emission features on either side of line center for the Ca II lines could in principle be produced in an isothermal atmosphere with coherent scattering in the line wings, as previously noted by Miyamoto,[24] or in a chromosphere if the whole line is formed in CRD. They were unable to choose among these two radically different explanations, although their calculations (Fig. 1) forcefully argue for the latter. <u>Thus a proper understanding and treatment of spectral line formation in the line wings is essential for the determination of the temperature structure in a stellar atmosphere.</u>

C. The standard partial redistribution problem

The PCS approximation assumes that the scattering process is coherent in the atom's rest frame and thus does not include physical processes in the atom's frame realistically. The next level of approximation, called partial redistribution (PRD), attempts to do so within a restricted set of assumptions. We start with the time-independent transfer equation for the unpolarized monochromatic specific intensity [Eq. (1)] and write the macroscopic absorption and emission coefficients for a two-level atom

$$k(\nu,\vec{n}) = \frac{h\nu}{4\pi} [n_L B_{LU} \phi(\nu,\vec{n}) - n_U B_{UL} \psi(\nu,\vec{n})] \qquad (12)$$

$$j(\nu,\vec{n}) = \frac{h\nu}{4\pi} [n_U A_{UL} \psi(\nu,\vec{n})] \quad . \qquad (13)$$

Here we treat stimulated emission as negative absorption (see Ref. 3 for an extended discussion of this point). Then

$$S(\nu,\vec{n}) \equiv \frac{j(\nu,\vec{n})}{k(\nu,\vec{n})} = \frac{A_{UL}}{\frac{n_L}{n_U} B_{LU} \frac{\phi(\nu,\vec{n})}{\psi(\nu,\vec{n})} - B_{UL}} \quad . \tag{14}$$

This expression shows that the source function becomes frequency independent in the CRD approximation ($\phi = \psi$). Oxenius[25] found a formal relation between ψ and ϕ for a two-level atom:

$$\psi(\nu,\vec{n}) = \frac{B_{LU} \bar{R}(\nu,\vec{n}) + C_{LU} \phi(\nu,\vec{n})}{B_{LU} \bar{J} + C_{LU}} \quad , \tag{15}$$

where

$$\bar{R}(\nu,\vec{n}) = \frac{1}{4\pi} \iint I(\nu',\vec{n}') R(\nu',\vec{n}';\nu,n) \, d\nu' \, d\Omega' \tag{16}$$

$$\bar{J} = \frac{1}{4\pi} \iint I(\nu',\vec{n}') \phi(\nu',\vec{n}') \, d\nu' \, d\Omega' \quad . \tag{17}$$

Omont, Smith, and Cooper[26] and subsequently Cooper, Ballagh, Burnett, and Hummer[7] and Cooper and Zoller[27] have questioned whether the heuristic semi-classical treatment of the radiative transfer equation and the use of macroscopic absorption and emission coefficients are justified from the perspective of quantum statistical mechanics. Starting from the general density matrix equation, they were able to derive statistical equilibrium and radiative transfer equations similar to those used above in the limit of dilute radiation fields and widely separated atoms. These limits are both valid for plasmas in stellar atmospheres.

Given this strong physical basis for the heuristic radiative transfer and source function equations, Hubeny[3] has defined what he calls the standard PRD problem by the following four assumptions: (1) photons in only one transition participate in a scattering process; i.e., the only correlated chain of radiative transitions considered is an absorption followed by a re-emission in the same line. (2) The stimulated emission is either negligible, or is treated as a negative absorption (i.e., with the absorption profile). (3) The velocity distributions of atoms in all energy states are assumed to be Maxwellian (more precisely, the velocity distribution of atoms in the lower level of a considered PRD transition is Maxwellian; the velocity distributions for other levels are irrelevant). (4) The atomic velocity is assumed unchanged during a scattering process.

A number of authors prior to 1962 have discussed frequency redistribution and redistribution functions within the context of these assumptions, but Hummer's[28] paper is an excellent starting point for a discussion of PRD radiative transfer. Hummer wrote redistribution functions for direct resonance scattering processes of the type i→j→i (for initial state i and excited state j), and all authors have generally followed his terminology, except that in subsequent work primed quantities refer to before scattering and unprimed quantities refer to after scattering. In what follows I adopt this revised convention. His terminology is summarized in Table 2. He defined the term redistribution function $R(\nu',\vec{n}';\nu,\vec{n})$ as the probability (as seen in the observer's frame) that a photon with initial frequency ν' and direction $\vec{n}'$ will be absorbed, leading to the re-emission in the line of a $(\nu,\vec{n})$ photon. He then defined four redistribution functions (see Table 3) for the cases of zero line width in the atom's frame (R_I), radiative damping of the upper state only (R_{II}), collisional and radiative damping of the upper state only (R_{III}), and radiative damping of both states (R_{IV}). Since it is difficult to solve the transfer equation including the full angle-dependent redistribution function, he wrote expressions for angle-averaged redistribution functions for the cases of the exact dipole phase function [e.g., $R_{II-B}(x',x)$] and the approximation of an isotropic scattering phase function [e.g., $R_{II-A}(x',x)$]. He then discussed some symmetry properties of the redistribution functions and showed that $R_{I-A}(x',x)$ and $R_{I-B}(x',x)$ are very similar but not identical. In most subsequent radiative transfer calculations, isotropic angle-averaged redistribution functions have been used.

Before proceeding further, I should mention that Heinzel[29] argued that Hummer's[28] R_{IV} function is based on an incorrect expression for scattering in the atomic frame. In its place, Heinzel derived a new redistribution function R_V for resonance scattering of subordinate lines assuming that both levels are radiatively broadened, and Heinzel and Hubený[30] generalized this redistribution function to include collisional broadening of both the upper and lower states. The R_V function includes all of the other redistribution functions as special cases. Computations of the angle averaged $R_{V-A}(x',x)$ are given by Heinzel and Hubený.[31]

It is important to understand what these redistribution functions mean and how they differ. Figure 3 depicts the physical processes described by the five different redistribution functions in terms of sublevel representations. The probability of re-emission at frequency x per absorption at frequency x' is shown in Figs. 4-6 for $R_{II}(x',x)$, $R_V(x',x)$, and $R_{III}(x',x)$. The important point to see in these figures is that for all redistribution functions there is a similar Doppler core for $|x| < 3$ and $|x'| < 3$, but different behavior in the line wings. For $R_{II}(x',x)$ the wing behavior is nearly coherent, i.e., $R_{II}(x',x) \sim \phi(x')$, and for $R_V(x',x)$ the wing behavior has both properties.

The first accurate numerical solutions of the radiative transfer equation including PRD to go beyond the schematic partially coherent scattering calculations of Jefferies and White are to be found in a paper by Hummer.[32]

These solutions, which assume constant plasma parameters, are for both plane-parallel and semi-infinite geometries. Hummer computed $S(\nu)$ for slabs with different line center thicknesses and ε values for $R_{I-A}(\nu',\nu)$ and $R_{I-B}(\nu',\nu)$. The latter differ only by a few percent, except near $x = 4$ for large optical depths T. He also computed $S(\nu)$ for $R_{II-A}(\nu,\nu')$ redistribution for several values of ε, A, and T, as well as for a semi-infinite geometry ($T = \infty$). He showed that $S(\nu)$ approaches arbitrarily close to the coherent scattering solution for sufficiently large values of ν, which means that the emergent intensity in the line wings can be substantially lower than the CRD solutions. These calculations demonstrated the effects of different assumed redistribution functions on the emergent line profiles and anticipated many of the results in later work.

The quantum mechanical calculations of Omont, Smith, and Cooper[26] showed that redistribution in the observer's frame can be written as the sum of natural broadening term and a collisional redistribution term that is essentially equal to CRD,

$$R(\nu',\nu) = \frac{\Gamma_R + \Gamma_I}{\Gamma_R + \Gamma_I + \Gamma_E} R_{II}(\nu',\nu) + \frac{\Gamma_E}{\Gamma_R + \Gamma_I + \Gamma_E} \phi(\nu')\phi(\nu)$$

$$= \gamma R_{II}(\nu',\nu) + (1-\gamma) \phi(\nu') \phi(\nu) \quad , \tag{18}$$

where Γ_R, Γ_I, and Γ_E are the broadening widths due to radiative decays, inelastic collisions (i.e., collisional de-excitation), and elastic collisions (i.e., Stark and van der Waals collisions). The quantity γ measures the probability that incident photons are coherently scattered in the atom's frame, and thus the probability for coherent scattering in the far line wings in the observer's frame. An analogous expression can be written for the case of natural broadening of both the lower and upper states by substituting $R_V(\nu',\nu)$ for $R_{II}(\nu',\nu)$.

Shortly thereafter, Milkey and Mihalas[33] proposed a PRD computational scheme in which they formally divided the upper state of a two-level atom into frequency substates and then solved the statistical equilibrium equations for these substates including appropriate collisional couplings. These statistical equilibrium equations were then coupled into the transfer equation using a complete linearization scheme and solved in the observer's frame. Essentially all stellar atmospheric computations to date which include PRD use Eq. (18) or an approximation to it and a computational scheme similar to that of Milkey and Mihalas.

Milkey and Mihalas solved the coupled statistical equilibrium-radiative transfer equations using the Vernazza, Avrett, and Loeser (VAL)[34] solar model and the fixed electron densities of the model. They did not attempt a detailed match to the solar $L\alpha$ profile by choosing an optimum turbulent velocity and temperature distribution. They did, however, arrive at several important results. (1) A fully self-consistent PRD calculation using depth-dependent coherence fractions computed from the broadening rates is feasible. (2) The computed line wings lie between the CRD and pure coherent scattering limits and are similar to the 93% coherency calculation of Vernazza.[35] (3) A proper inclusion of PRD effects in the wings of $L\alpha$ is needed to compute the electron density properly. This is because hydrogen is the dominant electron donor above 8000 K in the solar chromosphere, the ionization of hydrogen is controlled by photoionization from the second level, and the population of the second level depends on the $L\alpha$ radiation field, which at some locations in the chromosphere is controlled by the loss of photons in the line wings. In a subsequent paper Milkey and Mihalas[35] were able to match the shape of the observed $L\alpha$ profile by including a depth-dependent line profile function.

Despite the large improvement of the Milkey and Mihalas[33,36] Lα PRD calculations over the previous CRD calculations, Roussel-Dupré[37] called attention to two major problems in their method -- the Omont et al.[26] redistribution function used by Milkey and Mihalas is applicable only in the impact regime, which is not valid for the Lα wings, and it ignores the ℓ-degeneracy of the hydrogen energy levels. Both restrictions were relaxed by Yelnik et al.[38] Using their formulae for Γ_E, Roussel-Dupré[37] computed somewhat different wing intensities although similar wing shapes to those computed using the previous Omont et al.[26] formula for Γ_E. Thus the Milkey-Mihalas formulation of PRD was a major step forward, but it should not be viewed as the final answer to a complex radiative transfer problem.

Milkey and Mihalas[39] next applied their PRD formulation to the resonance lines of Mg II, treating each transition (the h and k lines) as equivalent to a two-level atom. An important aspect of these calculations is that they demonstrated the frequency dependence of $S(\nu)$ and its thermalization properties as a function of frequency. In the wings $S(\nu)$ behaves like the pure coherent scattering case with a monotonic decrease toward the surface since the chromosphere is optically thin in τ_ν (see Fig. 7, top). In the line core $S(\nu)$ is independent of ν and lies close to but slightly above the CRD case because thermalization involves redistribution to the optically thin wings where photons are lost from the atmosphere. Since redistribution to the line wings is made artificially large by the CRD assumption, there is less redistribution to the wings and a shorter thermalization length in the line core for PRD than CRD, so that $S_{PRD} > S_{CRD}$ in the line core. A second important point is that the h and k line profiles differ systematically in the sense that the h line wing lies above the k line wing and the h_1 minimum feature is brighter than k_2. Both effects naturally follow from the opacity in the h line being half that in the k line. The h line wing is brighter because at a given $\Delta\lambda$ optical depth unity occurs deeper in the photosphere where $B(T)$ is larger. The h_1 feature is brighter because the smaller h line opacity means that h_1 lies closer to line center than k_1, where the redistribution is less coherent. A final important point made in the paper is that coherent scattering depresses the intensities $I(h_1)$ and $I(k_1)$ (see Fig. 7, bottom), so that one cannot derive the temperature minimum located between the photosphere and chromosphere by setting the Planck function equal to the itensity at h_1 or k_1, i.e., $B_\nu(T_{min}) > I(h_1), I(k_1)$.

Milkey, Ayres, and Shine[40] then considered how stellar gravity might alter the Mg II line shapes. They computed PRD profiles of the Mg II lines for a solar model (log g = 4.44) and for a solar $T(\tau_{5000})$ distribution but log g = 2.0 to simulate an early G bright giant star. The effect of lowering the gravity is to decrease pressures in the photosphere. Since the major continuous opacity source (H^-) is proportional to P^2, while the Mg II optical depth scale is proportional to P, $\tau_{Mg\ II}(\tau_{5000})$ increases systematically with decreasing gravity. Since the temperature minimum occurs at larger $\tau_{Mg\ II}$ in a giant, the wavelengths of ($\Delta\lambda_{h_1}$, $\Delta\lambda_{k_1}$) of the h_1 and k_1 features increase. Ayres and coworkers[41] have used this argument to explain in part the Wilson-Bappu effect. In addition, the Mg II line wings darken with decreasing gravity since lower photospheric densities result in more nearly coherent scattering.

D. Partial redistribution for a multilevel atom

The first logical step beyond the standard PRD problem is to permit several transitions to participate in the scattering process. This step adds considerable complexity into the scattering process because now photons may be correlated in frequency. In other words, photons can remember the frequency of the previous photon and the process is non-Markovian in nature. At this point it is useful to introduce the concept of natural population[3,42]

> "An atomic level is said to be naturally populated if the probability of emitting (absorbing) a well-defined photon in a transition to a lower (higher) level, when averaged over an ensemble of identical atoms is independent of the previous history of the ensemble, i.e. the manner in which the level has been populated."

Natural population of a level occurs when the populating mechanism involves particles or photons whose energy spectra are broad compared to the width of a level so that the sublevels of an atomic level are uniformly populated. Examples of such processes are collisional excitation, spontaneous emission, ionization, and recombination. Deviations from natural populations occur as a result of processes that have an energy spectrum that changes across the width of a line. Such processes include absorptions and stimulated emissions when the specific intensity changes across the line profile.

Hubeny[43] introduced the concept of generalized redistribution functions (GRFs) analogous to the two-level atom redistribution functions. For a three-level atom with consecutive transitions $i \to j \to k$ starting from a naturally populated level i, the atomic frame GRF is $r_{ijk}(\xi_{ij}, \xi_{jk})$, where ξ_{ij} and ξ_{jk} are the frequencies of the two photons in the atomic frame. In general

$$r_{ijk} = \alpha \, p_V + \beta \, p_{III} \quad , \tag{19}$$

where $p_i (i = I-V)$ are analogous to the two-level atomic redistribution functions and the coefficients α and β indicate the relative importance of coherent scattering (with radiatively broadened upper and lower states) and collisional redistribution. This formalism can be extended to a chain of scatterings of arbitrary length.

The first and to my knowledge only use of GRFs for realistic computations was by Milkey, Shine, and Mihalas[44] for the case of atoms in which two allowed transitions have common upper states. The application they had in mind was the Ca II ion in which the 4^2P upper states radiatively decay to the 4^2S ground state via the H and K lines, and also radiatively decay to the metastable 3^2D states via the so-called infrared triplet lines. To include this additional decay and radiative excitation channel provided by the infrared triplet lines, they divided each upper state into substates associated with each spectral line as seen in the observer's frame. This leads to statistical equilibrium equations for each of these substates and to a new redistribution function

$$R(\nu',\nu) = \gamma' R^x(\nu',\nu) + (1-\gamma')\phi(\nu')\phi(\nu) \quad , \tag{20}$$

where $R^x(\nu',\nu)$, the cross-redistribution function, includes redistribution in the $i \to j$ transition via a third level k (i.e., $i \to j \to k \to j \to i$), and γ' is the branching ratio for this indirect process. This cross-redistribution provides additional noncoherence.

Hubený's[43] redistribution functions $p_i(i=1-5)$ in the atom's frame and $P_i(i=1-5)$ in the observer's frame, are analogous to the observer's frame redistribution functions $R_i(i=1-5)$ previously described but differ from $R_i(i=1-5)$ in that the final state may be different from the initial state. Thus R^x appears to be a special case of these generalized redistribution functions. Furthermore, Hubený, Oxenius, and Simonneau[42] have derived line profile coefficients for absorption and emission in multilevel atoms, taking into account generalized redistribution and the correlated re-emission that occurs when photoabsorption processes destroy the natural population of the excited state and elastic collisions tend to re-establish it.

In Paper V of their series, Shine, Milkey, and Mihalas[45] applied the cross-redistribution formulation to a five-level, five-transition (H, K, 8498 Å, 8542 Å, 8662 Å) Ca II ion. The main purpose of the paper was to consider whether in principle PRD effects can account for the many discrepancies between line profiles calculated assuming CRD and the observations. For these calculations they assumed the HSRA solar model and several microturbulent velocity distributions with the following results:

(1) The PRD line profiles (see Fig. 8) show limb darkening at all wavelengths, consistent with the data, whereas the CRD profiles limb-brighten only at H_3 and K_3. Athay and Skumanich[15b] could not account for the limb darkening of K_2 and H_2 assuming CRD.

(2) The decrease in intensity from K_2 to K_1 is abrupt in PRD but not in CRD, in accord with observations. This answered Wilson's[46] fundamental objections.

(3) The increase in line width is gradual toward the limb, as is observed.

(4) $I(H_1)/I(K_1) = 1.17$, as is observed.

(5) The infrared triplet line PRD and CRD profiles are nearly identical, implying that cross-redistribution is extremely important for these lines as one would expect on the basis that the A values for depopulating of the 4^2P states via the resonance lines are far larger than the corresponding rates for the infrared triplet lines.

(6) Computed values of $I(H_1)$ and $I(K_1)$ are much smaller than observed when they use the HSRA model which has $T_{min} \approx 4170$ K, but are consistent with observations when they raise T_{min} to 4450 K.

(7) For a schematic solar plage model they found that K_2 brightens and $\Delta\lambda_{K_1}$ increases as is observed, and the infrared triplet lines develop double reversals in their cores as is also observed.[47]

These successes provided strong evidence that the PRD approach includes much of the essential physics of the scattering process, however, even such approximate treatments of coherent scattering in the line wings as the partially coherent scattering model used by Vardavas and Cram[48] can qualitatively account for points (1), (2), (3), and (6). Also there are alternative methods of writing the transfer equation that include the same physical processes but in some cases may be easier computationally[49,50]

E. Further steps beyond the standard PRD problem

Hubeny[3] has discussed in detail the relaxation of other assumptions in the standard PRD problem (see above) and the reader is referred to his paper and the references he cites. Here I would like to just mention these generalizations and whether they are important in astrophysical radiative transfer calculation.

The radiative transfer equation is simplified greatly by ignoring stimulated emission or by treating it as negative absorption with the same profile as absorption [i.e., $\psi(\nu,\vec{n}) = \phi(\nu,\vec{n})$] and as stimulated emission in the statistical equilibrium equations. Cooper et al.[51] concluded after an extensive discussion that this is a valid approximation when spontaneous emission exceeds stimulated emission (i.e., $A_{UL} \gg B_{UL}\bar{J}$), which is generally valid for most astrophysical plasmas.

The standard PRD problem assumes that the velocity distribution of atoms in the lower state is at all times Maxwellian, whereas photoexcitations and collisional excitations tend to destroy this character. Also velocity-changing collisions (when the atom is excited during scattering process) can destroy photon correlations and produce CRD. These problems have been considered by Oxenius,[4,25] Cooper et al.,[7] and others. A satisfactory description of the latter problem has been given in the strong collision model of Cooper et al. in which elastic collisions either destroy photon correlations and produce a Maxwellian velocity distribution for the scattering atoms or leave the velocity unchanged and photons correlated.

III. APPLICATION OF PRD TECHNIQUES TO THE CONSTRUCTION OF STELLAR MODEL CHROMOSPHERES

A. Plane-parallel static models

Since 1972 a number of authors have used PRD techniques to compute profiles of the resonance lines of hydrogen, Ca^+, and Mg^+ formed in the chromospheres of the Sun and stars. These computations were done in order to derive models of these chromospheres (consisting of the distribution of temperature, pressure, and densities with height) by matching computed and observed line profiles. In general, these models assumed hydrostatic equilibrium and a plane-parallel geometry. This work is summarized by Linsky[5] and the reader is referred to this review of the topic. These computations are all within the constraints of the standard PRD problem and use the redistribution function $R(\nu',\nu)$ of Eq. (18) or in a few cases the cross redistribution function of Eq. (20).

B. The roles played by velocity fields and atmospheric extension

For many types of stars we have evidence that the atmospheres are not in hydrostatic equilibrium or plane-parallel (i.e. thickness $\ll$ stellar radius), but rather expand and are often geometrically extended. In addition, wave-like motions, circulation patterns, and inhomogeneity must occur in many stars as in the Sun. In two early papers Vardavas[52] and Cannon and Vardavas[53] calculated

PRD line profiles for an expanding atmosphere using the $R_I(x',x)$ redistribution function in the observer's frame. They found large differences between line profiles computed with CRD and PRD, which Magnan[54] argued must be incorrect. Mihalas et al.[55] argued that in a moving atmosphere accurate source functions can be computed in the observer's frame only when the full angular and frequency coupling is taken into account for the frequency redistribution, which is computationally very expensive. However, calculations in the comoving frame of the fluid are greatly simplified because then the scattering atoms see a nearly isotropic radiation field of limited frequency range, so that angle-averaged static redistribution functions are a good approximation. They then formulated the PRD transfer problem in the comoving frame and showed that for $R_I(x',x)$ redistribution line profiles for CRD and PRD are nearly identical. Vardavas[56] showed that accurate solutions to the PRD differential expanding atmosphere problem can be obtained using the comoving frame formulation for R_I and R_{II} redistribution in an isothermal atmosphere, and Vardavas and Cannon[57] did the same for a schematic chromosphere model with R_{II-A} redistribution.

In subsequent theoretical developments, solutions of the radiative transfer equation were generally formulated in the comoving frame for spherically symmetric geometries and they generally assumed CRD or, nearly equivalently, the Sobolev approximation, since the intended applications were for atmospheres of hot stars where the expansion velocities are much larger than the thermal and turbulent motions. To test the validity of the CRD approximation, Mihalas, Kunasz, and Hummer[58] solved the transfer equation for spherically symmetric flows with an angle-averaged redistribution function in the comoving frame using a variable-Eddington factor iterative scheme. They discussed the conditions for which the emergent line profiles for CRD and PRD can differ for

$R_I(x',x)$ and $R_{II}(x',x)$ redistribution. Mihalas[59] extended this work to include the full angle and frequency dependence of redistribution. Peraiah and Nagendra[60] have used the $R_{II-A}(x',x)$ redistribution function to interpret line profiles formed in the moving atmospheres of hot stars.

Since winds in late-type giants and supergiants have expansion velocities of only a few times the combined thermal-turbulent half-width, the CRD and the Sobolev approximations should not be accurate. However, there is ample evidence in the asymmetric Mg II h and k lines[61] that significant expansion velocities occur in the chromospheres of these stars. To explore the range of velocities and chromospheric geometric extents that should be representative for the late-type giants, Drake and Linsky[62] made a series of calculations using the comoving frame PRD code developed by Mihalas et al.[55] In their exploration of parameter space they found that with increasing flow speed and geometrical extension (see Fig. 9), the effect on the Mg II line profiles is to suppress the blue emission peak, shift the central minimum (k_3) to the blue, and enhance the red emission peak giving the Mg II lines a P Cygni character. For all cases in which the maximum flow speed is less than six times the Doppler width, the PRD profiles are very different from the CRD case. They also explored the effect of changing the chromospheric temperature gradient and the location of the chromosphere. Drake[63] has reported on his calculations to match the observed Mg II line profiles of Arcturus.

Avrett and Loeser[50] described an integral-equation method for solving the transfer equation in the comoving frame also including the effects of PRD, spherical geometry, and expansion. Using the partially coherent scattering formulation of $R_{II}(\nu',\nu)$ by Kneer,[20] they showed sample calculations for a constant property atmosphere and a velocity law with $v_{max}/\Delta v_D = 3$. Hartmann and Avrett[64] used this formalism to compute resonant line profiles for their

Alfvén-wave driven wind model of α Ori (M2 Iab). Differences between computed and observed line profiles then led them to propose a more complex velocity field in the atmosphere of this star than initially computed. These calculations highlight the role that observations have and should continue to play in stimulating progress in spectral line formation theory.

IV. CONCLUSIONS

As should be clear from the preceding discussion, the shape of a spectral line profile emergent from a stellar atmosphere depends in a complex way on the properties of the atmosphere (temperature-density structure, geometrical extent, inhomogeneity, velocity fields, and radiation fields in different lines and continua), as well as the radiative and collisional processes that lead to redistribution of photons during the scattering process. While the theory of PRD even for a multilevel atom is now becoming mature, we have no assurance that all of the important physical effects have yet been included properly. Nevertheless, the shape of spectral line profiles for optically thick resonance lines and perhaps also important subordinate lines depends critically on the proper inclusion of PRD, atmospheric extension, and expansion effects. Thus we cannot expect to derive the structure of a stellar atmosphere without properly including these effects, but at the same time we should not avoid using the powerful tool of spectral line profile analysis because of the mathematical complexity of the theory or possible uncertainties concerning its physical basis.

This work was supported in part by NASA grants NAG5-82 and NGL-06-003-057 through the University of Colorado. I would like to thank Drs. E. H. Avrett, T. R. Ayres, S. A. Drake, and D. G. Hummer for stimulating discussions.

REFERENCES

1. D. Mihalas, Stellar Atmospheres, 1st Ed. (W. H. Freeman, San Francisco, 1970).

2. R. N. Thomas, in Stellar Chromospheres, edited by S. D. Jordan and E. H. Avrett, NASA SP-317 (1973), p. 312.

3. I. Hubený, in Progress in Stellar Spectral Line Formation Theory, edited by J. E. Beckman and L. Crivellari (Reidel, Dordrecht, 1985), p. 27.

4. J. Oxenius, in Progress in Stellar Spectral Line Formation Theory, edited by J. E. Beckman and L. Crivellari (Reidel, Dordrecht, 1985), p. 59.

5. J. L. Linsky, in Progress in Stellar Spectral Line Formation Theory, edited by J. E. Beckman and L. Crivellari (Reidel, Dordrecht, 1985), p. 1.

6. I. Hubený, in Spectral Line Shapes, Proceedings of the Seventh International Conference on Spectral Line Shapes (1985).

7. J. Cooper, R. J. Ballagh, K. Burnett, and D. G. Hummer, Astrophys. J. **260**, 299 (1982).

8. D. Mihalas, Stellar Atmospheres, 2nd Ed. (W. Freeman, San Francisco, 1978), Ch. 13.

9. R. N. Thomas, Astrophys. J. **125**, 260 (1957).

10. C. Kenty, Phys. Rev. **42**, 823 (1932).

11. J. T. Jefferies and N. D. Morrison, in Stellar Chromospheres, edited by S. D. Jordan and E. H. Avrett, NASA SP-317 (1973), p. 3.

12. E. H. Avrett, in Stellar Chromospheres, edited by S. D. Jordan and E. H. Avrett, NASA SP-317 (1973), p. 27.

13. J. T. Jefferies and R. N. Thomas, Astrophys. J. **129**, 401 (1959); J. T. Jefferies and R. N. Thomas, Astrophys. J. **131**, 695 (1960).

14. E. H. Avrett, SAO Special Report No. 174, p. 101 (1965).

15. a) R. G. Athay and A. Skumanich, Solar Phys. **3**, 181 (1968).

    b) R. G. Athay and A. Skumanich, Solar Phys. **4**, 176 (1968).

16. J. L. Linsky, SAO Special Report No. 274 (1968).

17. J. L. Linsky and E. H. Avrett, Publ. Astron. Soc. Pacific **82**, 169 (1970).

18. R. A. Shine and J. L. Linsky, Solar Phys. **39**, 49 (1974).

19. J. T. Jefferies and O. R. White, Astrophys. J. **132**, 767 (1960).

20. F. Kneer, Astrophys. J. **200**, 367 (1975).

21. G. S. Basri, Astrophys. J. **242**, 1133 (1980).

22. T. R. Ayres, Astrophys. J., in press.

23. I. Hubený, Astron. Astrophys. **145**, 461 (1985).

24. S. Miyamoto, Publ. Astron. Soc. Japan, **5**, 142 (1953); S. Miyamoto, Publ. Astron. Soc. Japan **6**, 140 (1954).

25. J. Oxenius, J. Quant. Spectrosc. Radiat. Transfer **5**, 771 (1965).

26. A. Omont, E. W. Smith, and J. Cooper, Astrophys. J. **175**, 185 (1972).

27. J. Cooper and P. Zoller, Astrophys. J. **277**, 813 (1984).

28. D. G. Hummer, Monthly Notices Roy. Astron. Soc. **125**, 21 (1962).

29. P. Heinzel, J. Quant. Spectrosc. Radiat. Transfer **25**, 483 (1981).

30. P. Heinzel and I. Hubený, J. Quant. Spectrosc. Radiat. Transfer **27**, 1 (1982).

31. P. Heinzel and I. Hubený, J. Quant. Spectrosc. Radiat. Transfer **30**, 77 (1983).

32. D. G. Hummer, Monthly Notices Roy. Astron. Soc. **145**, 95 (1969).

33. R. W. Milkey and D. Mihalas, Astrophys. J. **185**, 709 (1973).

34. J. E. Vernazza, E. H. Avrett, and R. Loeser, Astrophys. J. **184**, 605 (1973).

35. J. E. Vernazza, Ph.D. Thesis, Harvard University (1972).

36. R. W. Milkey and D. Mihalas, Solar Phys. **32**, 361 (1973).

37. D. Roussel-Dupré, Astrophys. J. **272**, 723 (1983).

38. B. Yelnik, K. Burnett, J. Cooper, R. J. Ballagh, and D. Voslamber, Astrophys. J. **248**, 705 (1981).

39. R. W. Milkey and D. Mihalas, Astrophys. J. **192**, 769 (1974).

40. R. W. Milkey, T. R. Ayres, and R. A. Shine, Astrophys. J. **197**, 143 (1975).

41. T. R. Ayres, J. L. Linsky, and R. A. Shine, Astrophys. J. (Letters) **195**, L121 (1975); T. R. Ayres, Astrophys. J. **228**, 509 (1979).

42. I. Hubený, J. Oxenius, and E. Simonneau, J. Quant. Spectrosc. Radiat. Transfer **29**, 477 (1983); I. Hubený, J. Oxenius, and E. Simonneau, J. Quant. Spectrosc. Radiat. Transfer **29**, 495 (1983).

43. I. Hubený, J. Quant. Spectrosc. Radiat. Transfer **27**, 593 (1982).

44. R. W. Milkey, R. A. Shine, and D. Mihalas, Astrophys. J. **199**, 718 (1975).

45. R. A. Shine, R. W. Milkey, and D. Mihalas, Astrophys. J. **199**, 724 (1975).

46. O. C. Wilson, in <u>Stellar Chromospheres</u>, edited by S. D. Jordan and E. H. Avrett, NASA SP-317 (1973), p. 305.

47. R. A. Shine and J. L. Linsky, Solar Phys. **25**, 357 (1972).

48. I. M. Vardavas and L. E. Cram, Solar Phys. **38**, 367 (1974).

49. N. J. Heasley and F. Kneer, Astrophys. J. **203**, 660 (1976).

50. E. H. Avrett and R. Loeser, in <u>Methods in Radiative Transfer</u>, edited by W. Kalkofen (Cambridge University Press, Cambridge, 1983), p. 341.

51. J. Cooper, I. Hubený, and J. Oxenius, Astron. Astrophys. **127**, 224 (1983).

52. I. M. Vardavas, J. Quant. Spectrosc. Radiat. Transfer **14**, 909 (1974).

53. C. J. Cannon and I. M. Vardavas, Astron. Astrophys. **32**, 85 (1974).

54. C. Magnan, Astron. Astrophys. **35**, 233 (1974).

55. D. Mihalas, R. A. Shine, P. B. Kunasz, and D. G. Hummer, Astrophys J. **205**, 492 (1976).

56. I. M. Vardavas, J. Quant. Spectrosc. Radiat. Transfer **16**, 781 (1976).
57. I. M. Vardavas and C. J. Cannon, Astron. Astrophys. **53**, 107 (1976).
58. D. Mihalas, P. B. Kunasz, and D. G. Hummer, Astrophys. J. **210**, 419 (1976).
59. D. Mihalas, Astrophys. J. **238**, 1034 (1980).
60. A. Peraiah and K. N. Nagendra, Astrophys. Space Sci. **90**, 237 (1983).
61. R. E. Stencel and D. J. Mullan, Astrophys. J. **238**, 221 (1980); R. E. Stencel, D. J. Mullan, J. L. Linsky, G. S. Basri, and S. P. Worden, Astrophys. J. Suppl. **44**, 383 (1980).
62. S. A. Drake and J. L. Linsky, Astrophys. J. **273**, 299 (1983).
63. S. A. Drake, in <u>Progress in Stellar Spectral Line Formation Theory</u>, edited by J. E. Beckman and L. Crivellari (Reidel, Dordrecht, 1985), p. 351.
64. L. Hartmann and E. H. Avrett, Astrophys. J. **284**, 238 (1984)

Table 1. Properties of astrophysical plasmas

Region	$n_{TOT}(cm^{-3})$	n_e/n_H	$T_e(K)$	$L(cm)$
Interstellar Medium				
(a) Hot phase	10^{-3}	1	10^6	3×10^{20}
(b) Warm phase	10^{-1}	0.5	10^4	10^{19}
(c) Cold phase	$10^2 - 10^4$	10^{-4}	10^2	10^{16}
Stellar Atmospheres				
(a) Flares	$10^{11} - 10^{12}$	1	10^7	$10^6 - 10^8$
(b) Coronae	$10^8 - 10^{10}$	1	10^6	$10^{11} - 10^{12}$
(c) Chromospheres	$10^{10} - 10^{12}$	1	10^4	10^8
(d) Winds of cool supergiants	$10^7 - 10^9$	10^{-2}	4×10^3	$10^{15} - 10^{16}$
(e) Winds of young hot stars	$10^{10} - 10^{12}$	1	3×10^4	10^{15}
(f) Surface of a white dwarf	10^{19}	1	10^4	10^5

Table 2. Hummer's (Ref. 28) notation with indicies reversed

Quantity	Atom's frame	Observer's frame
Incoming frequency	ξ'	$\nu' = \xi' + (v_o/c)\vec{n}'\cdot\vec{v}$ $x' = (\nu'-\nu_o)/\Delta$
Outgoing frequency	ξ	$\nu = \xi + (v_o/c)\vec{n}\cdot\vec{v}$ $x = (\nu-\nu_o)/\Delta$
Absorption probability for photon $(\xi',\vec{n}')$	$f(\xi')\, d\xi'\, \dfrac{d\Omega'}{4\pi}$	
Probability of frequency redistribution	$p(\xi',\xi)d\xi$	
Probability of angle redistribution	$g(\vec{n}',\vec{n})d\Omega$	
Probability that an absorption at $(\xi',\vec{n}')$ leads to a re-emission at $(\xi,\vec{n})$	$f(\xi')p(\xi',\xi)g(\vec{n}',\vec{n})$ $\cdot \dfrac{d\Omega'}{4\pi}\, d\Omega d\xi' d\xi$	
(The redistribution function)		$R(\nu',\vec{n}';\nu,\vec{n})$

Table 3. Different types of redistribution functions

Type of scattering	Atom's frame	Observer's frame
Coherent in atom's frame	$p(\xi',\xi) = \delta(\xi'-\xi)$	$R(\nu',\vec{n}';\nu,\vec{n})$
Zero line width in atom's frame	$f(\xi')d\xi' = \delta(\xi'-\nu_0)d\xi'$ (isotropic) (dipole)	$R_I(\nu',\vec{n}';\nu,\vec{n})$ $R_{I-A}(x',x)$ $R_{I-B}(x',x)$
Radiation damping (upper state only)	$f(\xi')d\xi' = \dfrac{\delta}{\pi} \dfrac{d\xi'}{(\xi'-\nu_0)^2+\delta^2}$	$R_{II}(\nu',\vec{n}';\nu,\vec{n})$ $R_{II-A}(x',x)$ $R_{II-B}(x',x)$
Collisions and natural broadening (upper state only)	$p(\xi',\xi)d\xi = f(\xi)d\xi'$ $= \dfrac{\delta}{\pi} \dfrac{d\xi}{(\xi-\nu_0)^2+\delta^2}$	$R_{III}(\nu',\vec{n}';\nu,\vec{n})$ $R_{III-A}(x',x)$ $R_{III-B}(x',x)$
Resonance scattering (both upper and lower states broadened)	$f(\xi')p(\xi',\xi) =$ $\dfrac{\delta_i \delta_j}{\pi^2} \cdot \dfrac{1}{[(\xi'-\xi)^2+\delta_i^2]}$ $\cdot \dfrac{1}{[(\xi-\nu_0)^2+\delta_j^2]}$	$R_{IV}(\nu',\vec{n}';\nu,\vec{n})$ $R_{IV-A}(x',x)$ $R_{IV-B}(x',x)$ $R_V(\nu',\vec{n}';\nu,\vec{n})$ $R_{V-A}(x',x)$ $R_{V-B}(x',x)$

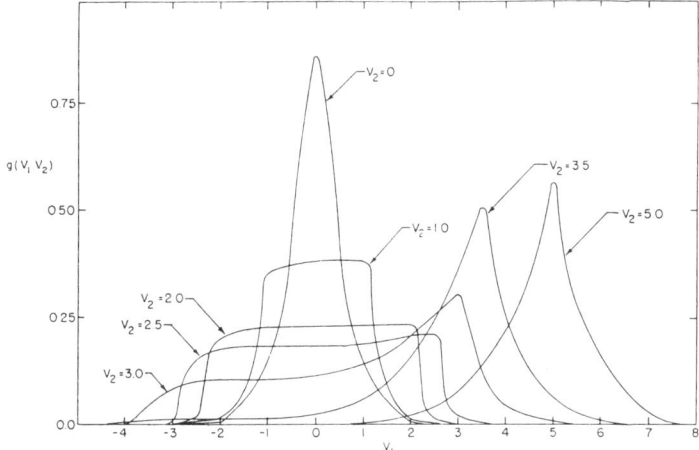

Fig. 1. The redistribution function for pure coherent scattering in the atom's rest frame $g(v_1,v_2) \equiv R_{I-A}(x',x)$ in the notation of this paper (from Jefferies and White, Ref. 19). Note that photons (in the observer's frame) at $v_2 < 3$ are non-coherently scattered and at $v_2 > 3$ are nearly coherently scattered about their initial frequency. Basri (Ref. 21) refers to the finite width of the redistribution function in the line wings ($v_2 > 3$) as Doppler drifting of "coherently" scattered photons.

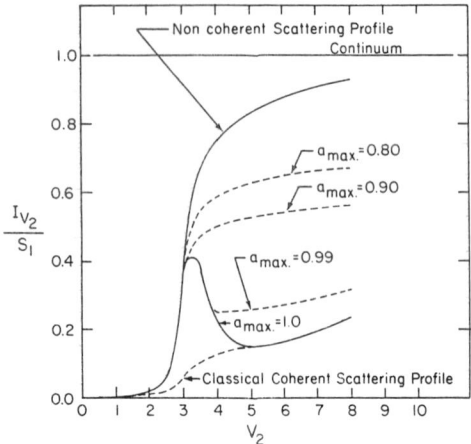

Fig. 2. Comparison of emergent line profiles for an isothermal atmosphere when the scattering is completely coherent in the atom's rest frame ($a_{max} = 1.0$), noncoherent ($a_{max} = 0$, upper solid line), and intermediate cases (from Ref. 19).

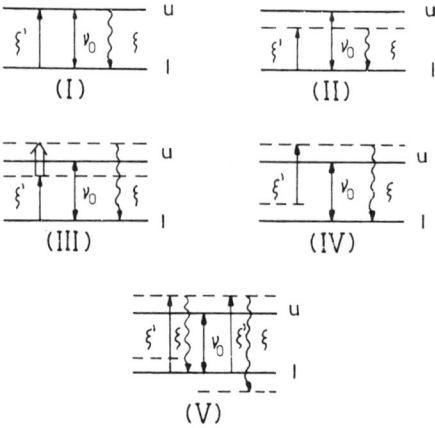

Fig. 3. Schematic sublevel representation of the scattering processes in a two-level atom (from Ref. 31). ξ' and ξ are the atomic-frame absorption and emission frequencies, respectively. ν_0 is the central frequency of the line. The double arrow indicates the collisional reshuffling of sublevels (for details refer to text).

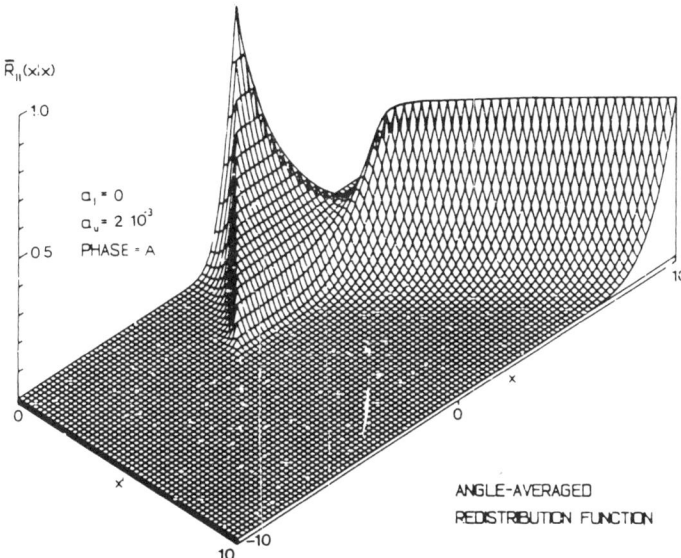

Fig. 4. Probability of re-emission at frequency x (in Doppler-width units) per absorption at frequency x', for the angle-averaged redistribution function $R_{II}(x',x)$. The Voigt damping parameters a_ℓ and a_u are indicated. The ordinate is $\bar{R}_{II}(x',x) = R_{II}(x',x)/\phi(x')$ (adopted from Heinzel and Hubený, Ref. 31).

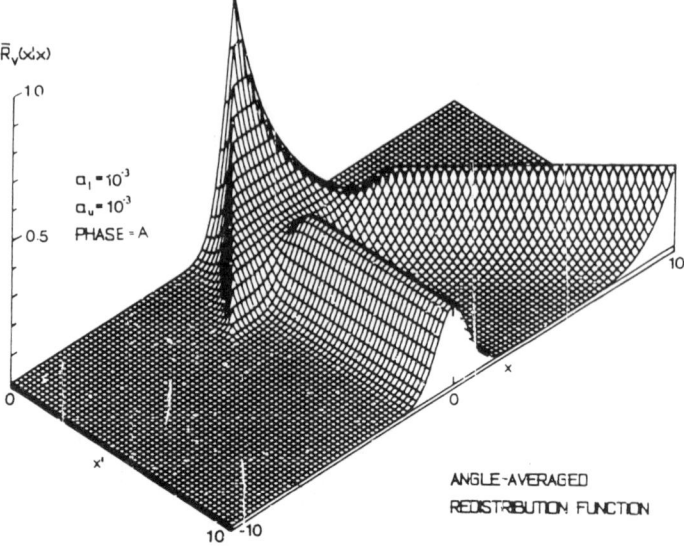

Fig. 5. As in Fig. 4, but for R_V (adapted from Ref. 31).

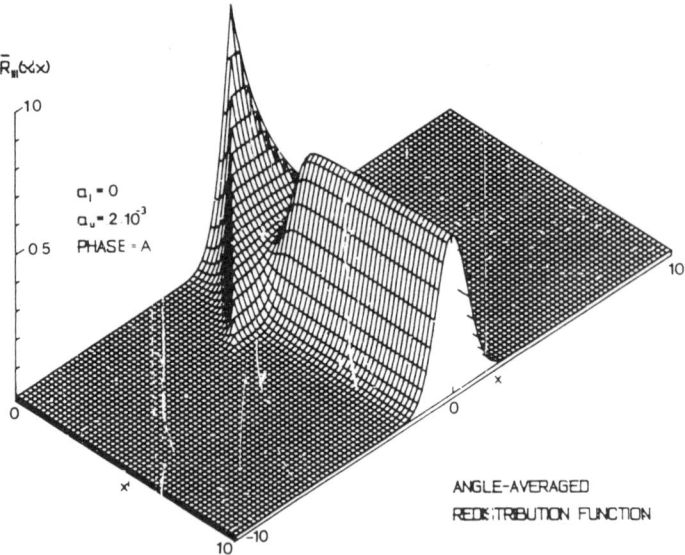

Fig. 6. As in Fig. 4, but for R_{III} (adapted from Ref. 31).

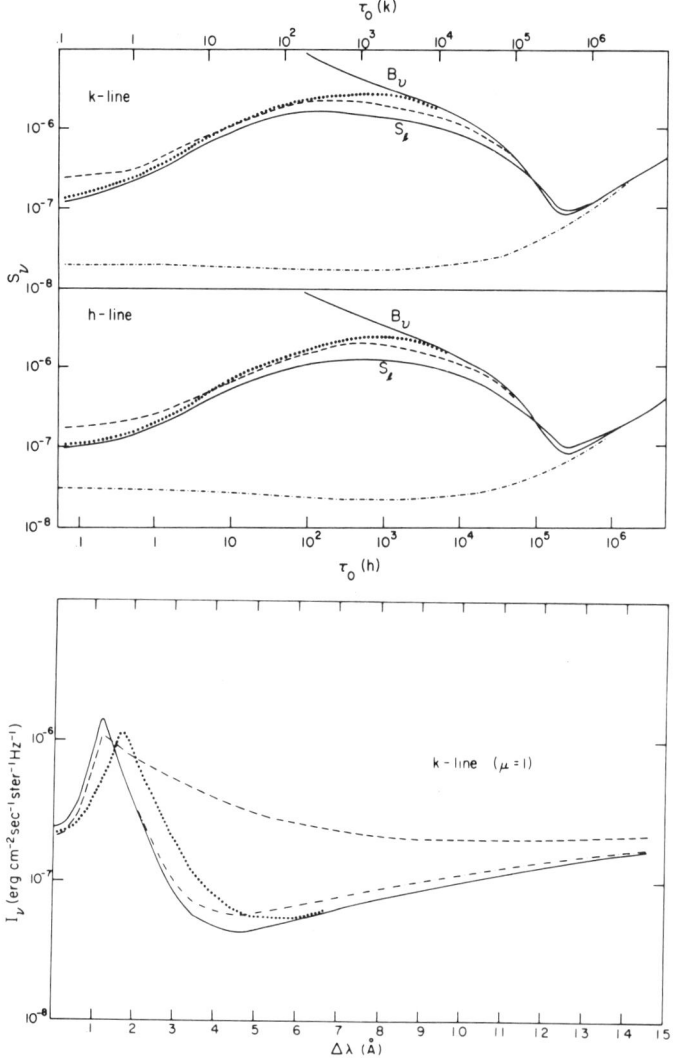

Fig. 7. Top -- Source functions (S_ν) for the Mg II h and k lines compared with B_ν: S_ℓ, complete redistribution; dots, line center ($\Delta\lambda = 0.109$ Å); dot-dashes, k_1 ($\Delta\lambda = 0.462$ Å), h_1 ($\Delta\lambda = 0.365$ Å) (from Ref. 38). Bottom -- Emergent Mg II k-line profiles for: dashes, complete redistribution; solid line, partial redistribution; dots, partial redistribution with enhanced turbulent velocities; dot-dashes, partial redistribution with enhanced Γ_{vdW} (from Ref. 38).

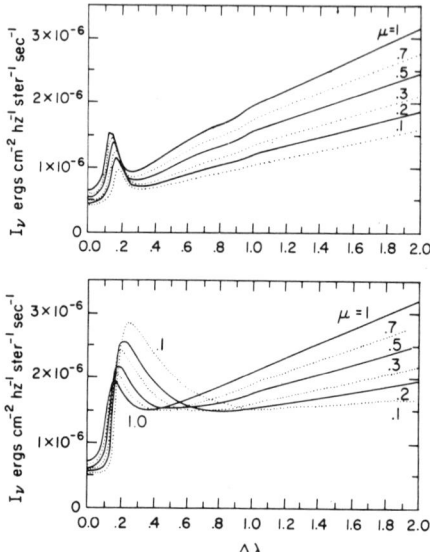

Fig. 8. Center-to-limb variation of the Ca II K line for a three-level atom with the HSRA model and the Linsky-Avrett velocity distribution. Top shows results for PRD; bottom shows results for CRD. Abscissa, wavelength separation from line center in Å; ordinate, specific intensity in absolute units. Curves are labeled with μ, the cosine of the angle from disk center (from Ref. 45).

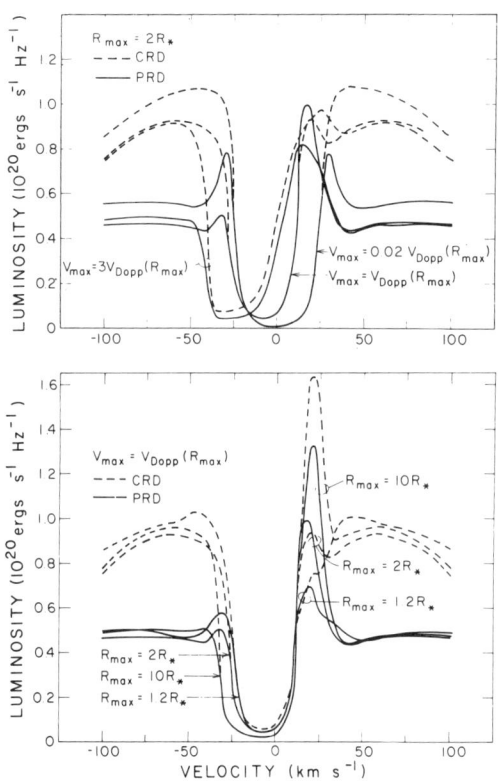

Fig. 9. (Top) Computed profiles of the Mg II k line for an early K giant star with $R_{max} = 2 R_*$. CRD and PRD profiles are given for three different velocity laws. (Bottom) Computed profiles keeping the same expansion velocity law but changing the maximum size of the chromosphere from 1.2 to 10 times the photospheric radius (from Ref. 62).

A FAST ACCURATE CODE FOR HYDROGEN AND HELIUM OPACITIES

T.R. Carson
Department of Astronomy and Astrophysics
University of St Andrews, Great Britain

Abstract

An opacity code, initially for hydrogen and helium mixtures, but which is readily extendable by the addition of more complex atomic ions, is presented. The ionization structure is computed by minimizing the Helmholtz free energy. The formation and dissociation of diatomic and triatomic molecules and radicals is also included. The atomic absorption data is provided, for the one-electron case, in the form of exact formulae, and for the two-electron case, in the form of accurate tables of spectroscopic energy levels and transition data, available in the literature. Perturbations due to adjacent particles can be treated either according to the confined atom model or according to the screened-Coulomb potential model. An additional feature of the code is the ability to include, in a routine fashion, the resonances in photoelectric absorption spectra. It is found that helium resonances make only a small contribution, at the one percent level, to the opacity.

1. Introduction

Stellar modelling, perhaps particularly hydrodynamical modelling, continues to point to the need for opacity tables with finer grids and higher accuracies, thus putting the emphasis on faster opacity codes. As computers become faster in operation and larger in memory, it becomes increasingly possible to store and process larger quantities of data specific to individual ions. After hydrogen, and similar one-electron systems, the two-electron systems, including neutral helium (as well as other members of its isoelectronic sequence), has been the most extensively studied both experimentally and theoretically to yield accurate energy level and transition data. Here we present detils of an opacity code, initially for hydrogen and helium mixtures, appropriate to Population III, extreme Population II stars and helium stars. The code can readily be extended to include more complex atomic ions, given the appropriate data.

Given the temperature, density and element abundances, the ionization equilibrium is computed by minimizing the Helmholtz free energy, including non-ideal terms, with respect to variations in the number densities of the different species - electrons and ions. Among the non-ideal terms are terms to represent the free energy due to the Coulomb interactions between charged species, as well as terms which describe the effects of interactions on the internal energy levels, and through them on the internal partition functions, of composite species. All the one-electron energy levels and transition data are computed from the exact (non-relativistic) wavemechanical formulae by means of specially designed fast subroutines, with memory storage and scaling to avoid repetition. The two-electron energy levels are obtained from spectroscopic data, augmented by accurate perturbative calculations, while transition data is taken from the growing body of high accuracy calculations. Perturbations can be treated according either to the confined atom model or the screened-Coulomb (Debye or Yukawa) potential model. In both cases a substantial gain in simplicity and speed of execution is achieved by reducing the minimization of the free energy to one with respect to a single variable, namely the free electron number density or the Debye length, according to the model chosen. The formation and dissocation of diatomic and triatomic molecules and radicals is included, using partition functions computed directly with the aid of spectroscopic data.

Included in the calculations are the usual scattering, free-free, bound-free and bound-bound radiative absorption processes involving neutrals, positive and negative atomic ions, and molecules, with the exception of molecular band spectra. In order to investigate the possible importance to opacity of resonance structure in photoelectric absorption spectra, provision is made for the routine treatment of resonance data, using a parameterization such as that of Fano. It is shown, however, that the inclusion of helium resonances leads to only a small increase, of at most about one percent, in the opacity. Both the Planck and Rosseland mean opacities are calculated.

2. Opacity and equation of state

In any application involving the transport of energy by radiation it may be possible to describe the transfer process in terms of a mean (or frequency averaged) coefficient of absorption. Thus we may define the opacity or mean mass absorption coefficient (absorption cross-section per unit mass)

$$k = \int W(\nu) k(\nu) d\nu$$

where ν is the frequency of the radiation, $k(\nu)$ the frequency-dependent (or monochromatic) mass absorption coefficient, and $W(\nu)$ is an appropriate weighting function of frequency. Two special cases have been identified, namely the Planck mean k_P and the Rosseland mean k_R appropriate to the cases of optically thin and optically thick radiative transfer, respectively. These are given by

$$k_P = \int P(\nu) k(\nu) d\nu \; ; \quad P(\nu) = B(\nu,T)/B(T)$$

and

$$k_R^{-1} = \int R(\nu) k^{-1}(\nu) d\nu; \; R(\nu) = (\partial B(\nu,T)/\partial T)/(\partial B(T)/\partial T)$$

where $B(\nu,T)$ and $B(T)$ are the frequency dependent and frequency integrated Planck blackbody specific intensity functions. In terms of the reduced frequency $u = h\nu/kT$ these weighting functions become

$$P(u) = 15\pi^{-4} u^3 (e^u - 1)^{-1}$$

and

$$R(u) = (15/4) \pi^{-4} u^4 e^u (e^u - 1)^{-2}$$

The absorption coefficient $k(\nu)$ is that for pure absorption, corrected in the optically thick case by the stimulated emission factor $(1-e^{-u})$, plus a contribution from scattering (Carson,[1]). It may be expressed in terms of the atomic radiative cross-sections via the relation

$$\rho k(\nu) = \mu(\nu) = N\sigma(\nu)$$

where ρ is the mass density, and $\mu(\nu)$ the volume absorption coefficient (absorption cross-section per unit-volume), N the number density of absorbers and $\sigma(\nu)$ the radiative absorption cross-section at frequency ν. Since a photon of frequency ν may be absorbed in any of a usually large number of specific transitions, $\mu(\nu)$ may be written as

$$\mu(\nu) = \sum_a N_a \sum_b \sigma_{ab}(\nu)$$

where a and b refer to the initial and final levels and N_a is the number density of occupation of the initial level a. The determination of the N_a for all the quantum levels (discrete and continuous) is the same problem that arises in the determination of the equation of state and all the other thermodynamic functions in an equilibrium state.

3. The Helmholtz free-energy minimization

The thermodynamic state of a system of particles of various species i with particle numbers N_i confined to a volume V at temperature T is described by the Helmholtz free energy

$$F = F[T,V,\{N_i\}] = -kT \ln Z[T,V,\{N_i\}]$$

where Z is the partition function for the system. The equilibrium state is that for which F is a minimum with respect to variations in the N_i, that is

$$\delta F = \sum_i (\partial F/\partial N_i) \delta N_i = \sum_i \mu_i \delta N_i = 0$$

where the μ_i are the chemical potentials. The minimization is to be carried out subject to the constraints of conservation appropriate to the situation. Where only chemical reactions, dissociation and ionization, are permitted then electrons and nuclei are conserved. The method can be used equally well where nuclear reactions are allowed conserving nucleons, or where elementary particle processes take place with the conservation of baryons and leptons. In the case of chemical equilibrium the free energy minimization method forms the basis for the mathematical formulation of the Law of Mass Action. For more modern applications see Harris[2] and Harris and Trulio.[3]   More recently still the method has been used for the calculation of thermodynamic properties of matter where there may be large departures from the ideal gas, such as in stars and planets. For details of those applications, see for example the work of Graboske et al. (Grossman and Graboske,[4] and earlier references) and of Hubbard et al. (Hubbard and De Witt,[5] and earlier references).

Confining our attention for the moment to ionization, let N_{ij} be the number of atoms of type (element) i in ionization stage j, N_i be the total number of atoms of type i, and N_e be the number of electrons. Then the minimization of F gives, together with the subsidiary conservation conditions,

$$\delta F = \sum_{ij} \mu_{ij} \delta N_{ij} + \mu_e \delta N_e = 0$$

$$\delta N_i = \sum_j \delta N_{ij} \qquad = 0 \quad \text{for all i}$$

$$\delta N_e = \sum_{ij} q_{ij} \delta N_{ij}$$

where the q_{ij} are the net positive charge of the ion indicated by the subscripts.[ij] Introducing Lagrange multipliers λ_i (all i) and λ_e we obtain

$$-\lambda_i = (\mu_{ij} + \mu_e q_{ij})$$

for all i and all j relating to that value of i. In general therefore any two adjacent states of ionization j and j+i of an atom i are related by

$$\mu_{ij} + \mu_e q_{ij} = \mu_{i\,j+1} + \mu_e q_{i\,j+1}$$

or

$$\mu_{ij} - \mu_{i\,j+1} = \mu_e$$

Analogous equations can be written down for dissociation equilibrium.

4. Evaluation of the chemical potentials

The evaluation of the chemical potentials $\mu_i = \partial F/\partial N_i$ requires a knowledge of $F = F(T,V,\{N_i\})$. The development of an expression for F is facilitated by certain assumptions regarding the independence of various degrees of freedom of the system, the additivity of the energies associated with different degrees of freedom, and the related factorization of Z, i.e. the additivity of $\ln Z$. Then we may write F as a sum of parts deriving from the translational, internal and configurational degrees of freedom

$$F = F_{trans} + F_{int} + F_{con}$$

For a Maxwell-Boltzmann gas of identical particles i

$$F_{i,trans} = -N_i kT[1 + \ln\{(V/N_i)G_i(T)\}]$$

giving

$$\mu_{i,trans} = -kT \ln[(V/N_i)G_i(T)]$$

where $G_i(T)$ is given by

$$G_i(T) = g_i(2\pi m_i kT/h^2)^{3/2}$$

wherein g_i and m_i are the intrinsic statistical weight and mass of particle i. For a Fermi-Dirac gas, such as the electron gas

$$F_{e,trans} = -N_e kT[U_{3/2}(\Lambda)/U_{1/2}(\Lambda) - \ln\Lambda] = -N_e kT[(2/3)I_{3/2}(\eta)/I_{1/2}(\eta) - \eta]$$

where $\eta = \ln \Lambda$, $I_n(\eta) = \int x^{-n}[\exp(x-\eta)+1]^{-1}dx$, $U_n(\Lambda) = I_n(\eta)/\Gamma(n)$

giving

$$\mu_{e,trans} = \eta kT$$

If $Z_{i,\text{int}}$ is independent of the $\{N_i\}$, as for an ideal gas then

$$F_{i,\text{int}} = -N_i kT \ln Z_{i,\text{int}}$$

and

$$\mu_{i,\text{int}} = -kT \ln Z_{i,\text{int}}$$

Thus for the moment ignoring F_{con}, the ionization equilibrium condition becomes

$$(N_{i\ j+i}/N_{ij}) = (Z_{i\ j+i}/Z_{ij})(G_{i\ j+i}/G_{ij}) \exp[-I_{ij}/kT - \eta]$$

where I_{ij} is the ionization potential, giving the generalization of the Saha equation for arbitrary (non-relativistic) electron degeneracy.

5. The non-ideal gas

The inclusion of interaction between particles, usually ignored in the ideal gas approximation, introduces a dependence of the internal energy levels and hence also of the internal partition function on the particle numbers. The internal contribution to the chemical potential then becomes

$$\mu_{i,\text{int}} = -kT[\ln Z_{i,\text{int}} + \sum_j N_j \partial(\ln Z_{j,\text{int}})/\partial N_i]$$

For ions with an infinite number of discrete levels, at least in the isolated case where $N_j = 0$, the internal partition function sum itself has always required to be converged by truncation or other device to simulate the interactions with neighbouring ions. This problem is now compounded by the appearance of the second term on the right which involves the derivative of the internal partition function with respect to the particle numbers. It is of course possible to truncate the partition function sum in a manner which involves only V and/or ρ but not the N_j individually, so that the required derivative vanishes leaving again only the first term. The more consistent approach, and the one we adopt, is to include the perturbations of all the energy levels, and thus automatically truncate the partition function sum as well as improving the accuracy of the sum itself. The derivative term, however, remains to be evaluated. While most workers have been content to carry out the minimization process numerically, Däppen[6] chose to represent the internal partition functions by approximate but differentiable analytical formulae. We however prefer to evaluate the level perturbations and the partition functions numerically thus retaining the integrity as well as the accuracy of the physical model. At the same time by ensuring that the partition function depends on only a single variable, the minimization process now effectively takes place in a one-dimensional rather than a multi-dimensional space. We now consider how to implement this scheme for the two most widely used atomic models, the confined atom model and the screened-Coulomb potential model.

In the confined atom model, each ion i is enclosed in a spherical box of radius r_i. The mere existence of a finite boundary is sufficient to perturb the energy levels of an electron in the otherwise unperturbed potential within the sphere, and to truncate the partition function sum, and this approach has been adopted by some workers (see Däppen, loc. cit.). Of course this ignores the real physical interactions which also perturb the potential within the sphere and the energy levels. One way to include these perturbations is to ascribe them to the field of the free-electron cloud which uniformly fills the space assigned to each ion, viz

$$\Delta V_i(r) = (3n_i e^2/2r_i)[1 - \frac{1}{3}(r/r_i)^2]$$

where $n_i = N_e v_i$, $v_i = (4/3)\pi r_i^3$ being the volume of the sphere. There is no unique prescription for the determination of the r_i or v_i. It may however be shown that all linear relations between v_i and the charge q_i on each ion, that is of the form

$$v_i = c(a + bq_i)$$

where a, b and c are constants, lead to the result

$$V = \sum_i N_i v_i = c(aN_a + bN_e)$$

where N_a is the total number of atoms/ions. From this it follows that the v_i or r_i and therefore the level perturbations and the internal partition functions are functions of N_e only. For $b = 0$ all ion spheres have the same volume; for $a = 0$ the ion sphere volume is proportional to q_i (giving zero volume for neutrals) while for $a = Z_i$ (the nuclear charge on the ion) and $b = -1$ the ion sphere volume is proportional to the number of bound electrons in the ion (giving zero for completely stripped ions).

In the screened-Coulomb potential model, the otherwise (approximate) Coulomb potential energy of the outer or valence electron

$$V_i(r) = z_i e^2/r$$

where $z_i = q_i + 1$, is replaced by a Debye-Hückel or Yukawa potential

$$V_i(r) = (z_i e^2/r)\exp(-r/\lambda)$$

where λ is the Debye length, given by

$$\lambda^{-2} = (4\pi e^2/kTV)\sum_c N_c z_c^2 \theta_c$$

where the subscript c now indicates all charged species, ions and electrons. The factor θ_c represents a correction for degeneracy effects, for $c \neq e$ $\theta_c = 1$ while for $c = e$ $\theta_c = I_{-\frac{1}{2}}(\eta)/I_{+\frac{1}{2}}(\eta)$. The bound eigenstates of the screened-Coulomb potential has been studied by numerous workers. The most extensive tabulations of the numerical solutions has been given by Rogers et al.[7] The work of Dickinson[8] shows that the semi-classical expectations as to the number of bound states supported by this potential are in very close agreement with the numerical results. Thus it follows also in the case of this model that the level perturbations and the internal partition functions are functions of the single variable λ only.

For both models the problem of calculating derivatives of the internal partition functions with respect to the particle numbers N_c is reduced to the determination of a derivative with respect to one variable only. In the confined atom case this is $(\partial Z_{int}/\partial N_e)$ and is readily obtained numerically. In the screened-Coulomb case we write

$$\partial Z_{int}/\partial N_c = (\partial \lambda/\partial N_c)(dZ_{int}/d\lambda)$$

in which the first factor on the right can be written down in analytical form, while the second factor is obtained numerically. Thus in neither model is it necessary to resort to a multi-dimensional completely numerical minimization of the internal partition functions.

In a highly ionized environment the most important configuration free energy term is that due to the Coulomb interactions of charged species. An expression for F_{con} in terms of the particle numbers N_c has been given by de Witt,[9] and need not be repeated. All the derivatives $\partial F_{con}/\partial N_c$ can be obtained analytically, though tediously, and will not be written down here. A particularly convenient property of all the expressions for the configuration free energy and its derivatives is that they involve only the charged particle numbers and the squares of their charges, regardless of the nuclear charge, so that at most a one-dimensional array of quantities is required to generate all the required derivatives. In an environment where there is a substantial number of neutral species, one should also include in F_{con} terms to represent neutral-charged and neutral-neutral interactions.

All of the non-ideal terms in the free energy, whether internal or configurational, thus effectively lead to corrections to the chemical potentials for all species, neutral and charged heavy ions and electrons, thus modifying each chemical potential term of the equilibrium equation.

6. Atomic Data

For one-electron systems, the free-free, bound-free and bound-bound cross-sections are expressed in terms of Gaunt factors and oscillator strengths which are generated from exact formulae. For this purpose it was deemed advantageous to derive hydrogenic formula specific to each lower level up to principal quantum number $n = 5$ and orbital angular momentum quantum number $l = 4$, for arbitrary upper level, both bound and free. Specially designed fast subroutines are then used to construct tables for multiple use, using scaling laws where applicable, or to calculate quantities as required.

For two-electron systems the ever-growing amount of accurate results both experimental and theoretical is utilized. In the case of H^- the analytic fits of Tsuji[10] for free-free and bound-free cross-sections are used. For HeI spectroscopic data for levels, both the singlet and triplet series, is obtained from the compilation of Moore,[11] and for lines, including oscillator strengths, from Wiese, Smith and Glennon.[12] Bound-free transition data is taken from Bell and Kingston,[13,14] Jacobs and Burke,[15] Bell, Kingston and Taylor,[16] Jacobs,[17] and Stewart.[18,19]

Where necessary all the data is extended beyond what is available by various techniques, perturbation theory, quantum-defect theory, interpolation and extrapolation. In some cases data for the same process from different sources are spliced together in a complementary fashion.

7. Resonances

Resonances occur in photoionization absorption spectra when either (i) the initial ion has virtual discrete states lying in the continuum or (ii) the final ion has real discrete states lying in the continuum. In helium the lowest resonance levels are (i) the doubly excited (2s, 2p) level of HeI at 4.42 rydbergs above the $1s^2$ ground level and at 2.91 rydbergs above the (1s, 2s) first excited level, and (ii) the singly excited (2s or 2p) level of HeII at 4.81 rydbergs above the $1s^2$ level and at 3.29 rydbergs above the (1s, 2s) level. These resonances give rise to almost discontinuous features in the photoionization cross-section, frequently represented as Fano profiles where the oscillator strength density is given in parametrized form

$$df/d\varepsilon = B(\varepsilon+q)^2/(1+\varepsilon^2)$$

where B is the background, q is a measure of the resonance 'strength', and ε is a reduced energy variable given by

$$\varepsilon = 2(E-E_r)/\Gamma_r$$

E being the photon energy, E_r its value at the resonance and Γ_r the resonance width. In terms of these quantities the integrated oscillator strength is

$$f_r = (\pi \Gamma_r/2) B(q^2-1)$$

Values of the parameters for helium resonances have been computed by Burke,[20] Norcross,[21] Jacobs,[17] McGreevy and Stewart,[22] and Stewart.[23] Generally speaking the resonance width is narrow, but in any case the resonance profile must be convolved with the profile for the initial level. In the case of a Cauchy profile the resulting profile can be written down as a simple analytical expression, but for other cases, such as a Gauss or Gamma profile, a numerical quadrature is necessary.

8. Results and discussion

A sample of results is presented for a mixture consisting of 75% hydrogen and 25% helium by mass. Both the confined atom (CA) model and the screened-Coulomb (SC) model have been used. Figure 1 shows the absorption spectrum for the case log T = 4.7, log ρ = -7.0 on the basis of the confined atom model. The upper curve (a) shows the continuous spectrum together with the helium resonances, while the lower curve (b) shows the continuous plus line spectrum. The temperature is chosen so as to display all the main photoelectric edges and line series for the ions of hydrogen and helium. On the reduced energy scale $u = h\nu/kT$, 1 Rydberg corresponds to u = 3.15. At low density, such as the case illustrated, the confined atom model and the screened Coulomb model produce virtually identical spectra. Figures 2 and 3 respectively show the Planck and Rosseland mean continuous opacities as a function of temperature for several densities. In each case the upper curves are for the confined atom model and the lower curves are for the screened Coulomb model. It is seen that at the lower densities the two models give results which are almost identical. However at higher densities the results for the two models diverge at low temperature while remaining in close agreement at high temperature. Along a line in the temperature-density plane appropriate to stellar conditions there is very little difference between the absorption spectra and the mean opacities produced by the two models.

---

[1] T.R. Carson, in Stellar Evolution, edited by H-Y. Chiu and A. Muriel, Ch. 15, p.427 (The MIT Press, Cambridge, Ma., 1972).
[2] G.M. Harris, J. Chem. Phys. 31, 1211 (1959).
[3] G.M. Harris and J. Trulio, J. Nucl. Energy C 2, 224 (1961).
[4] A.S. Grossman and H.C. Graboske, Jr., Astrophys. J. 164, 475 (1971).
[5] W.B. Hubbard and H.E. De Witt, Astrophys. J. 290, 388 (1985).
[6] W.Däppen, Astron. Astrophys. 91, 212 (1980).
[7] F.J. Rogers, H.C. Graboske, Jr, and D.J. Harwood, Phys. Rev. A 1, 1577 (1970).
[8] A.S. Dickinson, J. Phys. B 4, L116 (1971).
[9] H.E. De Witt, in Low Luminosity Stars, edited by S. Kumar, p.211 (Gordon and Breach, New York, 1969).
[10] T. Tsuji, Publ. Astr. Soc. Japan 18, 127 (1966).
[11] C.E. Moore, Atomic Energy Levels, Vol.I (NBS, 1949)
[12] W.L. Wiese, M.W. Smith and B.M. Glennon, Atomic Transition Probabilities, Vol.I (NBS, 1966).
[13] K.L. Bell and A.E. Kingston, J. Phys. B 3, 1433 (1970).
[14] K.L. Bell and A.E. Kingston, J. Phys. B 4, 1308 (1971).
[15] V.L. Jacobs and P.G. Burke, J. Phys. B 5, L67 (1972).
[16] K.L. Bell, A.E. Kingston and I.R. Taylor, J. Phys. B 6, 2271 (1973).
[17] V.L. Jacobs, Phys. Rev. A 9, 1938 (1974).
[18] A.L. Stewart, J. Phys. B 11, L431 (1978).
[19] A.L. Stewart, J. Phys. B 12, 401 (1979).
[20] P.G. Burke, in Proc. 5th ICPEAC, edited by L.M. Branscomb, p. 128 (1967).
[21] D.W. Norcross, J. Phys. B 4, 652 (1971).
[22] E. McGreevy and A.L. Stewart, J. Phys. B 10, L527 (1977).
[23] A.L. Stewart, J. Phys. B 11, 2449 (1978).

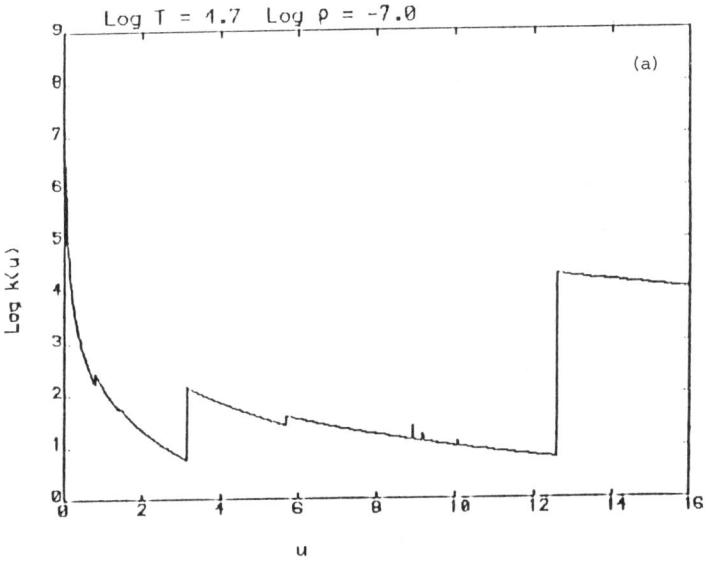

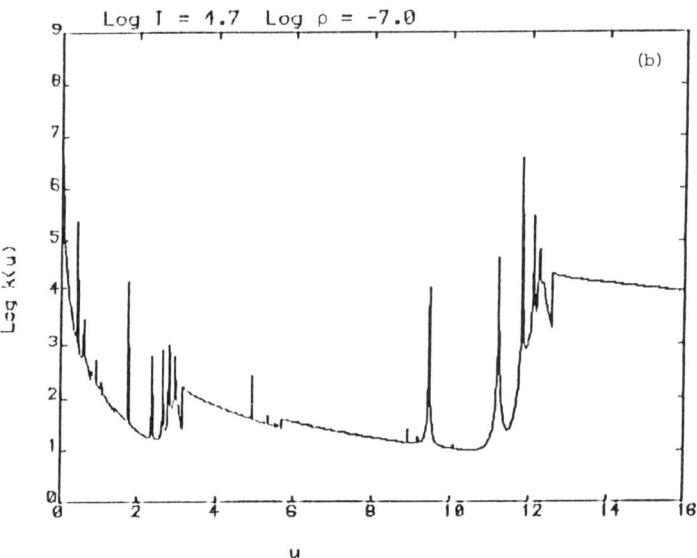

Figure 1. Absorption coefficient
(a) continuous spectrum   (b) line spectrum

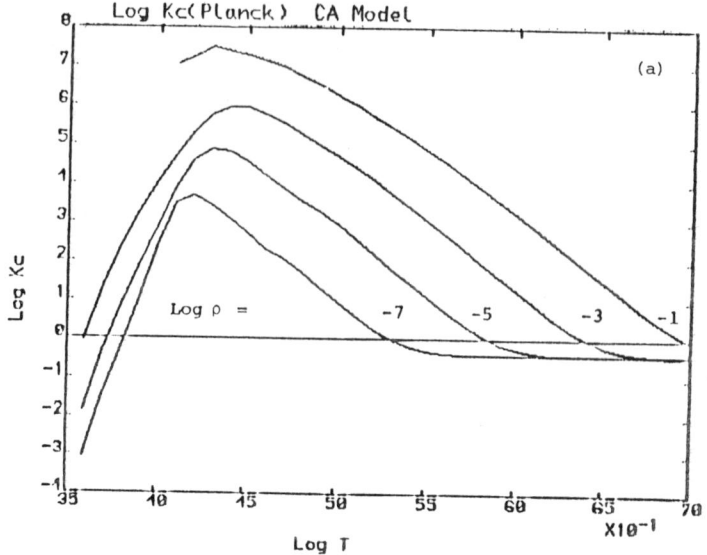

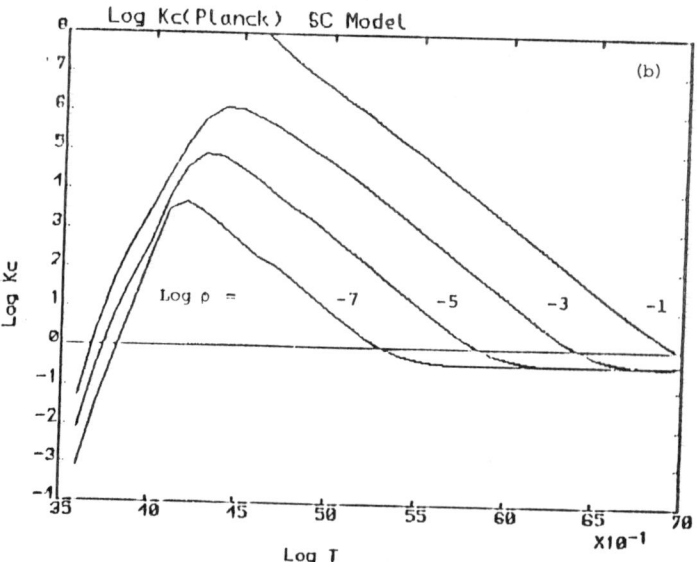

Figure 2. Planck continuous opacity
 (a) CA model (b) SC model

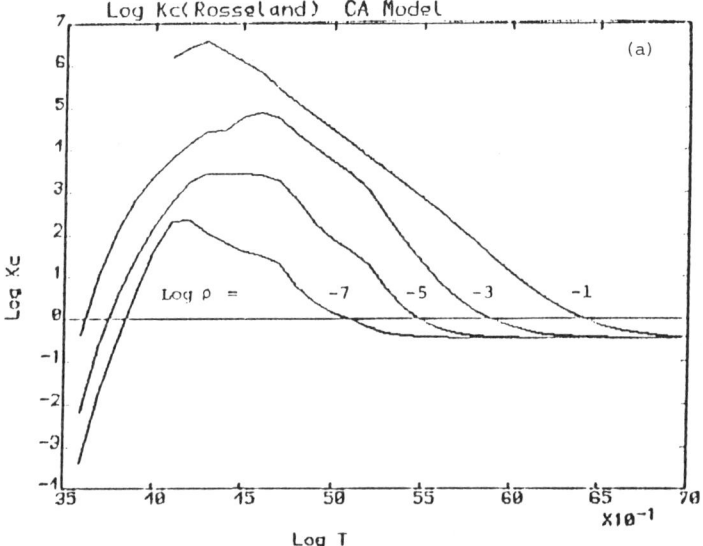

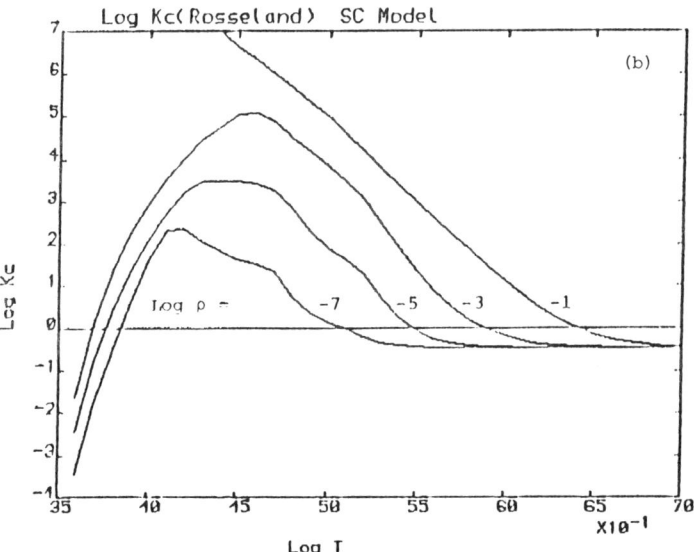

Figure 3. Rosseland continuous opacity
(a) CA model (b) SC model

IV. EQUATION OF STATE, RATE EQUATIONS AND ELECTRODYNAMICS

Electrical Conductivity of Dense Plasmas at Temperatures
of .5 to 1.0 eV*

S. W. Daniels, H. R. Griem and A. D. Krumbein**
Laboratory for Plasma and Fusion Energy Studies
University of Maryland
College Park, Maryland 20742

Abstract

The electrical conductivity of a plasma containing a mixture of light elements at densities near 1 g/cm^3 is studied at temperatures from 0.5 to 1 eV. This regime, encountered in chemical explosives, has received little attention because it is at the lower bound of interest for plasma physics and the upper bound of interest for chemistry. The dissociation equilibria of molecules are treated using Vardya's model[1] for pressure dissociation. The Saha equation is used to form a system of equations which is solved numerically to determine equilibrium populations. Effects which are considered include the existence of negative ions, reduction of ionization potentials in a plasma, and pressure ionization. It is found that H^- can be a primary negative carrier. The electrical conductivity is then calculated using an approximate electron-atom cross section. The results for hydrogen are compared with a number of published measurements and calculations.

* Sponsored by NAVSEA-SYSCOM

** On leave from the Soreq Nuclear Research Center, Yavne, Israel.

Introduction

We have studied high density, low temperature plasmas consisting of mixtures of H, C, N, and O in a preliminary theoretical attempt to find the electrical conductivity. Although it seems reasonable to expect condensed matter physics and quantum mechanical principles to become important in this regime, these are difficult to apply to this problem. We have therefore, used a plasma physics approach, which allows for a more tractable problem, in the hope of elucidating the physical situation.

Dissociation of Molecules

First we consider the molecules. We use the pressure dissociation model of Vardya[1] which describes collisions of hydrogen atoms with hydrogen molecules. This model will give us a lower limit on the dissociation of hydrogen.

We will use the law of mass action

$$K_p = P_H^2/P_{H_2}$$

where P_H is the pressure of hydrogen atoms and P_{H_2} is the pressure of molecular hydrogen. Also, K_p is known as the equilibrium constant and is, at low densities, only a function of the temperature,

$$\ln K_p = -\frac{D_{eff}}{RT} + \frac{5}{2} \ln T - \ln Q_o + 11.346$$

where T is in °K, D_{eff} is the effective dissociation energy, Q_o is the internal partition function, and R is the gas constant.

We find D_{eff} by considering the change in the potential energy during the atom-molecule interactions. The model of Vardya assumes that dissociation is caused by a hydrogen atom striking a hydrogen molecule along the line connecting the centers of the atoms in the molecule. When the vibrational and rotational energy in the molecule exceeds D_{eff} then dissociation has occurred. According to this model

$$\Delta D = |43.8106[(2.7504r - 1.04) \exp(- 2.0628r) \sinh(.6876r) - .52\exp(- 1.3752r)]|$$

were r is in angstroms and ΔD is in eV. ΔD is then defined as

$$D_{eff} = D_e - \Delta D$$

and D_e is the dissociation energy measured above the lowest energy with no perturbation.

The radius r is now set equal to the average distance between atoms,

$$r = (\frac{4\pi}{3} N)^{-1/3}$$

where N is the density of atoms. The quantity Q_o in the K_p equation is obtained from tables[1].

Using the ideal gas law and the conservation of mass we get an equation for the pressure of atomic hydrogen

$$P_H = -\frac{K_p}{4} + \sqrt{\frac{K_p}{16} + \frac{NRT\,K_p}{2}}$$

The results of this model for a 1 g/cm^3 hydrogen gas range from 82% to over 93% dissociated for temperatures from 0.5 eV to 1.0 eV. The dissociation of the other diatomic molecules is of the same order of magnitude and so we consider them to be as dissociated as hydrogen. This high degree of dissociation means that we can now consider this a gas of single atoms and go on to look at the results of applying ionization equilibrium relations to these atoms.

Ionization and Negative Ion Formation

The ionization of the atoms is governed by the Saha equation and the quasi-neutrality condition. The Saha equations are,

$$\frac{H^+ N_e}{H_o} = S_H(T) = \frac{2\,Z^{H^+}}{Z^{H_o}} \left(\frac{mkT}{2\pi\,h^2}\right) \exp\left(-\frac{E_I - \Delta E}{kT}\right)$$

$$\frac{C^+ N_e}{C_o} = S_C(T)$$

$$\frac{N^+ N_e}{N_o} = S_N(T)$$

$$\frac{O^+ N_e}{O_o} = S_O(T)$$

where E_I is the ionization energy from the ground state, Z is the partition function of the species, and N_e, H^+, H_o, etc. are the densities of the electrons, hydrogen ions, hydrogen neutrals, etc.

The change in the ionization energy, ΔE, reflects the energy released on immersing an electron ion pair in a plasma[2]. It has the effect of lowering the ionization potential that the neutral atom sees. It also sets a limit on the otherwise infinite summation in the partition function. Because of the very short Debye length in this dense plasma we were unable to obtain a meaningful calculation of ΔE. We arbitrarily chose about 4 eV because this is about the largest number that we could justify using as a perturbation.

In this model we also allow negative hydrogen ions. A semi-empirical Thomas-Fermi calculation[3] yields about 1.5 free electrons per atom in this regime. We feel that this is an overestimate which comes from trying to use a many electron atom approach for a low Z atom. We take the opposite approach and assume very few free electrons per atom so that the free electrons interact mostly with neutral atoms. The electron affinity of neutral hydrogen is lower than that of carbon, nitrogen, or oxygen but at these densities the size of the negative ion becomes a factor and the hydrogen ion is the smallest of the four[4]. The negative hydrogen ion introduces a Saha-like equation of the form[2]

$$\frac{H^o N_e}{H^-} = S_{H^-}(T) = 4\left(\frac{mkT}{2\pi h^2}\right)^{3/2} \exp\left[-\frac{E_A}{kT}\right]$$

where E_A is the electron affinity of hydrogen. There is no change in the ionization energy because there is no change in the charge of the plasma whether it is H^- or H_o and a free electron.

We also assume quasi-neutrality and conservation of mass. These generate

$$N_e + H^- = H^+ + C^+ + N^+ + O^+$$

$$D_H = H_o + H^+ + H^-$$

$$D_C = C_o + C^+$$

$$D_N = N_o + N^+$$

$$D_O = O_o + O^+$$

where the D's are the mass densities of each species, which we assume are given.

This non-linear system of equations can be reduced to a sixth order polynomial in the electron density which can then be solved numerically using Newton's Method. The solution gives the number density of each species in the plasma. From this we can generate the number of negative charge carriers, electrons and negative hydrogen ions, per atom. The results show that we have only a very small amount of ionization. This is shown in Figs. 1 and 2. Figure 2 shows a hydrogen plasma which will be our sole concern for the rest of the analysis here.

Electrical Conductivity

Armed with the electron density, hydrogen neutral density, and hydrogen ion densities in this hydrogen plasma we can go on to calculate the conductivity. Consider the conductivity given by[5]

$$\sigma = \frac{N_e}{H_o} \frac{e^2}{M_e} \frac{1}{\overline{V} \, \overline{\sigma}_{e \to H}}$$

where $\overline{V}$ is the average velocity and $\overline{\sigma}_{e \to H}$ is the effective collision cross section between electrons and hydrogen neutrals.

We use

$$N_e = (e^- + H^-)$$

because the density is so high that an electron associated with H^- will be so close to an H_o that it could easily hop over to the H_o and so seem essentially free.

The average thermal velocity of these electrons is given as

$$\overline{V} = \sqrt{\frac{3 \, kT}{m}}$$

We then have for the conductivity

$$\sigma = \frac{(e^- + H^-)}{H_o \sqrt{kT}} \frac{e^2}{\sqrt{3m} \, \sigma_{e \to H}}$$

and using $\sigma_{e \to H}$ as 10^{-15} cm^2 we can generate Fig. 3.

Discussion

The electrical conductivity results for hydrogen, as shown in Fig. 3, have been compared with measurements[6,7] and with a recent calculation[8]. At 0.85 eV (9900° K), where our calculations overlap the measurements, our calculations are lower than the measured values by about a factor of 2. The calculations of Ref. 8 overlap our calculations to a larger extent. Here our results are consistently higher by a factor of 3. It is, perhaps, fortuitous that our calculations are closer to experiment than those of Ref. 8, since we have used the simplest theories available to us. Reference 8 does not assume Local Thermal Equilibrium (LTE) whereas we do. However, in the experiments it was necessary to use LTE to analyze the data, which may help explain the more favorable comparison obtained. Use of a non-LTE theory becomes increasingly more essential as plasma temperatures fall below 1 eV at low mass densities.

A comparison with the SESAME tables[9], shown in Fig. 4, which are based on the Thomas-Fermi model, shows a difference of several orders of magnitude and a very different temperature dependence. One must conclude that the theory upon which the SESAME calculations are based does not appear to be applicable to the plasma discussed in this paper. Apparently condensed matter and quantum mechanical principles are very difficult to implement at this level.

Given the uncertainty in calculating the electrical conductivity in this high density, low temperature regime even for pure hydrogen, it would appear to be advantageous to attempt to determine the electrical conductivity experimentally for the systems of interest. We are now in the process of designing such an experiment.

References

1. M. S. Vardya, Monthly Notices of the Royal Astronomical Society, 129, 345 (1965).

2. H. R. Griem, Plasma Spectroscopy, McGraw-Hill Inc., 1964.

3. R. More, Atomic Physics in ICF, UCRL-84991 (1981) (unpublished).

4. B. L. Christensen-Dalsgaard, J. Phys. B: At. Mol. Phys. 18 (1985), p. L407 - L411.

5. L. A. Arzimovich, Elementary Plasma Physics, Blaisdell Pub. Co., 1965.

6. J. C. Morris, R. P. Rudis, and J. M. Yos, Phys. Fluids 13, 608 (1970).

7. K. Behringer and N. van Cung, Appl. Phys. 22, 373 (1980).

8. D. A. Erwin and J. A. Kunc, Phys Fluids 28, 3349 (1985).

9. Equation-of-State and Opacity Group, SESAME '83: Report on the Los Alamos Equation-of-State Library, LALP-83-4 (1983) (unpublished).

Figure Captions

Fig. 1. Number of free electrons per atom as a function of temperature for a mixture of hydrogen, carbon, nitrogen and oxygen at a density of 1 gm/cm^3.

Fig. 2. Number of free electrons per atom as a function of temperature for pure hydrogen at a density of 1 gm/cm^3.

Fig. 3. Electrical conductivity as a function of temperature for pure hydrogen. This is a comparison of our calculatons with results of other investigators.

Fig. 4. Electrical conductivity as a function of temperature for pure hydrogen at a density of 1 gm/cm^3 - SESAME tables.

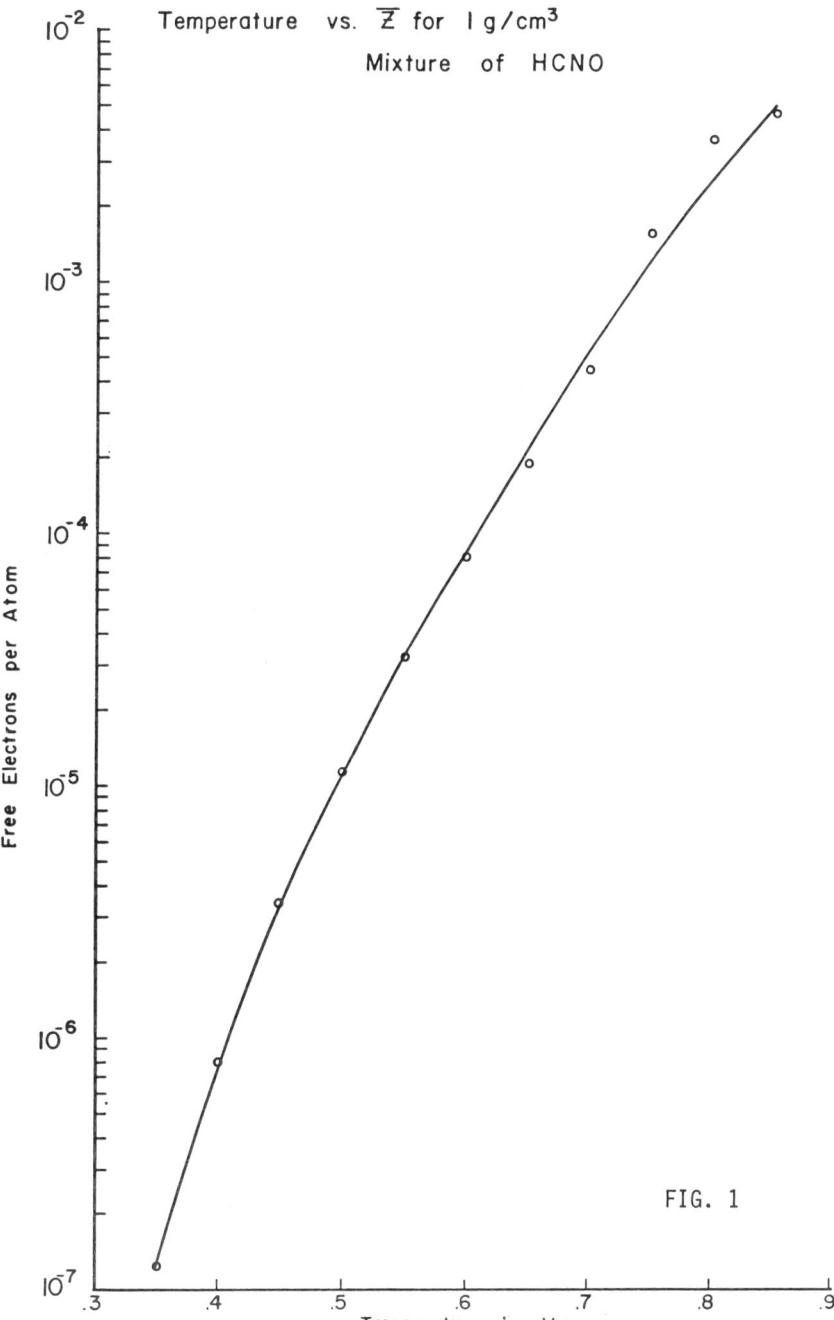

FIG. 1

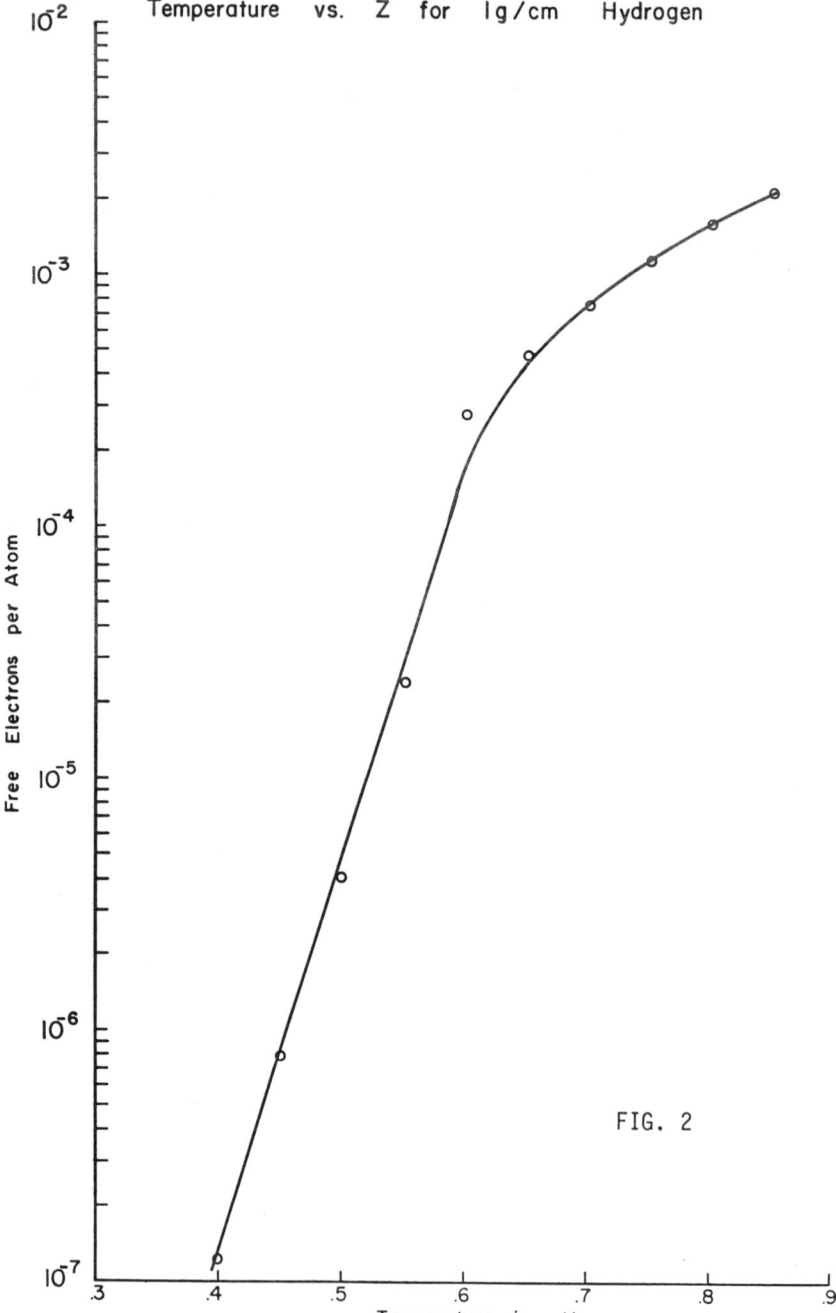
FIG. 2

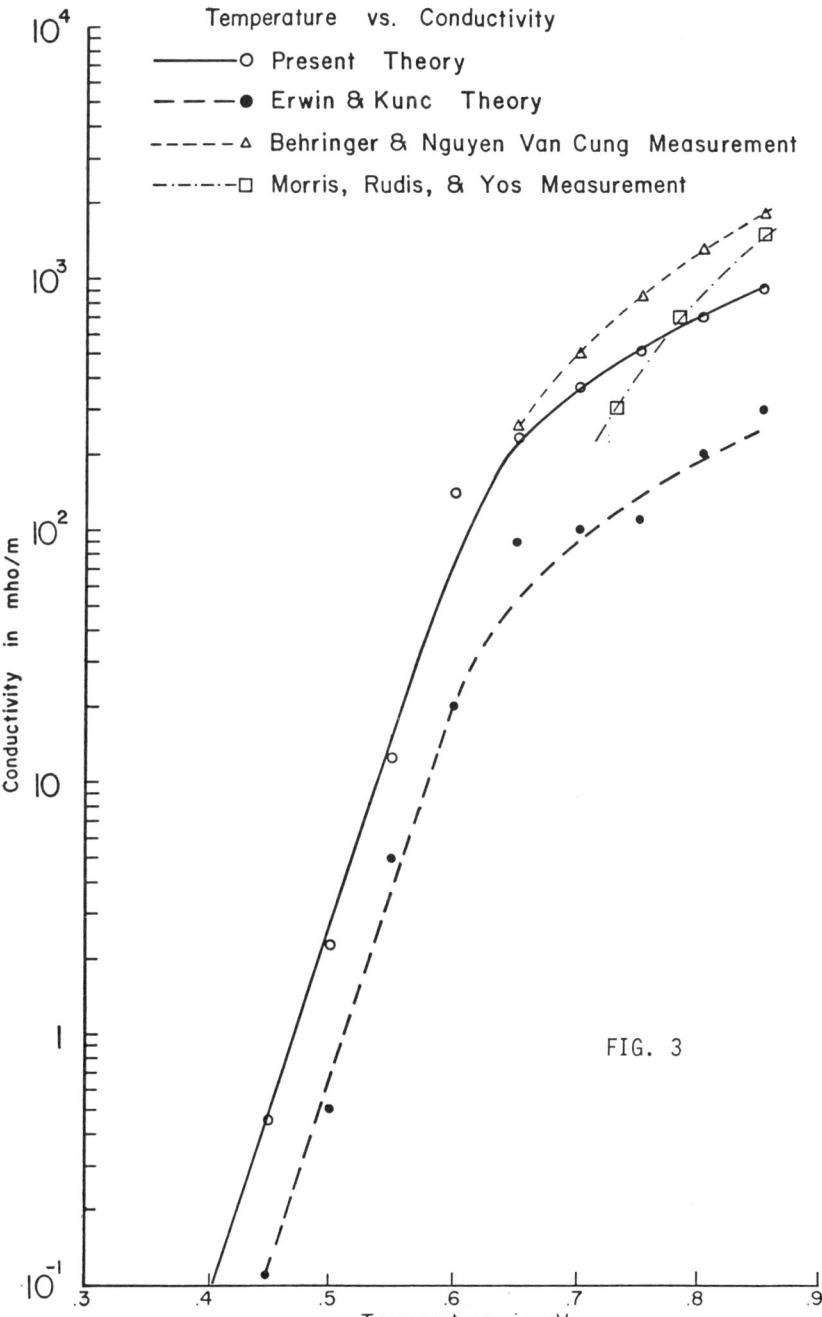

FIG. 3

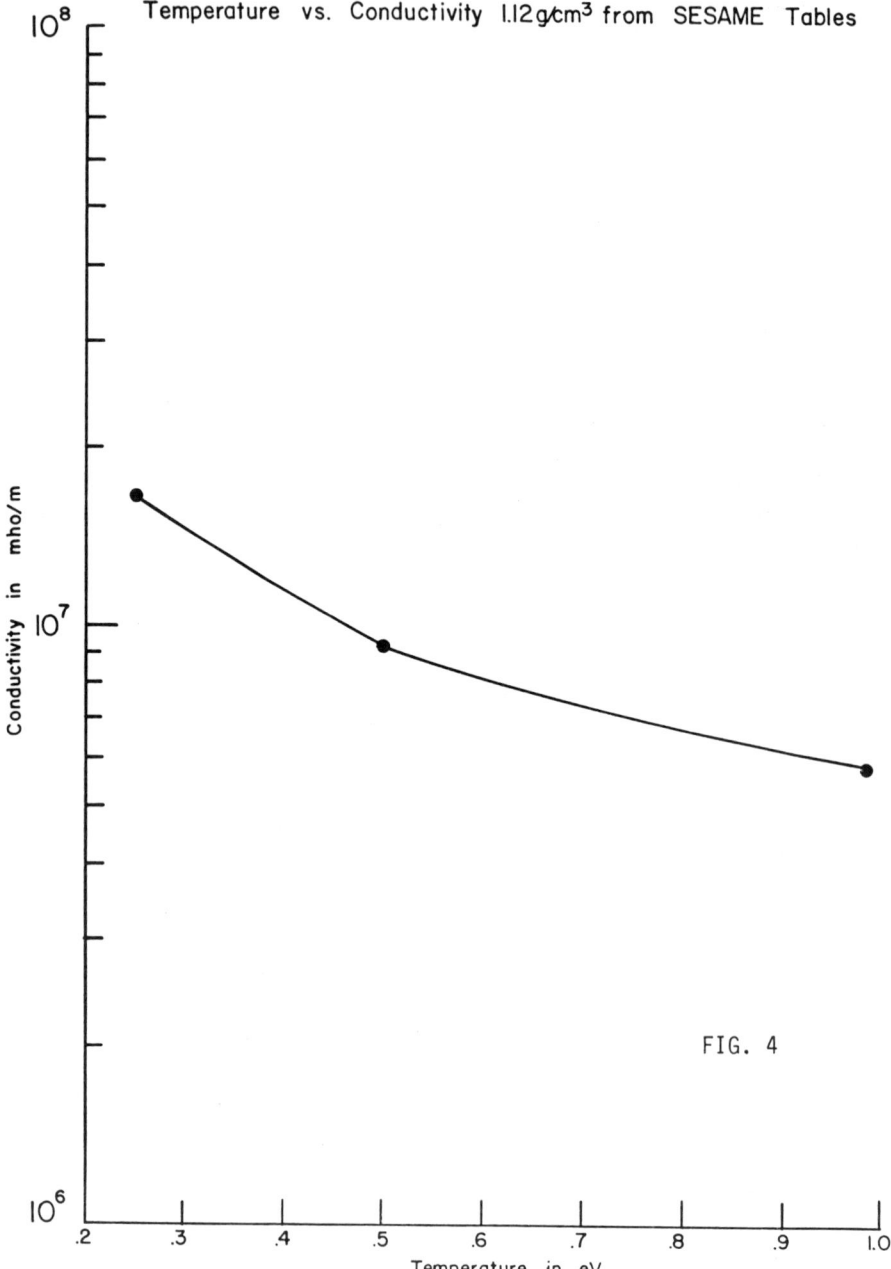

FIG. 4

Density Fluctuations And Channel Mixing Effects
In The Absorption Of Light By Dense Plasmas

François Grimaldi
Service de Physique et Techniques Nucleaires
Centre d'Etudes de Bruyeres Le Châtel
B.P. 12, 91680 Bruyeres Le Châtel, France

and

Annette Grimaldi-Lecourt
Centre d'Etudes de Limeil-Valenton
B.P. 27, 94190 Villeneuve-St-Georges, France

and

M.W.C. Dharma-wardana
Division of Physics
National Research Council, Ottawa, Canada K1A 0R6

Abstract

The independent particle picture of light absorption is well known to be quite inadequate for a proper description of the absorption cross section of light by atoms due to the screening of the incident electric field arising from the relaxation of the system via density fluctuations, channel mixing, exchange and correlation effects etc. We exploit the simplicity and the proven success of time dependent local density approximation (TDLDA) and density functional theory to introduce these many body effects into calculations of light absorption by plasmas. Calculations for an iron plasma at normal density and at a temperature of 100 eV are presented together with details of computational aspects necessary for the implementation of the method.

405

I. Introduction

The calculation of radiative properties (e.g., opacities) of hot plasmas involves a calculation of absorption due to bound-bound, bound-free and free-free transitions. The standard models used in plasma physics and astrophysics are based on those of atomic physics modified to take into account the presence of the plasma environment.[1] But, due to the complexity of the problem these models tend to sacrifice a lot of necessary physics for the sake of computational simplicity. In this respect, we may note that channel mixing should be even more important in a plasma than in a free atom, since, unlike in a free atom at zero Kelvin, all energy levels are only partially occupied. Such channel mixing effects, electron correlation effects etc. can have drastic consequencies on the spectrum, even to the extent of complete disappearance of expected lines, or appearance of unexpected resonances, absorption windows etc.[2] It is possible to study such effects (without the complications due to the plasma) in the photoionization spectrum of complex isolated atoms. The calculation of accurate photoionization cross sections in atoms has proved to be very demanding and extensive and elaborate calculations based on many body perturbation theory (MBPT)[3] or the random phase approximation with exchange (RPAE)[4] have been necessary.

The use of such MBPT or RPAE is entirely impractical as far as atomic processes in plasmas are concerned. The best practical description of such plasmas has to depend on density functional theory (DFT) using a local density approximation (LDA) as the calculational method[5,6] which leads to an average atom (AA) model description of the one particle states and energies in the plasma. It is well known that such LDA models give very bad energy eigenvalues. But one may ask if these models at least give the qualitative shape of the absorption cross section correctly. In fig. 1 we show a photoabsorption cross section calculated for an isolated Xe atom (at 0 Kelvin) using DFT-LDA, and using the Hartree-Fock (H-F) calculations for Xe, using the length formula (L) and the velocity formula (V) for the dipole transition matrix elements. In LDA both the L and V calculations agree. In H-F these differ, as is well known. But the important point is that all three calculations <u>differ dramatically</u> from the experimental cross section.

Fig. 1 clearly suggests that there is little reason to trust the bound-bound, bound-free and free-free cross sections of LDA or H-F calculations for plasmas, even if the latter were practicable.

In Fig. 1 we also show a calculation marked in triangles and based on time dependent channel mixing (TDCM) methods. This method, more precisely known as the time dependent local density approximation, TDLDA, has been introduced by Zangwill and Soven (ZS)[7,8] and shown to give excellent results in predicting the absorption cross section $\sigma(\omega)$ of isolated atoms with only modest computational effort. The essential feature of the method is the replacement of the external electric field term $\vec{r} \cdot \vec{E}$ in the Hamiltonian by an effective driving field term $\vec{r}(\vec{x},\omega) \cdot \vec{E}$ where $\vec{x}$ is the position and ω the frequency. The new effective interaction arises from electron redistribution. That is, the electron density fluctuations induce a field and adjust themselves to be consistent with the total effective field. This can be interpreted in terms of exchange and correlation, channel mixing etc. The effective driving field displays a phase lag with respect to the external field. It is complex and strongly dependent on the dielectric response of the system to the external field. In the new theory transition matrix elements of the type $\langle i | \vec{r}(\vec{x},\omega) | j \rangle$ appear instead of the more usual $\langle i | \vec{r} | j \rangle$ of the standard theory.

The great merit of TDCM is its numerical simplicity and practicality in comparison to MBPT or RPAE. Thus we have generalized[9] Zangwill and Soven's method to (i) include bound-bound, and free-free processes, in addition to the bound-free processes originally considered by ZS (ii) include finite temperature and "free" electron effects, lifetime effects as well as ion-density profile effects as is required for plasmas. In ref. 9 a detailed study of an iron plasma at 100 eV and normal compression using TDLDA was reported. In this paper we complement ref. 9 by providing more computational and other details necessary for implementing the method. We follow ref. 9 and omit the free-free processes from this discussion.

II. Density functional formulation of the light absorption cross section

We consider a charge neutral plasma having an average free electron density $\bar{n}$ at a temperature $T = 1/\beta$ (in energy units). The nuclear charge of the ions is Z and each ion will support a number of bound states, giving rise to an effective ionic charge $\bar{Z}$. In DFT we consider the plasma environment around a central nucleus, within a sufficiently large sphere of radius R larger than any of the characteristic lengths of the plasma. The potentials for r > R are taken to be zero and all particle densities take on their mean values.

Since light absorption processes and electronic transitions tend to be much faster than the characteristic time scales of the ion subsystem, it is well accepted that a <u>quasi-static ion distribution</u> can be used for the calculation of the electronic configuration of the radiating ion. Like all other atomic plasma models, DFT uses a <u>static</u> ion distribution. Since the plasma is already assumed to be in collisional equilibrium the ions are most likely to be in configurations which are statistically close to the average ion distribution. Hence the "zeroth order" electronic states of the radiator are taken to be those of an ion placed in the most probable static ion distribution around the radiator and the effect of ion-distribution fluctuations are relegated to a separate problem. In other words, in this study we have calculated the absorption due to electrons in <u>a chosen</u> static ion distribution but the averaging of the absorption over an ensemble of quasi-static distributions has <u>still to be</u> carried out. The latter can be treated via, say, an ion microfield approach or using suitable ion-electron vertex insertions in the electron response function. A more serious difficulty arises in the neglect of electron density fluctuation effects since we use an average atom model where even the electron distribution is treated using average occupation numbers given by the fermi factor $f(\varepsilon)$ for the electronic level of energy ε, rather than 0 or 1 as would be the case in the configurations actually sampled by the light. The fluctuations in $f(\varepsilon)$ have the same time scales as optical transitions and hence should be included in the theory.[10] Some allowance for this can be made via self-energy corrections to the zeroth order model which is the average atom model. For the present we work with an average atom model where the electron distribution $n(r)$ is determined from the DFT-equations and the levels are corrected by the lowest order self-energy insertions.

The DFT equations[5-6] for the electrons and ions are obtained from the variational property that the thermodynamic potential $\Omega[n,\rho]$ is minimized for the exact electron distribution $n(r)$ and the exact ion distribution $\rho(r)$, while being subject to the constraints of charge neutrality and the Friedel sum rule. The equation $\delta\Omega/\delta n$ leads to a Schrödinger equation with an effective one-electron potential, viz., the Kohn-Sham potential V_{K-S}. The solution of this Kohn-Sham equation provides a set of eigenvalues ε_i, and eigenfunctions ϕ_i, corresponding to bound and free electron states. Although there is no strict justification for identifying them as the single-electron states of the system they can be used with complete rigor to calculate single particle properties of the system since the exact electron density distribution $n(r)$ is correctly given by the set ϕ_ν. Thus

$$n(r) = \sum_i f_i |\phi_\nu(r)|^2$$

where f_i is the Fermi factor defining the occupation probability of the level $i \equiv n,\ell,m,s$ for bound states, and $i = k,\ell,m,s$ for continuum states with the energy $\varepsilon_k = k^2/2$ in Hartree atomic units. It should of course be emphasized that DFT is used in an __approximation__ where the correlation and exchange effects which enter into the effective one-electron potential are calculated using the local density approximation (LDA). That is, results from the finite temperature electron gas problem[11] are invoked to define $V_{XC}(r)$, viz., the exchange-correlation potential, as being that corresponding to the local electron density $n(r)$.

The most probable static-ion density distribution $\rho(r)$ around a given ion (taken to be the origin) is defined by the variational equation $\delta\Omega[n,\rho]/\delta\rho = 0$, and leads to an equation which has to be solved self-consistently, coupled to the Kohn-Sham Schrödinger equation. The details of these procedures are given in ref. 2. As far as this study is concerned, the ion density profile may be considered as providing an external potential due to the most probable configuration of ions around the central ion whose photoabsorption is to be evaluated. Thus we neglect the very small coupling of the ion distribution with the radiation field and study how the electrons (as described by the Kohn-Sham LDA) polarize to generate a frequency dependent induced dipole moment. The photoabsorption will be evaluated following reference 7.

If the electrons in the system were assumed to be independent, the induced density in the presence of the external radiation field $\phi_{ext}(r,\omega)$ can be written as

$$\delta n(\underline{r},\omega) = \int \chi_0(\underline{r},\underline{r}')\phi_{ext}(\underline{r}',\omega)d\underline{r}' \qquad (2.1)$$

where $\chi_0(\underline{r},\underline{r}')$ is the single particle response function. This is essentially the Fourier transform of the retarded density-density correlation function for non-interacting particles moving in the potential $V_{K-S}(r)$. Using the complete set of functions $\phi_i(\underline{r})$ obtained from the LDA-DFT calculation we have

$$\chi_o(\underset{\sim}{r},\underset{\sim}{r}') = \sum_{i,j} \frac{(f_i - f_j)\rho_{ij}(\underset{\sim}{r})\rho_{ij}^*(\underset{\sim}{r}')}{\omega + \varepsilon_{ij} + i\delta} \qquad (2.2)$$

where

$$\varepsilon_{ij} = \varepsilon_i - \varepsilon_j$$

$$\rho_{ij}(\underset{\sim}{r}) = \phi_i^*(\underset{\sim}{r})\phi_j(\underset{\sim}{r})$$

$$f_i = [1 + \exp\{\beta(\varepsilon_i - \mu)\}]^{-1}$$

and δ is a positive infinitisimal. μ is the chemical potential which appears in the DFT equations. Note that if $\phi_i(\underset{\sim}{r})$ be replaced by plane waves, viz. $\phi_k(\underset{\sim}{r}) \propto e^{-i\underset{\sim}{k}\cdot\underset{\sim}{r}}$, with $\varepsilon_i = k^2/2m$, equation (2.2) reduces to the form of the Lindhard function.

In a real plasma the electrons interact and hence the induced density $\delta n(\underset{\sim}{r},\omega)$ creates an induced Coulomb potential

$$\delta V_C(\underset{\sim}{r},\omega) = \int \frac{\delta n(\underset{\sim}{r}',\omega)}{|\underset{\sim}{r} - \underset{\sim}{r}'|} dr' \qquad (2.3)$$

and also an exchange-correlation contribution

$$\delta V_{XC}(\underset{\sim}{r},\omega) = \frac{\delta V_{XC}(n)}{\delta n}\bigg|_{n=n(\underset{\sim}{r})} \delta n(\underset{\sim}{r},\omega) \ . \qquad (2.4)$$

Equation (2.4) is the linear correction to the LDA exchange-correlation potential produced by the density change $\delta n(\underset{\sim}{r},\omega)$. Hence we define the induced potential $V_{ind}(\underset{\sim}{r},\omega)$ and the self-consistently determined total driving potential $\phi_{SCF}(\underset{\sim}{r},\omega)$ by

$$V_{ind}(\underset{\sim}{r},\omega) = \delta V_C(\underset{\sim}{r},\omega) + \delta V_{XC}(\underset{\sim}{r},\omega)$$

$$\delta n(\underset{\sim}{r},\omega) = \int \chi_0(\underset{\sim}{r},\underset{\sim}{r}')[\phi_{ext}(\underset{\sim}{r}'\omega) + V_{ind}(\underset{\sim}{r}',\omega)]d\underset{\sim}{r}'$$

$$= \int \chi_0(\underset{\sim}{r},\underset{\sim}{r}')\phi_{SCF}(\underset{\sim}{r}',\omega)d\underset{\sim}{r}' \ . \qquad (2.5)$$

In effect the response of the interacting system converts the external potential $\phi_{ext}(\underset{\sim}{r},t)$ to an effective driving potential determined self-consistently from the equation

$$\phi_{SCF}(\underset{\sim}{r},\omega) = \phi_{ext}(\underset{\sim}{r},\omega) + \int K(\underset{\sim}{r},\underset{\sim}{r}')\chi_0(\underset{\sim}{r}',\underset{\sim}{r}'',\omega)\phi_{SCF}(\underset{\sim}{r}'',\omega)d\underset{\sim}{r}'d\underset{\sim}{r}'' \qquad (2.6)$$

where the kernel $K(\underset{\sim}{r},\underset{\sim}{r}')$ is defined by

$$V_{ind}(\underset{\sim}{r},\omega) = \int K(\underset{\sim}{r},\underset{\sim}{r}')\delta n(\underset{\sim}{r}',\omega)d\underset{\sim}{r}' \ . \qquad (2.7)$$

As in ZS, ref. 7, $K(\underset{\sim}{r},\underset{\sim}{r}')$ is assumed to be independent of the frequency and $\delta n(\underset{\sim}{r}',\omega)$ carries all the time dependent effects of the induced field.

We take the external electric field to be of magnitude ε_{ext} directed along the z axis, with

$$\phi_{ext}(\underset{\sim}{r},\omega) = \varepsilon_{ext} Z. \qquad (2.8)$$

Then we use the Golden rule formula

$$\sigma(\omega) = 4\pi(\omega/c)\operatorname{Im}\alpha(\omega) . \qquad (2.9)$$

where $\alpha(\omega)$ is the polarizability defined to be the ratio of the induced dipole moment to the external field, viz.

$$\alpha(\omega) = -\frac{1}{\varepsilon_{ext}} \int Z \delta n(\underset{\sim}{r},\omega) d\underset{\sim}{r} .$$

Then it can be established that (see ref. 7)

$$\begin{aligned}\sigma(\omega) &= -\frac{4\pi\omega}{c\varepsilon_{ext}^2} \operatorname{Im}\int \phi_{ext}(\underset{\sim}{r},\omega)\chi_0(\underset{\sim}{r},\underset{\sim}{r}',\omega)\phi_{SCF}(\underset{\sim}{r}',\omega)d\underset{\sim}{r}d\underset{\sim}{r}' \\ &= -\frac{4\pi\omega}{c\varepsilon_{ext}^2} \int \phi_{SCF}(\underset{\sim}{r},\omega)\operatorname{Im}\chi_0(\underset{\sim}{r},\underset{\sim}{r}',\omega)\phi_{SCF}^*(\underset{\sim}{r}',\omega)d\underset{\sim}{r}d\underset{\sim}{r}' \qquad (2.10)\end{aligned}$$

Equation (2.10) is the essential equation for the calculation of light absorption by the system at frequency ω inclusive of relaxation effects. The velocity of light c appearing in (2.9) is that appropriate to the medium and can be calculated self-consistently. But for simplicity, in the present study we neglect this renormalization. If $\chi_0(\underline{r},\underline{r}',\omega)$ is replaced by (2.2) we have, instead of (2.10), the equation

$$\sigma(\omega) = 4\pi^2(\omega/c) \sum_{ij} (f_i - f_j) |<i|Z_{SCF}(\underline{r},\omega)|j>|^2 \delta(\omega+\varepsilon_i-\varepsilon_j) \qquad (2.11)$$

where it should be noted that $Z_{SCF}(\underline{r},\omega)$ has the same symmetry as the external field (directed along z) and is defined by

$$\phi_{SCF}(\underline{r},\omega) = - \varepsilon_{ext} Z_{SCF}(\underline{r},\omega) \; . \qquad (2.12)$$

Since the Fermi-factor difference $(f_i - f_j)$ appearing in equation (2.11) be written as

$$f_i - f_j = f_i(1-f_j) - f_j(1-f_i)$$

we immediately see that (2.10)-(2.11) gives the <u>net absorption</u> from the processes $i \to j$ and $j \to i$. If the plasma is in local thermodynamic equilibrium (LTE) these two processes are coupled by the conditions of detailed balance. Hence, for LTE plasmas, if only the process $i \to j$ is to be calculated, irrespective of induced emission, then the $(f_i - f_j)$ factors appearing in (2.10) and (2.11) should be replaced by $f_i(1-f_j)$. However, we note that the form (2.10) may be applied to non-LTE situations if the response functions reflect the non-LTE aspects of the system. In this paper we restrict ourselves to LTE plasmas only.

Equation (2.10) is more general than (2.11) since (2.10) can be used even when the response function $\chi_0(\underset{\sim}{r},\underset{\sim}{r}',\omega)$ becomes modified to a form $\chi(\underset{\sim}{r},\underset{\sim}{r}',\omega)$ where collisions and lifetime effects are taken into account. The effect of the collisions etc. can be approximately treated by writing

$$\chi(\underset{\sim}{r},\underset{\sim}{r}',\omega) = \sum_{ij} \frac{(\tilde{f}_i - \tilde{f}_j)\rho_{ij}(\underset{\sim}{r})\rho_{ij}^*(\underset{\sim}{r}')}{(\omega + \varepsilon_{ij} + \text{Re}\tilde{\Gamma}_{ij}) + i\text{Im}\tilde{\Gamma}_{ij}} \qquad (2.13)$$

where $\tilde{\Gamma}_{ij}$ is a complex local field whose imaginary part can be identified with the damping of the transition (ij). It is useful to artificially separate out the contributions to σ although only the total absorption is observable. That is, in analogy with equation (11) we write equation (10) in the form

$$\sigma(\omega) = -4\pi\left(\frac{\omega}{c}\right)\text{Im}\sum_{ij}(f_i-f_j)<i|Z_{SCF}(\underset{\sim}{r},\omega)|j>^2\left[\omega+\varepsilon_i-\varepsilon_j+i\tilde{\Gamma}_{ij}(\omega)\right]^{-1}. \qquad (2.14)$$

Finally we note that (2.10) will give the absorption corresponding to bound-bound, bound-free and free-free processes depending on the value of ω chosen and the existence of transitions corresponding to ω. In the single particle theory where there is no channel mixing or self-energy effects we have $Z_{SCF}(\underset{\sim}{r},\omega) \rightarrow Z$ (i.e., the dipole operator) and $\tilde{\Gamma}_{ij} \rightarrow +\delta$ where δ is a small positive infinitisimal. To examine the contributions to σ we define, for the unrelaxed model

$$\sigma^0(\omega) = \sigma^0(\omega)_{\text{line}} + \sigma^0(\omega)_{\text{ph}} + \sigma^0(\omega)_{\text{brem}}$$

and similarly, from (2.14)

$$\sigma(\omega) = \sigma(\omega)_{line} + \sigma(\omega)_{ph} + \sigma(\omega)_{brem}$$

where σ_{line} is obtained by restricting the i,j sums to bound states only. σ_{ph} is the "photoabsorption" contribution obtained by restricting i to bound states and j to continuum (i.e. "free") states. The bremstrahlung term, σ_{brem} arises from free-free contributions. The reported absorption cross section does not include $\sigma(\omega)_{brem}$ but is the sum of σ_{ph} and σ_{line}. That is, in this study we have not included transitions where both i and j are in the continuum. In the calculation of ϕ_{SCF} this amounts to the neglect of relaxation effects induced in the system by purely free-free (Brehmstrahlung like) transitions. In general these effects would manifest themselves via plasmon-like processes but their effects on polarization processes involving atomic electrons is expected to be quite small. These effects would be examined in greater detail in a future publication.

III. Calculations for an iron-plasma

We study an iron-plasma at a temperature of 3.6751 Hartree atomic units (100 eV) and at a free electron density such that the electron sphere radius r_s = 1.206 a.u. ($\bar{n}$ = 1.35 × 10^{24} electrons per cm^3). The chemical potential calculated from DFT is -6.7232 a.u. and the bound energy levels are given in table 1 (note that all energies are in Hartrees).

The bound electron energies, wavefunctions and the self-consistent one-electron potential V_{K-S} contained in the Kohn-Sham equation were used as inputs to the present calculation. The necessary free electron (i.e. continuum) wavefunctions were generated <u>in situ</u> from a Schrödinger code using the given V_{K-S}. The iterative procedure for calculating $\phi_{SCF}(\underline{r},\omega)$ consisted in assuming $\phi_{SCF}^{(1)}$ to be ϕ_{ext} in the first iteration (see eqn. 2.5) and then $V_{ind}(\underline{r},\omega)$ was calculated. A straightforward iterative procedure is to mix the induced potential, $V_{ind}(\underline{r},\omega)$ of two successive iterations to obtain the current value at any given iteration. This method was quite successful but failed at frequencies near the line centres. A modified method, based on the direct calculation of bound-bound matrix elements of $\phi_{SCF}(\underline{r},\omega)$, rather than $\phi_{SCF}(\underline{r},\omega)$ itself, with the photo-contribution held fixed, was implemented and found to be much more stable and efficient. (This will be discussed in section IV.) Once a self-consistent $\phi_{SCF}(\underline{r},\omega)$ was obtained, $\sigma(\omega)$ could be calculated from (2.9) or (2.10). Note that all the frequency dependent quantities (e.g. $\phi(\underline{r},\omega)$) become complex after the first iteration. In effect, the driving field becomes out of phase with the applied external field.

In figs. 2(a) and 3(a) we show the real part of the effective (radial) driving potential $\phi_{SCF}(r,\omega)$, together with the external potential $\phi_{ext}(r)$ set equal to r, for two different values of ω for illustration. The imaginary parts are illustrated in figs. 2(b) and 3(b). In fig. 4 the photoabsorption cross section $\sigma_{ph}(\omega)$ for $22 < \omega < 34$ is shown. The net photoabsorption calculated from the usual single particle theory (without relaxation) is denoted by $\sigma_{ph}^{o}(\omega)$. This has been checked against standard calculations for an iron-plasma and provided a useful verification of the computer coding. The self-consistent results, viz. $\sigma_{ph}(\omega)$ are seen to be significantly different due to channel mixing. For more details and explanation of the observed features see ref. 9. The full absorption curve is given in fig. 5.

IV. Calculation of the effective driving potential $\phi_{SCF}(r,\omega)$

The main difference between the time dependent model and the usual single particle model is the use of an effective driving potential $\phi_{SCF}(r,\omega)$ instead of the unrelaxed external potential ϕ_{ext} of the usual theory. In this section we give some details necessary for the numerical implementation of the method.

In the original Zangwill-Soven formulation the evaluation of the response function was carried out in terms of Green functions evaluated directly from the Kohn-Sham, Schrödinger equation. We will instead directly sum over the pairs of states i,j of the Kohn-Sham basis where, for example, $i = n_i, \ell_i, m_i$ for bound states and $i = \varepsilon, \ell_i, m_i$ with $\varepsilon = k^2/2m$ for continuum states. When the angular momentum sums are carried out using standard vector-coupling methods it turns out that these i,j summations pose no major difficulty. There is also the formal advantage that the induced potential, viz. $V_{ind}(r,\omega)$, can be written as separate contributions from bound-bound (i.e., line), bound-free (i.e., photo) and free-free (i.e., Bremstrahlung) processes. Thus

$$V_{ind}(r,\omega) = V_{ind}^L(r,\omega) + V_{ind}^P(r,\omega) + V_{ind}^B(r,\omega) \ .$$

Here V_{ind}^L, V_{ind}^P, V_{ind}^B contain i,j summations where (i,j) are bound,bound), (bound,free) and (free,free) states corresponding to line-, photo- and Bremstrahlung processes respectively. In a rigorous sense it is no longer meaningful to clearly separate out these three physical processes due to channel mixing. But this formal separation has a physical meaning in the single particle picture although here it merely indicates a division of the sum into three types of terms. As already stated, in our calculations we ignore the V^B term which arises when both i and j are continuum states. The calculation of V^B is reported elsewhere.[12] It is neglected in this study.

From eqn. (2.1) to (2.6) we have

$$\phi_{SCF}(\underline{r},\omega) = \phi_{ext}(\underline{r},\omega) + V_{ind}(\underline{r},\omega)$$

with

$$V_{ind}(\underline{r},\omega) = \iint \chi_0(\underline{r}',\underline{r}'',\omega)\phi_{SCF}(\underline{r}'',\omega)\left[\frac{1}{|\underline{r}-\underline{r}'|} + \frac{\delta V_{XC}(n(\underline{r}'))}{\delta n}\delta(\underline{r}-\underline{r}')\right]d\underline{r}'d\underline{r}''. \quad (4.1)$$

Writing for the radial part,

$V_{ind}(r,\omega) = V_{ind}^L(\sim,\omega) + V_{ind}^P(r,\omega)$ is is easy to show that for either of them

$$V_{ind}(r,\omega) = \sum_{i<j} \theta_{ij} D_{ij} \phi_{ij} V_{ij}(r) \quad (4.2)$$

where θ_{ij} arises from the angular integrals in (4.1)

$$\theta_{ij} = 2(f_i - f_j)(2\ell_i + 1)(2\ell_j + 1) \frac{1}{3} \begin{pmatrix} \ell_i, 1, \ell_j \\ 0, 0, 0 \end{pmatrix}^2 \quad (4.3)$$

where i denotes the degenerate levels n_i, ℓ_i, and similarly for j. The Fermi factors f_i, f_j arise from the numerator of $\chi_0(\underline{r}', \underline{r}'', \omega)$. Its denominator is included in D_{ij} which has a real and imaginary part. Thus, writing $\alpha = (i,j)$ we have

$$D_{ij}^L(\omega) = D_\alpha^L(\omega) = \frac{(\omega - \varepsilon_\alpha) - i\Gamma_\alpha(\omega)}{(\omega - \varepsilon_\alpha)^2 + \Gamma_\alpha^2(\omega)} - \frac{(\omega + \varepsilon_\alpha) - i\Gamma_\alpha(\omega)}{(\omega + \varepsilon_\alpha)^2 + \Gamma_\alpha^2(\omega)} \quad (4.4)$$

$$\varepsilon_\alpha = \varepsilon_j - \varepsilon_i \quad (4.5)$$

where each pair $\alpha(=i,j)$ appears in the sum only once. Here $\Gamma_\alpha(\omega)$ is the width associated with the transitions (i,j). This is evaluated from the <u>level</u> widths γ_i and γ_j as described in the appendix of ref. 9. A similar expression is obtained for $D_{ij}^P(\omega)$ with the difference that Γ_{ij}, i.e. $\Gamma_{i,\varepsilon\ell m}$ involves only the width of the bound state i. That is, the width of the continuum state is neglected. The integration over the continuum states is replaced by a Gaussian quadrature and hence the coding is very similar for both bound states and continuum states.

The term ϕ_{ij} in (4.2) arises from the radial integrals. These are evaluated numerically via a Simpson type quadrature. ϕ_{ij} is given by

$$\phi_\alpha = \phi_{ij}^{SCF}(\omega) = \int R_i(r) R_j(r) \phi_{SCF}(r, \omega) r^2 dr \ .$$

The radial functions $R_i(r)$ are the eigenfunctions obtained by solving the radial Kohn-Sham Schrödinger equation for the Fe-ion in the plasma. In the first iteration $\phi_{SCF}(r,\omega)$ is simply the external potential and $\phi_{ij}(\omega)$ reduces to the dipole matrix element $<i|r|j>$ for dipolar coupling of light to the atomic states. In subsequent iterations $\phi_{ij}(\omega)$ becomes complex with

$$\phi_\alpha(\omega) = \phi_\alpha(\omega)_{Re} + i\, \phi_\alpha(\omega)_{Im} \ .$$

Finally, the term $V_{ij}(r)$ contained in (4.2) includes the Coulomb and exchange-correlation potential

$$V_{ij}(r) = r \int_r^\infty dr' \left(\frac{1}{r'}\right)^2 R_i(r')R_j(r') + \frac{1}{r^2}\int_0^r dr'\, r'\, R_k(r')R_\ell(r') + \frac{3}{4\pi}\frac{\delta V^{XC}}{\delta n} R_i(r)R_j(r)$$

Iterative procedure

Initially the induced potential $V_{ind}(r,\omega)$ is zero and $<i|\phi_{SCF}(r,\omega)|j>$ simply leads to the usual $<i|r|j>$ dipole matrix element if the magnitude of the external field is set to unity for convenience. In calculating $V_{ind}(r,\omega)$ we found it numerically more stable to treat $V_{ind}^L(r,\omega)$ and $V_{ind}^P(r,\omega)$ separately. That is, we use a two step iterative process where each of them is modified in turn while the other is held fixed, until convergence in the total induced potential $V_{ind}^L + V_{ind}^P$ is obtained. The method was found to work in most circumstances. The two steps in each iteration are as follows.

Step 1: V^L_{ind} is held fixed and only the given V^P_{ind} is iterated to obtain the self consistent value of V^P_{ind} from

$$\phi^P_{ind}(r,\omega) = \sum_{\substack{i=n_i,\ell_i \\ j=\varepsilon,\ell_i\pm 1}} \theta_\alpha D_\alpha \phi_\alpha V_\alpha(r) \qquad (4.7)$$

where

$$\phi_\alpha = \phi_{ij}(\omega) = \int R_i(r) R_j(r)(r + V^L_{ind} + V^P_{ind}) dr \;.$$

This involves integrations over the bound states for i, and for the continuum states for $j = \varepsilon, \ell_i \pm 1$.

Step 2: $V^L_{ind}(r)$ is obtained from a direct determination of the matrix elements $\phi_\alpha(r)$ where $\alpha = (i,j)$ is restricted to bound-bound i,j pairs. We have

$$\phi_\alpha(\omega) = r_\alpha + V^P_\alpha + V^L_\alpha \;. \qquad (4.9)$$

Here the matrix element $V^L_\alpha = V^L_{ind,\alpha} = <R_i(r) |V^L_{ind}(r,\omega)| R_j(r)>$. If we introduce the full expression for $V^L_{ind}(r)$ in (4.9) we have

$$\phi_\alpha(r) = r_\alpha + V^P_\alpha + \sum_\beta \theta_\beta D_\beta W_{\alpha\beta} \phi_\beta \qquad (4.10)$$

where $\beta = (k,\ell)$ is a pair of bound states. $W_{\alpha\beta}$ is given by

$$W_{\alpha\beta} = \int r^2 dr\, r'^2 dr'\, R_i(r) R_j(r) \left(\frac{r_<}{r_>^2}\right) R_k(r') R_\ell(r') \qquad (4.11)$$

where $r_<$ and $r_>$ are the smaller and the larger of r and r' respectively. Since ϕ_α is a complex quantity

$$\phi_\alpha = \phi_\alpha^{SCF}(r) = \phi_{\alpha,Re}(r) + i\,\phi_{\alpha,Im}(r)$$

eq. (4.10) gives us a set of 2N linear equations where N is the number of pairs of bound states, for the radial matrix elements $\phi_{\alpha,Re}$ and $\phi_{\alpha,Im}$.

$$\left[1 - \begin{pmatrix} a_{\alpha\beta} & -b_{\alpha\beta} \\ b_{\alpha\beta} & a_{\alpha\beta} \end{pmatrix}\right] \begin{pmatrix} \phi_{\alpha,Re} \\ \phi_{\alpha,Im} \end{pmatrix} = \begin{pmatrix} V_{\alpha,Re}^P + r_\alpha \\ V_{\alpha,Im}^P \end{pmatrix} \qquad (4.12)$$

where $a_{\alpha\beta}$ and $b_{\alpha\beta}$ are given by

$$a_{\alpha\beta} + i\,b_{\alpha\beta} = W_{\alpha\beta}\theta_\alpha D_\alpha^L(\omega) \qquad (4.13)$$

with $D_\alpha^L(\omega)$ defined by eq. (4.4).

Once $\phi_{\alpha,Re}$ and $\phi_{\alpha,Im}$ are obtained by these two steps which involved the relaxation of V_α^L and V_α^P separately, with one or the other held fixed, a new $V_\alpha^L(r)$ is constructed and introduced in eq. 4.7 of step 1 and a new $V_{ind}^P(r)$

is determined. Convergence is checked on the modulus of ϕ^{SCF}. The method does not converge if the matrix defined by eq. (4.12) becomes singular.

Note that the matrix method could be used in principle even for $V_{ind}^P(r,\omega)$ but in practice this is unwieldy since there are a large number of continuum states and hence the number of linear equations to be solved becomes large.

Conclusions

We have shown how the relaxation of an atomic system in a plasma and the consequent channel mixing arising during the action of an external field can be taken into account by a self consistent calculation of the induced fields in the system. Unlike the original Zangwill-Soven method for isolated atoms at T = 0 Kelvin, we do not use a Green function solution of the Kohn-Sham equation to fold the summations over the intermediate states. Instead we directly use the spectral resolution and hence we can separately deal with what may be termed bound-bound, bound-free and free-free contributions to the total absorption. These contributions are those of a model system (Kohn-Sham basis) and do not necessarily correspond to an actual physical separation of these processes. In the true physical system line-, photo- and free-free processes are completely intermixed due to channel mixing and only the total absorption coefficient $\sigma(\omega)$ has a true physical meaning. Within the framework used $\sigma(\omega)$ is simply the sum of what we have called the line-, photo- and free-free contributions to $\sigma(\omega)$, calculated using the Kohn-Sham basis.

Unlike MBPT or RPAE, the time dependent-DFT method given here is eminently practical for plasmas and it is hoped that light absorption cross sections for plasmas would be calculated with the kind of improvement shown in fig. 1 for atoms.

References

1. R.M. More in Atomic and Molecular Physics of Controlled Thermonuclear Fusion, edited by C.J. Joachin and D.E. Post (Plenum, New York, 1983).

2. G. Wendin, in Photoionization and other probes of many-electron interactions, edited by F.J. Wuilleumier (Plenum, NY 1976).

3. H.P. Kelly and S.L. Carter, Phys. Scrip. (Sweden) 21, 448 (1980).

4. M.Ya. Amusia, in Advances in Atomic and Molecular Physics, Vol. 17, edited by D. Bates and B. Bederson (Academic, NY 1981).

5. M.W.C. Dharma-wardana and F. Perrot, Phys. Rev. A26, 2096 (1982), F. Perrot and M.W.C. Dharma-wardana, Phys. Rev. A29, 1378 (1984).

6. F. Perrot and M.W.C. Dharma-wardana, Phys. Rev. A31, 970 (1985).

7. A. Zangwill and P. Soven, Phys. Rev. A21, 1561 (1980).

8. A. Zangwill, Ph.D. Thesis, University of Pennsylvania 1981.

9. F. Grimaldi, A. Grimaldi-Lecourt and M.W.C. Dharma-wardana, Phys. Rev. A32, 1063 (1985).

10. F. Grimaldi and A. Lecourt, Proc. 2nd Int. Conf. on Rad. properties of hot dense matter (World Scientific Publishing Co., Singapore 1985).

11. F. Perrot and M.W.C. Dharma-wardana, Phys. Rev. A30, 2619 (1984).

12. F. Grimaldi, A. Grimaldi-Lecourt and M.W.C. Dharma-wardana (unpublished).

Figure Captions

Fig. 1 The 4d photoabsorption of atomic xe at zero temperature. Solid line: (LDA). Dashed line: Hartree-Fock length (L) and velocity (V) formulae. Dotted line: experiment and triangles TD-LDA. See ref. 8.

Fig. 2 Panel (a) shows the real part of the self-consistent (radial) driving potential $\phi_{SCF}(r,\omega)$ for $\omega = 1.8$ a.u. The dashed line is the external potential which is the unrelaxed dipole factor r. Panel (b) shows the imaginary part of $\phi_{SCF}(r,\omega)$. The dashed (horizontal) line shows that $Im\phi_{ext} = 0$.

Fig. 3 Same as fig. 2 but for $\omega = 27.1$ a.u. The effective field is enhanced inside the atom and out of phase with the external field, as implied by the large imaginary part.

Fig. 4 Comparison of $\sigma_{photo}(\omega)$ and the usual unrelaxed result, $\sigma^o_{photo}(\omega)$ for $22 < \omega < 34$ a.u. The new features in σ_{photo} correspond to admixture of line processes.

Fig. 5 The total photoabsorption cross section. The short dashed line is the contribution from σ_{line}. The long dashed line in the left panel is σ_{photo}. This is denoted as triangles in the right panel and dominates for large ω.

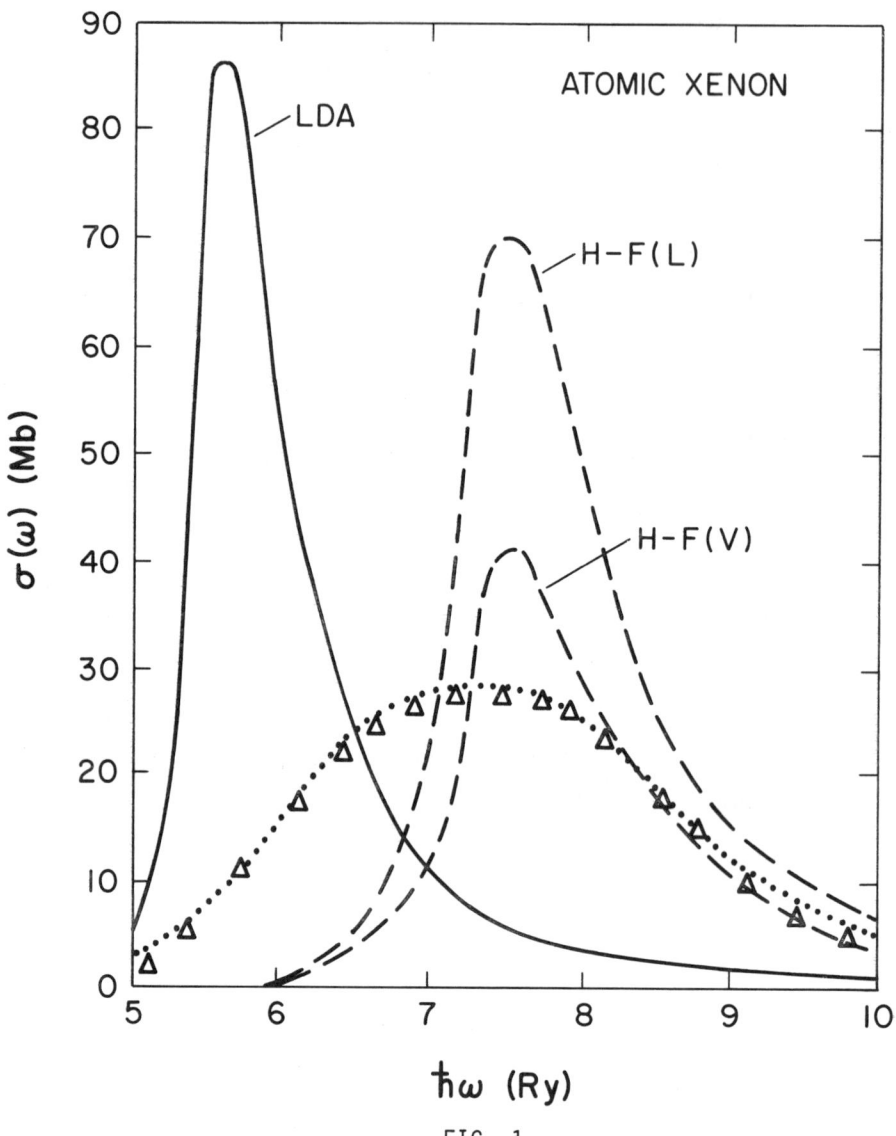

FIG. 1

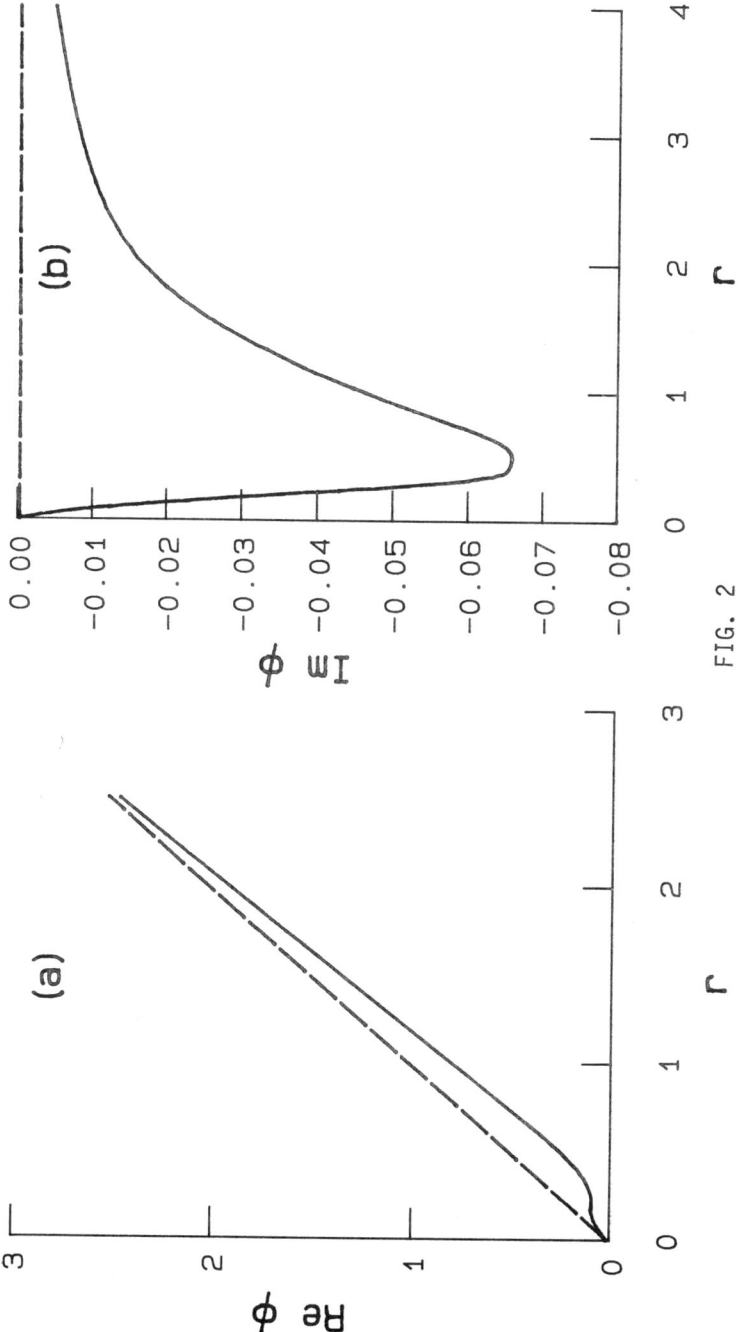

FIG. 2

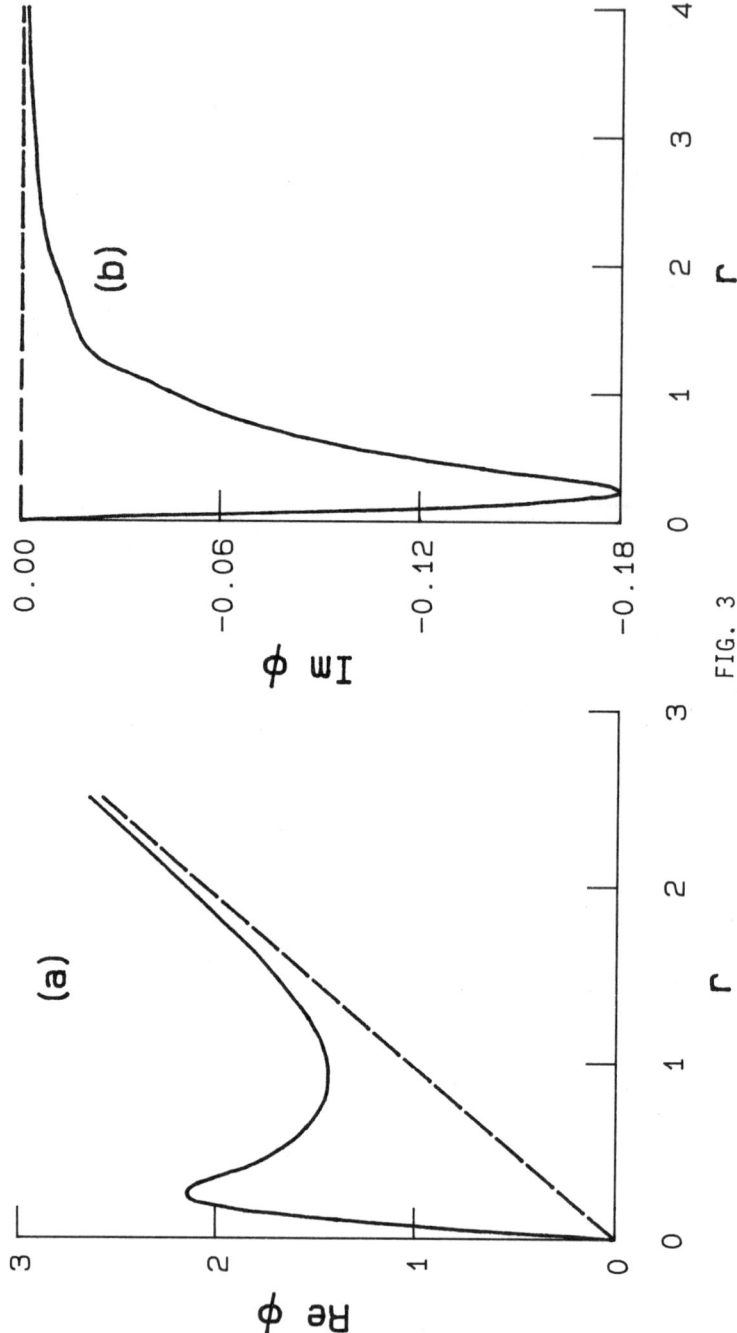

FIG. 3

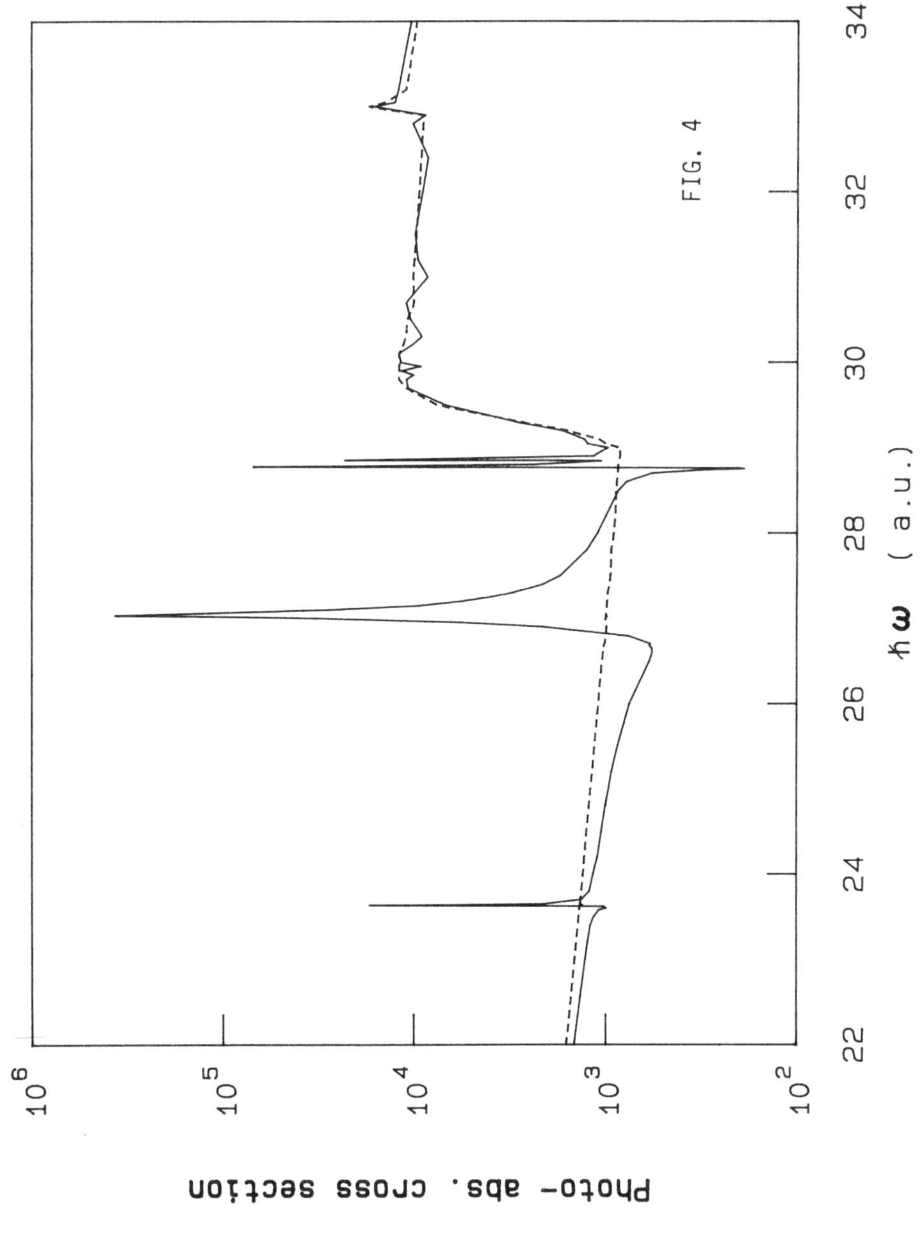

FIG. 4
Photo- abs. cross section
10^6
10^5
10^4
10^3
10^2
ħω (a.u.)
22
24
26
28
30
32
34

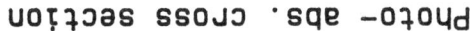

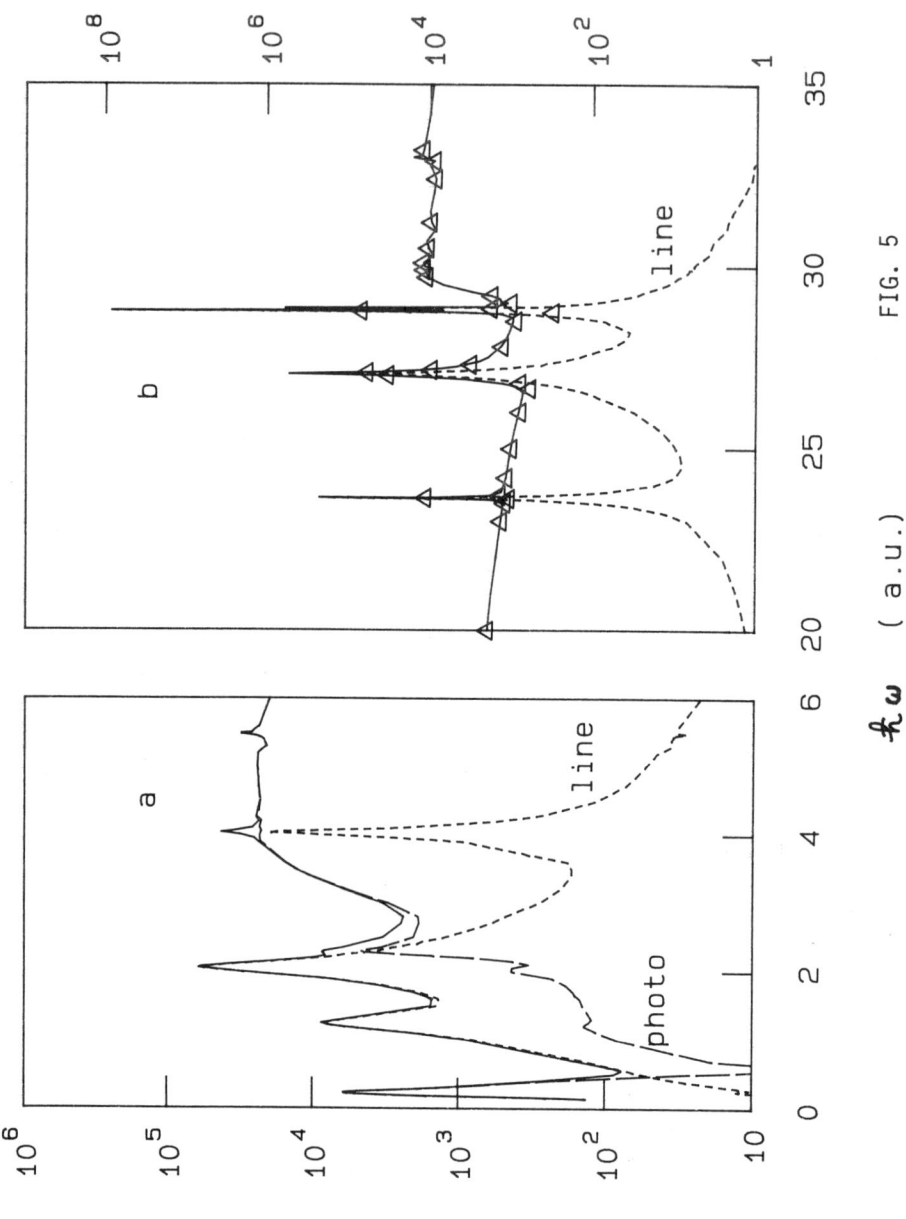

FIG. 5

Atomic Process Calculations In Hot Dense Plasmas Using Average Atom Models

G. Velarde, J. M. Aragones, L. Gamez, J. J. Honrubia,

J. M. Martinez-Val, E. Minguez, J. L. Ocaña, J. M. Perlado,

J. F. Serrano

Instituto de Fusion Nuclear
Universidad Politecnica de Madrid
E.T.S. Ingenieros Industriales
pº Castellana, 80 - 28046 Madrid, España

1. INTRODUCTION

For the analysis of the physics and performance of ICF targets, it seems very necessary to dispose of a computer code, which includes the relevant physical processes. To this end only a few computer codes have been developed in the world (Yamanaka, 1983).

During the past years, an important effort has been devoted in our Institute to develop the NORCLA code, which in the first version (Velarde et al. 1978, Velarde et al. 1980a; Velarde et al. 1980b) was characterized by the following features: one-dimensional lagrangian mesh; equilibrium between radiation, ion and electron species; local alpha energy deposition; neutron transport by the discrete ordinates method and analytical equation of state, opacities and conductivities.

In the succesive versions of NORCLA, EOS and electron conductivities were modified by the pressure ionization and degeneracy corrections (Leira et al. 1981); a module was also developed for computing the energy deposition of the incident ion beams coupled to the energy equation, and a code to calculate the alpha particle transport and energy deposition (Velarde et al. 1984; Honrubia & Aragonés 1986). Recently, a 3T version of the NORCLA code, with tabular EOS, opacities and conductivities, laser ray tracing and suprathermal electrons transport has been produced (Velarde et al. 1986).

In this article, the atomic physic models developed to determine more accurate the atomic data, such as EOS and opacities will be explained in the Section 2, giving a brief description and a comparison of them.

As a result of this development, a DENIM Atomic Data Library is being generated, taking some data and procedures from the SESAME Library. In section 3 this library will be presented, including a comparison of the opacity data for aluminium and iron at different densities and temperatures.

Finally, conclussions about this work will be presented, and the ongoing developments will be summarized in order to follow this work.

2. ATOMIC PHYSICS MODELS DEVELOPED

The detailed knowledge of the atomic structure for the different constituents is required both for Equation of State (EOS) and opacity calculations. In the most general case we should calculate for each component of the plasma the population of its various ionic states and then, for each degree of ionization, the relative abundances of the different energy states defined by a set of electronic energy levels and populations. Given that for many-electron atoms there are millions of excited states the aforementioned scheme is not longer valid for the most practical cases and some simplified descriptions have to be used. The most widely used approximation is the so-called average atom (AA) (Rozsnyai 1972; Rozsnyai 1982) model wherein the atomic structure of the plasma is defined via a fictitious, with fractional electronic occupation numbers, configuration which represents an average over all the real species in the plasma.

Following the average description, we have developed different models in order to determine several atomic data in the high density and temperature region assuming LTE conditions. On the other hand, a general NLTE-AA code, which will improve our simulation capabilities in the low density region, as well as in soft X-ray production, is now under development. In both LTE and NLTE cases we are also working in different splitting schemes to obtain from the average the dominant configurations and their abundances.

For the high density and high temperature regime, where the electron collisions with the plasma constituents are dominant and radiation has a negligible effect on electron population, we have developed two different models. The first one, PANDORA, is screened hydrogenic average whereas the second, ADELA, gives a better quantum treatment by solving the Dirac equation for different descriptions of the central potential V(r).

2.1. PANDORA CODE

The atomic data such as energy levels, E_{nlj}, electronic populations, P_{nlj}, and the average ionization, Z, for LTE plasmas as a function of density and temperature of the medium are obtained from an iterative solution of the following equations (Lokke & Grasberger, 1977).

$$P_{nlj} = \frac{g_{nlj}}{1+\exp(\alpha + E_{nlj}/T)} \qquad (1)$$

$$E_{nlj} = E_{nl}^o + \Delta E \pm \frac{1}{2}\Delta E_{LS} \qquad (2)$$

$$Z = Z - \sum_{nlj} P_{nlj} \qquad (3)$$

where:

- T = material temperature
- α = degeneracy parameter
- g_{nlj} = multiplicity of the level
- P_{nlj} = level population
- Z = atomic number

The degeneracy parameter is known when the Fermi-Dirac integral $F_{1/2}(\alpha)$ is given by means of several well-known expressions depending of the α value. We distinguish between strong, weak and partial degeneracy, being the intermediate range calculated via an analytical expression fitted from the tabular results by Cox.(1965)

For an isolated hydrogenic ion the bound electron energy levels are given from the Dirac equation as:

$$E^o_{nl}(eV) = 511 \times 10^3 \left\{ \left[1 + \left(\frac{\beta Z_n}{n - (l+1) + \sqrt{(l+1)^2 - \beta^2 \overline{Z}_n^2}} \right)^2 \right]^{-1/2} - 1 \right\} \qquad (4)$$

being β the fine structure constant and Z_n the net nuclear charge as seen by the electrons in the nth level. This effective nuclear charge is calculated as a function of the level occupations through a set of screening constants.

At high densities, the interaction with the neighbour ions and free electrons changes the former energy levels (and therefore the degree of ionization) obtained for the isolated ion. In this way the correction term ΔE was included in eq. (2) following the simple theory of continuum lowering given by Stewart & Pyatt (1966). In addition, to correctly describe the pressure ionization in the high density region we use a corrected degeneracy (Zimmerman & More 1980) of the level as:

$$g_{nlj} = \frac{2j+1}{1 + (a \frac{r}{R_o})^b} \qquad (5)$$

where r and R_o are the radius of the electronic orbital and the ion-sphere radius respectively. The constants a and b are fitted from the Thomas-Fermi results and provide a reasonable smoothing of the ionization curves. Finally, the ΔE_{LS} term in eq. (2) corresponds to the LS coupling.

In figures 1 to 4 the average ionization for Aluminium calculated with PANDORA is compared with the results of the SESAME Library (Cooper, 1983) for the density range of 10^{-6} up to 10^3 g/cm^3 and temperatures between 200 eV and 5 keV.

2.2. ADELA CODE

The model used in this code is based on the calculation of a self-consistent central potential from the local electron density which includes both bound and free electron contributions. The potential has the usual form with a term from the nucleus, another from the electron charge distribution and a Slater type exchange term (Liberman 1979). This potential will be used self-consistently to calculate the electron density in an iterative scheme until convergence be reached.

When dealing with high-Z materials, the former scheme can be simplified by using an externally calculated Thomas-Fermi field. In this case no iterations are done and shorter computational times can be achieved.

Under LTE conditions the radial bound electron density is expressed as:

$$n_b(r) = \frac{1}{4\pi r^2} \sum_i \frac{2j+1}{1+exp[(E^i-\mu)/KT]} [R^i(r)]^2 \qquad (6)$$

where "i" represents a spin-orbital defined by a set of quantum numbers (n,l,j). E^i is the corresponding eigenvalue and $R^i(r)$ the radial wave function obtained by solving the radial Dirac equation for the i'th orbital.

Then the population of the i'th level is given by

$$P^i = \frac{g_i}{1+exp[(E^i-\mu)/KT]} \qquad (7)$$

being g_i the level degeneracy and μ the chemical potential.

The one-electron orbital functions used in the model are obtained following the work of Liberman et al. (1971). The expression for the free electron density given by the Fermi-Dirac distribution is:

$$\rho_f(r) = \int_{P_o(r)}^{\infty} \frac{8\pi p^2}{h^3} \frac{dp}{1 + exp[(\frac{p^2}{2m} - eV(r) - \mu)/KT]} \quad (8)$$

$$\frac{P_o^2(r)}{2m} = eV(r)$$

and therefore:

$$n_f(r) = 4\pi r^2 \rho_f(r) \quad (cm^{-1}) \quad (9)$$

Finally the total radial electron density will be $n(r) = n_b(r) + n_f(r)$, being the chemical potential, μ, determined from the closure relation:

$$Z = \int_o^{R_o} n(r) dr \quad (10)$$

which enforces the electrical neutrality of the ion sphere of radius R_o.

2.3. NUMERICAL RESULTS AND COMPARISON

In Table I the energy levels for iron at 200 eV and 7.59 g/cm³ obtained with several different versions of our two average atom models are shown. We have used as reference the results given by More (1981) from a work of Liberman (1979).

The ADELA code carries out the numerical computation of the model just proposed. When the self-consistent solution is used we describe it as SCP while the notation TF indicates that the simpler Thomas-Fermi version has been used. The results obtained with our first analytical model, PANDORA (Velarde et al. 1984), and those from the improved version, PANDORA-S, which includes LS coupling are also presented. The table is completed with the average ionization, ZBAR, the degeneracy parameter, $\alpha = -\mu/KT$, and some characteristic computing times (on a CYBER-835).

From these results we can conclude the convenience of the use of the simpler models when calculating the plasma ionization or degeneracy. In these magnitudes we obtain very similar results with much less expensive codes. On the other hand, if we are interested in defining the atomic structure (as in the opacity calculations), the more elaborated model is required because the observed differences on the energy levels can produce important discrepancies in the calculated opacities.

In figure 5 the radial electron densities, $n(r) = 4\pi r^2 \rho(r)$, for the aforementioned iron case calculated with the two versions of the ADELA code are compared. For lower radii less important differences are observed whereas for radii greater than 0.5 a.u., where the free electron density becomes dominant, both results are very close. At higher temperatures and matter densities the agreement between both models will be better. Then, for a wide range of temperatures and densities, the simpler version could be used for EOS or stopping power calculations where a very detailed knowledge of the atomic structure is not necessary.

Figure 6 shows the same quantity, $n(r)$, for iron at density 7.59 g/cm^3 and two different temperatures: 200 eV and 1 KeV. The tails of the curves come from the non uniform free electron density distribution (Rozsnyai, 1972). These results have been obtained with the SCP version.

The density effect on the upper states with large Bohr radii is shown in figures 7 and 8, where the bound and total electron densities are plotted. The solid line corresponds to iron at 50 eV and 7.59 g/cm^3 while the dashed line has been calculated at 75.9 g/cm^3. Note the two different sphere radii and how the n = 3 level has been pressure ionized in the second case.

Finally in fig. 9 the local effective (screened) nuclear charge $Z^*(r) = -rV(r)$, for the iron case at 200 eV and 7.59 g/cm^3 is shown. The different behaviour at $r = R_o$ is due to the fact that in the TF case no exchange term was included in the local potential.

3. ATOMIC DATA DETERMINATIONS FOR ICF TARGET CALCULATIONS

3.1. DENIM/AT-I LIBRARY

Directly driven ICF targets can be simulated using the one dimensional NORCLA code (Velarde et al., 1986). Atomic data for the materials used in these targets are generated with DENIM/AT-I Library, fully coupled with the NORCLA code.

This library contains mainly EOS, electron conductivities and mean opacities, generated by our own methods. The structure of this library is basically the same than SESAME one, because some data are taken from it, and its retrieval programs are extended to our library.

In the figure 10, the structure of this library with auxiliary programs to generate all data is shown. At present time, a testing of data for some materials is done, and new versions are now under development in order to extend the library to treat indirectly driven targets, essentially those with X-ray generation mechanism.

3.2. OPTICAL PROPERTIES CALCULATIONS

Although details of the GEMINIS code (Minguez et al., 1984) have been already reported, a short review of the present state of the calculation model is now given:

i) Bound-bound transition simulation includes hydrogenic oscillator strength, a Voigt profile with natural and Doppler line broadenings and transitions between levels (no transitions between sublevels of the same level are calculated).
ii) Bound-free and free-free transitions are evaluated from Kramer's formulae with Gaunt's quantum and degeneracy correction factors.
iii) Thompson scattering including collective effects and a Klein-Nishima correction factor is considered.

These frequency dependent opacities weighted with the Planck function yield the Rosseland mean opacities.

Three atomic physics models are coupled with GEMINIS: PANDORA (P), PANDORA splitting (P-SP) and ADELA (A). The first two ones are very cheap in computing time. P only gives energy levels and electron populations for each n-level whereas P-SP can obtain a set of energy levels and populations for each n_{lj} level. On the other hand, A consumes more computer time but provides a wide set of data for each n_{lj} orbital.

The calculation model here proposed is an average atom model in a LTE condition whose range of validity according to the numerical simulation of direct irradiated targets is limited to the range of high densities.

Aluminium and iron were selected as low-Z materials in this analysis because they are included in the Astrophysical Library (Ast. Lib.) (Huebner et al. 1977).

Although the range of densities and temperatures analyzed was longer: 10^{-3} to $10^3 g/cm^3$ and 50 eV to 5000 eV; only four temperatures (50, 100, 300 and 1000 eV) and a large range of densities for aluminium are reported here, because conclusions may be well established and generalized for the whole range and extended to iron plasma.

The first part of the analyzed cases consists in the calculation with PANDORA-GEMINIS (PG) of Rosseland mean opacities for aluminium in the range of densities and temperatures abovementioned, assuming the following:

i) Hydrogenic oscillator strength.
ii) Natural line broadening.
iii) Energy levels and populations for each n level (no splitting).
iv) No degeneracy correction in the bound-free transitions.

The second part of the calculations considers the following effects in the Rosseland mean opacities:

i) Splitting of energy levels (nlj level) and also degeneracy correction in the bound-free transition (PSP-GEMINIS).
ii) Energy levels and populations from ADELA (A-GEMINIS).
iii) Doppler line broadening.

For this analysis, aluminium at 300 eV, 0.1 g/cm^3 and iron at normal density (7.59 g/cm^3), 200 eV were calculated.

In figures 11 to 13, total and continuous Rosseland mean opacities from the P-G model and the Ast. Lib. are plotted. At 50eV and 300 eV line transitions are important as it is shown in the Ast. Lib. results, but this effect cannot be simulated with P-G because only natural line broadening was taken into account and no splitting of energy levels was assumed in the calculation. When continuous transitions are the dominant effect, a good agreement between Ast. Lib. and P-G is found.

The splitting effect of energy levels and its use in the opacity calculation yields in a weak increase in the Rosseland mean opacities, even when considering only natural line broadening (Tables II and III) because more transitions were calculated.

Results via ADELA-GEMINIS are more sensible, not only in the continuous transitions but also in line transitions, because of differences (~10%) in energy level intensities from ADELA-GEMINIS and those from PANDORA-SP-GEMINIS.

The inclusion of the Doppler effect increases the Rosseland mean opacities and low deviations were calculated. As it can be shown, continuous transition deviation from ADELA-GEMINIS are in a good agreement with Ast. Lib.

4. SUMMARY AND FINAL REMARKS

Two average ion models have been explained, and finally tested solving a case of iron at 200 eV and normal density. The results show that PANDORA computating times are much shorter than those of ADELA. The former can be used except in the calculation of energy levels (for instance, for opacity calculations) where important deviations can be obtained. These deviations can be neglected for Z, and other atomic physics data.

As a conclusion, PANDORA will be improved in the sense of taking into account screening constants for each sublevel and for each material, and continuous lowering will be better analyzed in order to explain differences in the ionization, specially to high densities and low temperatures.

Detailed configuration for high-Z materials will be used, which are now under development.

Optical properties calculations for low-Z materials have also been presented, following the developed atomic models, coupled to GEMINIS that contains analytical formulas for atomic transitions the following conclusions were reached:

i) With very simple atomic physics models, such as PANDORA, without splitting and using Kramer's formulae, Rosseland mean opacities can be well calculated only when the plasma is almost fully ionized because continuous transitions are fairly dominant.

ii) When the plasma is not fully ionized, line transitions are an important effect, and it is necessary to employ a more sophisticated atomic physics model such as ADELA.

iii) For low density plasmas NLTE-AA models should be used. In both cases, LTE and NLTE conditions, the opacity results will be improved if the mean configuration is resolved into its components.

iv) In the frame of AA models less deviations were reported with the use of Doppler line broadening and following the ADELA-GEMINIS scheme. This model is expensive in computer time and thus, PANDORA-SP-GEMINIS will be fitted with some correction parameters in order to get approximate results with less calculation time.

In a future work, less deviations for low-Z opacity calculations will be obtained if the oscillator strength is calculated via orbital functions (dipole strength) from ADELA code and if detailed configurations are taken into account.

All these ongoing developments will be used for calculating the opacities of high-Z materials, solving at the same time a Non-LTE model, when LTE conditions are not valid.

References

COOPER, N.G., ed. 1983 "Equation of State and Opacity Group Sesame' 83".Los Alamos LALP-83-4.

COX, A. 1965 "Stellar Absorption Coefficients and Opacities in Stellar Structure"; Aller, L.H., Mc Laughlui, D.B., Fols. Univ. of Chicago Press.

HONRUBIA, J. & ARAGONES, J.M. 1986 Nucl. Sci. Eng. To be published.

HUEBNER, W.F. et al. 1977. Los Alamos National Laboratory. Report No. LA-6760-M.

LEIRA, G., PERLADO, J.M., VELARDE, G. et al. 1981 Proceedings of the 10th European Conference on Controlled Fusion and Plasma Physics. Moscow.

LIBERMAN, D.A., CROMER, D.T. & WABER, J.J. 1971 Comp. Phys. Comm. **2**, 107.

LIBERMAN, D.A. 1979 Phys. Rev. B**20**, 12.

LOKKE, N.A. & GRASBERGER, W.H. 1977. Lawrence Livermore National Laboratory Report No. UCRL-52276.

MINGUEZ, E., SERRANO, J.F. & CABEZUDO, C. 1984, Annual Research Report DENIM-025.

MORE, R.M. 1981 Lawrence Livermore National Laboratory. Report No. UCRL-84991 Part I and II. Preprint.

ROZSNYAI, B.F. 1972, Phys. Rev. A5,3.

ROZSNYAI, B.F. 1982, J. Quant Spectros. Radiat. Transfer 27, 211.

STEWART, J. & PYATT, K. 1966, Astroph. Journal **144**, 1203.

VELARDE, G. et al., 1978 Atomkernenergie/Kerntechnik **32**, 58.

VELARDE, G. et al., 1980a Atomkernenergie/Kerntechnik **35**, 40.

VELARDE, G. et al., 1980b Atomkernenergie/Kerntechnik **36**/213.

VELARDE, G. et al. 1984. Atomkernenergie/Kerntechnik **44**, 3.

VELARDE, G., MINGUEZ, E., SERRANO, J.F. et al. 1984 Report DENIM. DENIM-010.

VELARDE G. et al., 1986 Submitted to Laser and Particle Beams.

YAMANAKA, C. 1983 Nuclear Fusion 23, 108.

ZIMMERMAN, G.B. & MORE, R.M. 1980 J.Q.S.R.T. 23, 517.

TABLE I.- Energy levels in Fe^{26} at 200 eV and 7.59 g/cm^3 from different models

J	LIBERMAN * E(EV)	ADELA (SCP) E(EV)	ADELA (TF) E(EV)	PANDORA-S E(EV)	PANDORA E(EV)
$1S_{1/2}$	-7284	-7280	-7108	-7516	-7522
$2S_{1/2}$	-1063	-1062	-1031	-1078	-1075
$2p_{1/2}$	- 961	- 960	- 925	-1075	-1075
$2p_{3/2}$	- 948	- 947	- 913	-1068	-1075
$3S_{1/2}$	- 238	- 240	- 225	- 177.9	- 178
$3p_{1/2}$	- 206	- 207	- 192	- 177.5	- 178
$3p_{3/2}$	- 203	- 205	- 190	- 176.6	- 178
$3d_{3/2}$	- 154	- 155	- 143	- 176.9	- 178
$3d_{5/2}$	- 153	- 154.8	- 143	- 176.6	- 178
	α	2.805	2.752	2.963	2.966
	ZBAR	14.535	14.66	14.66	14.61
	RUN TIME	1100sec	60 sec	.13 sec	.1 sec

Table II. Rosseland mean opacities comparison for aluminium at 300 eV and 0.1 g/cc. (In K_C no lines are included)

	ASTROPHYSICAL LIBRARY	PANDORA	PANDORA SP	ADELA	PANDORA DOP.	PANDORA SP.-DOP.	ADELA DOP
K_T	13.54	7.181	8.77	9.49	12.6	12.94	14.15
K_C	5.787	5.12	5.13	5.58	5.12		

Table III. Rosseland mean opacities comparison for iron at 200 eV and normal density (In K_C no lines are included)

	ASTROPHYSICAL LIBRARY	PANDORA	PANDORA SP	ADELA	PANDORA DOP.	PANDORA SP.-DOP.	ADELA DOP
K_T	2679.	579.	591.	1014.	739.	759.	2200.
K_C	988.	561.	563.	862.	561.	563.	862.

Figure Captions

Fig. 1. Average ionization vs. density for Aluminium plasma at 200 eV (——Sesame; --- Pandora).

Fig. 2. Average ionization vs. density for Aluminium plasma at 500 eV (——Sesame; --- Pandora).

Fig. 3. Average ionization vs. density for Aluminium plasma at 1 KeV (——Sesame; --- Pandora).

Fig. 4. Average ionization vs. density for Aluminium plasma at 5 KeV (——Sesame; --- Pandora).

Fig. 5. Total electron radial density vs. atomic radius for iron at 200 eV and normal density. Effect of the potential used in the Adela code (——SCP, self consistent potential; --- TF).

Fig. 6. Effect of the temperature in the total electron radial density for iron at normal density (——T = 200 eV; --- T = 1KeV).

Fig. 7. Effect of the density in the bound electron radial density for iron at 50 eV.

Fig. 8. Effect of the density in the total electron radial density for iron at 50 eV.

Fig. 9. Local effective nuclear charge vs. the potential for iron at 200 eV and normal density.

Fig. 10. Scheme calculation for the DENIM Atomic Tables.

Fig. 11. Rosseland mean opacities from Astrophysical Library and Geminis code, vs. density for aluminium at 50 eV.

Fig. 12. Rosseland mean opacities from Astrophysical Library and Geminis code, vs. density for aluminium at 300 eV.

Fig. 13. Rosseland mean opacities from Astrophysical Library and Geminis code, vs. density for aluminium at 1 KeV.

Figure 1

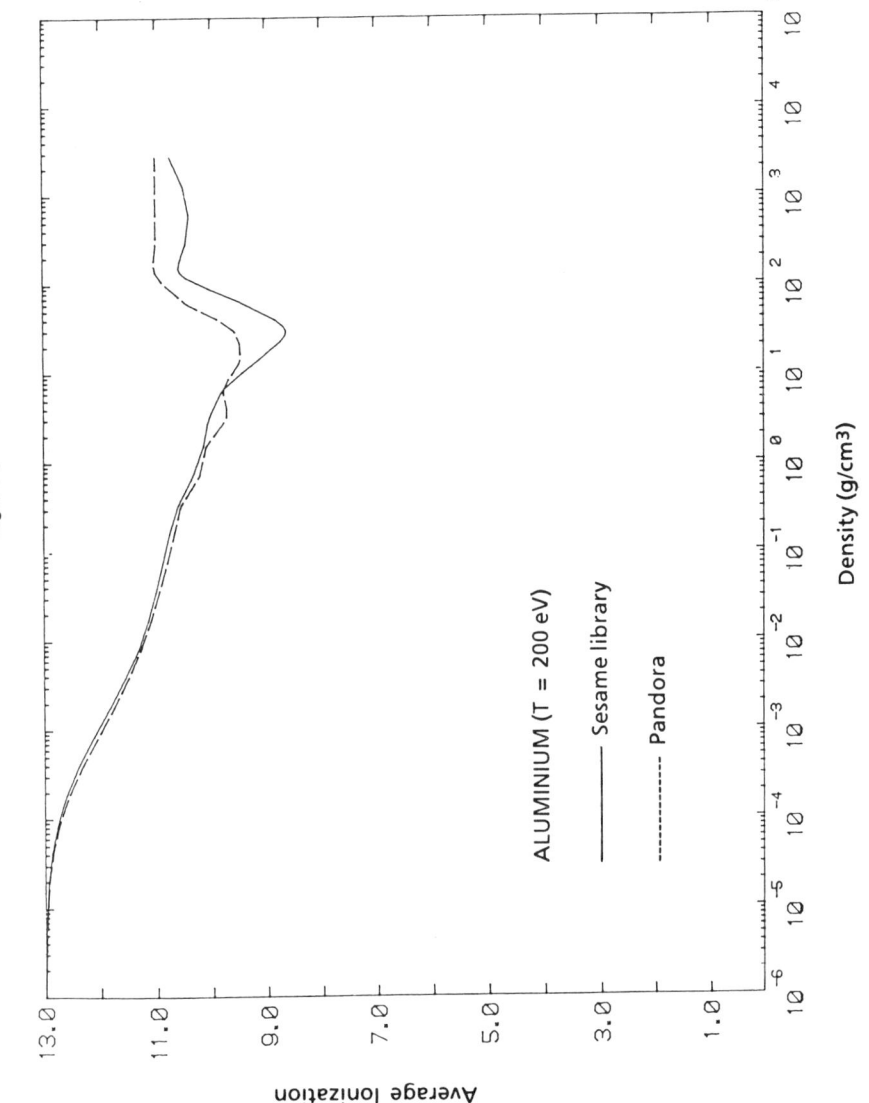

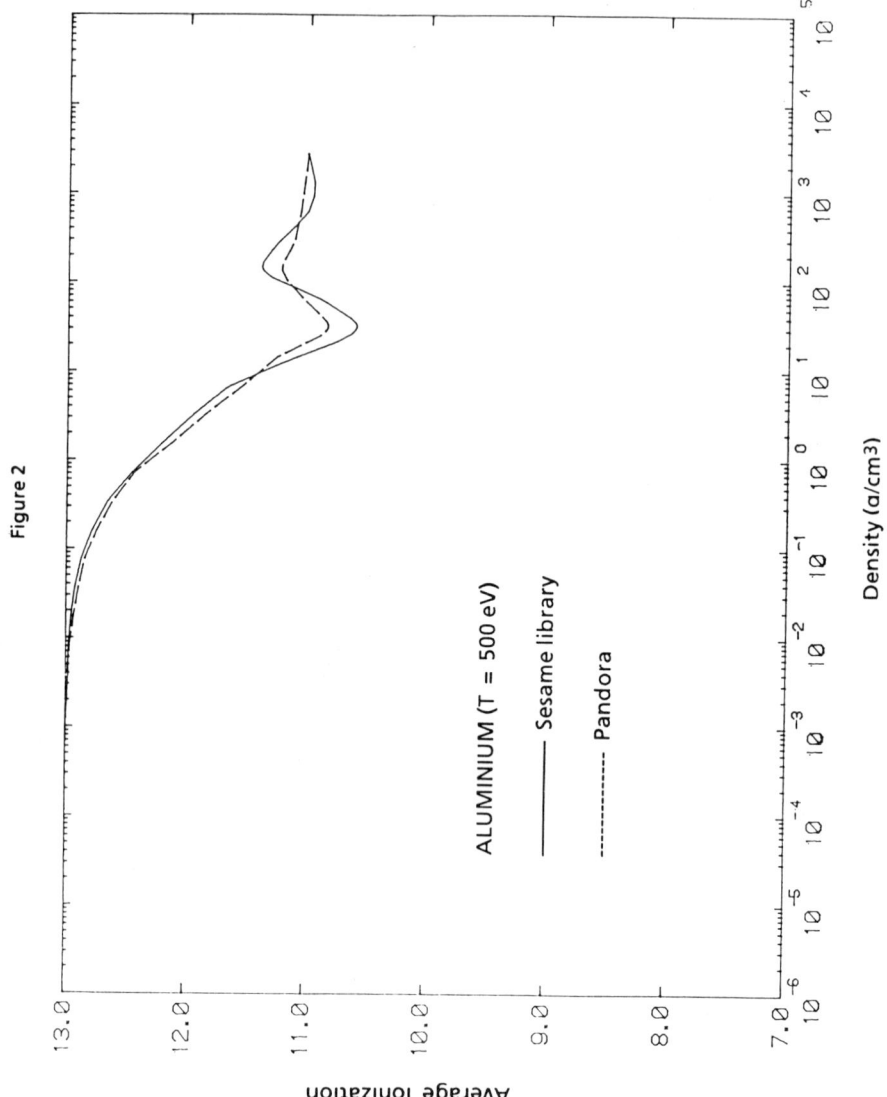

Figure 2

Figure 3

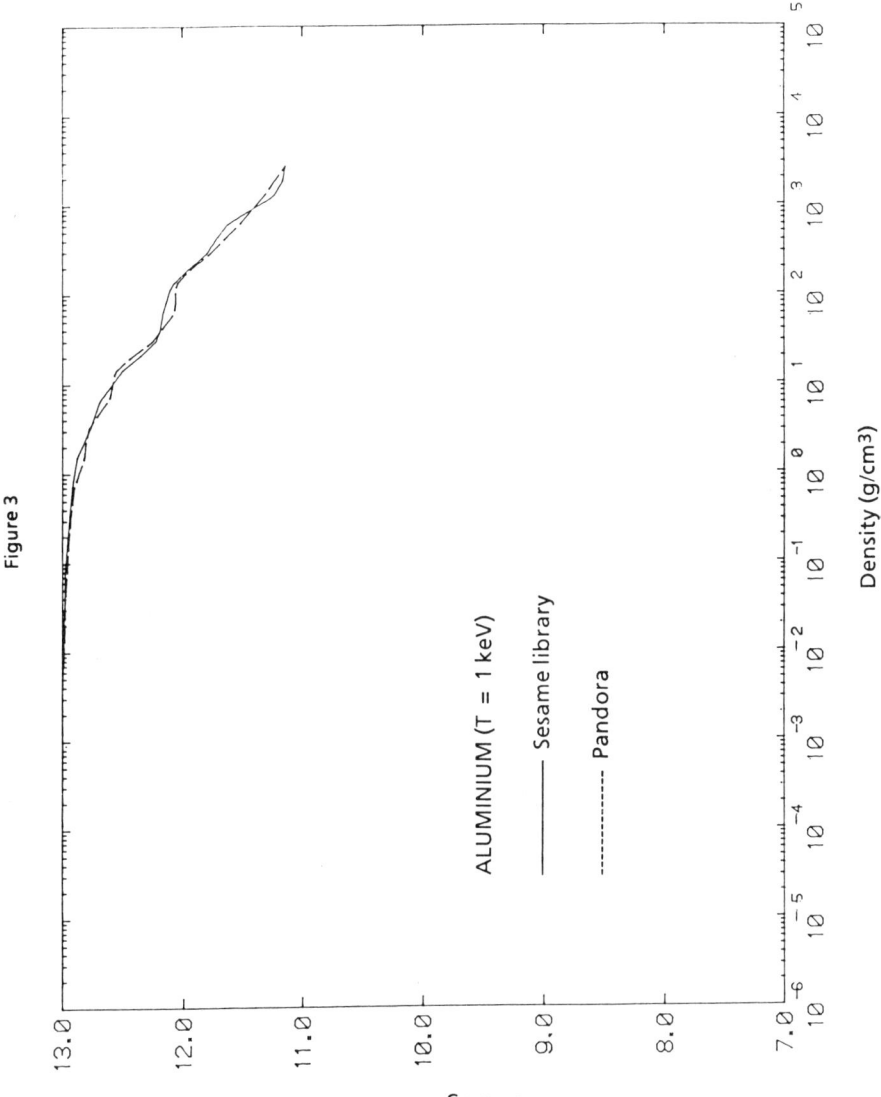

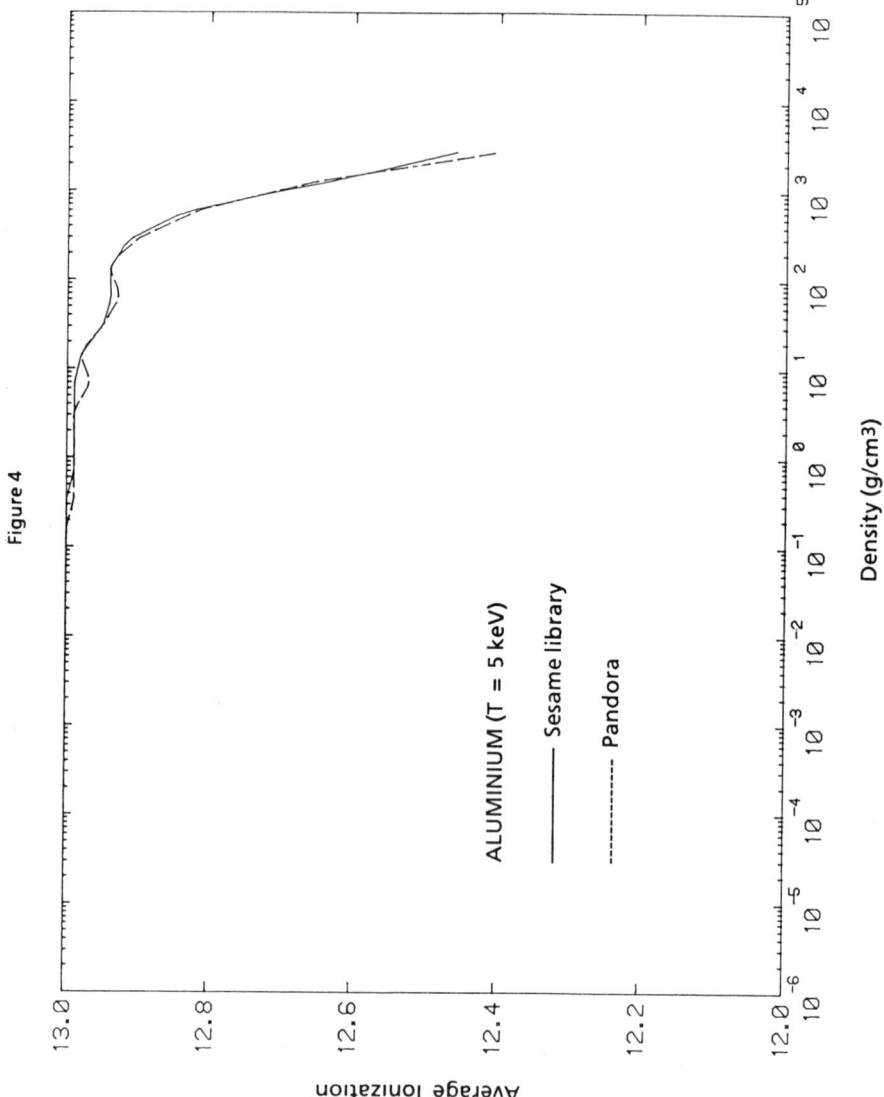

Figure 4

Figure 5

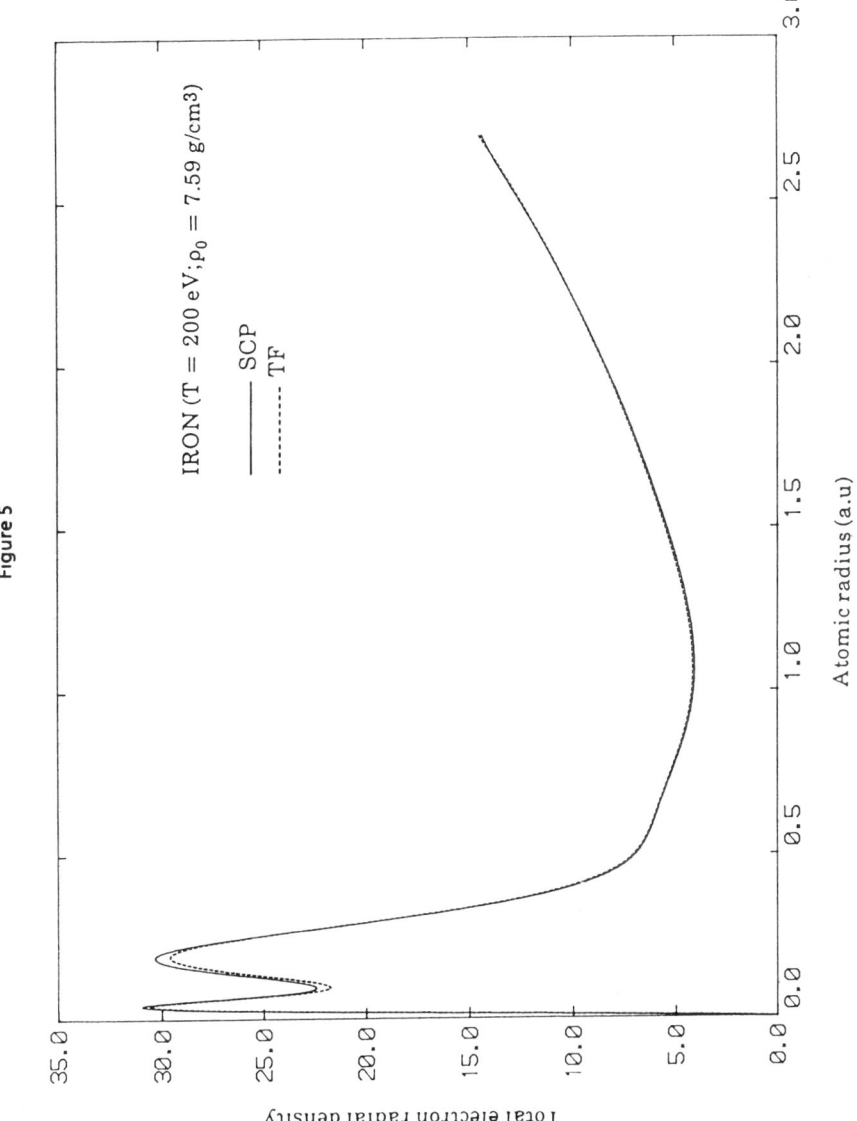

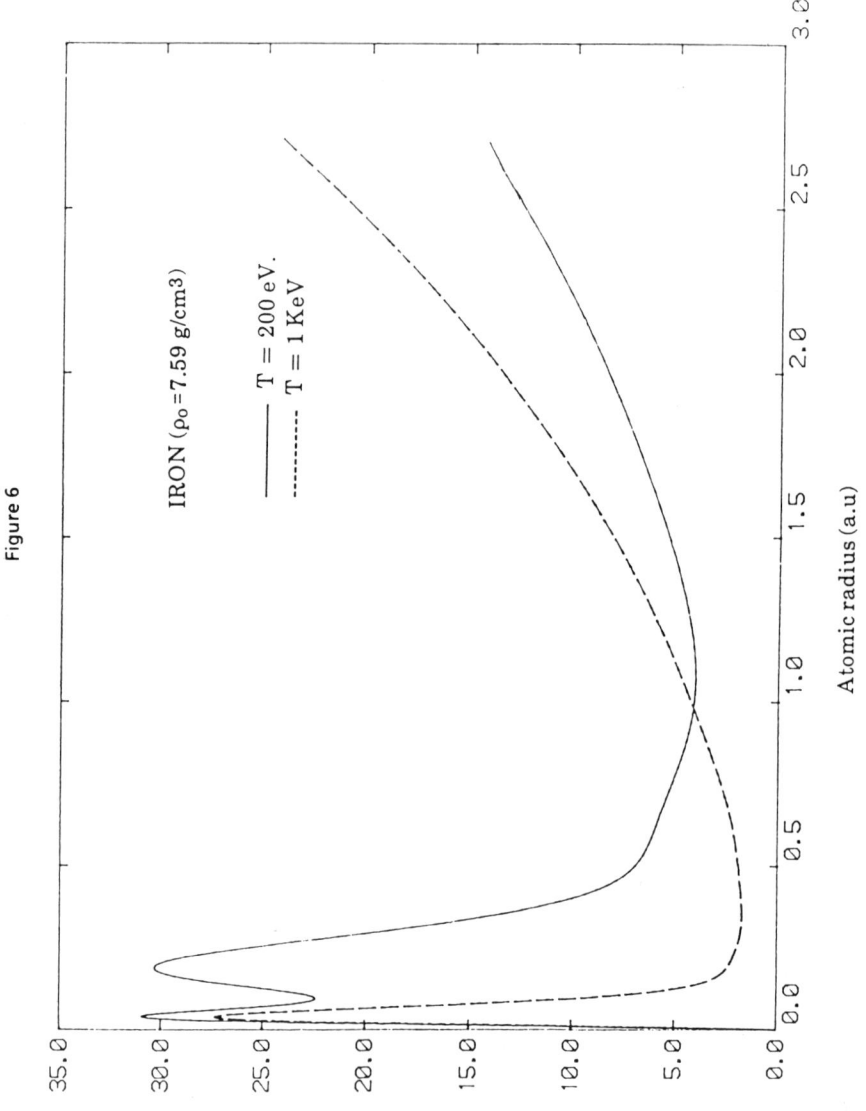

Figure 6

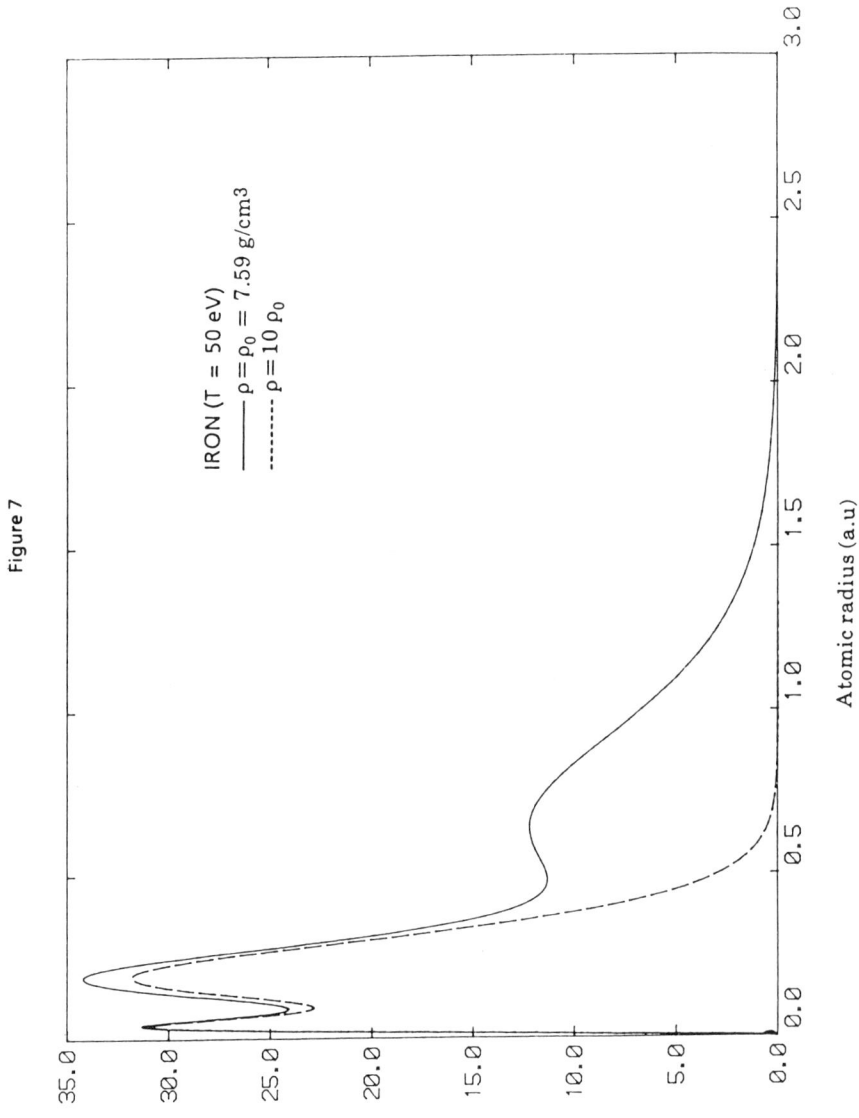

Figure 7

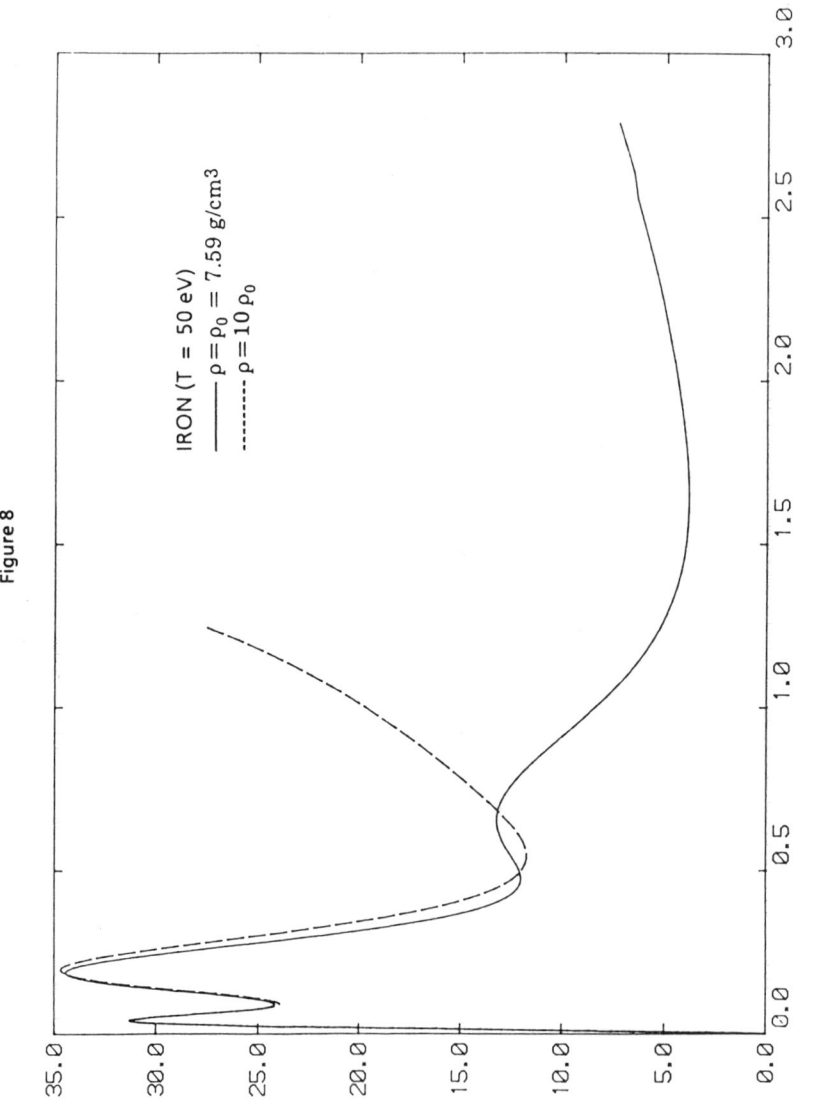

Figure 8

Figure 9

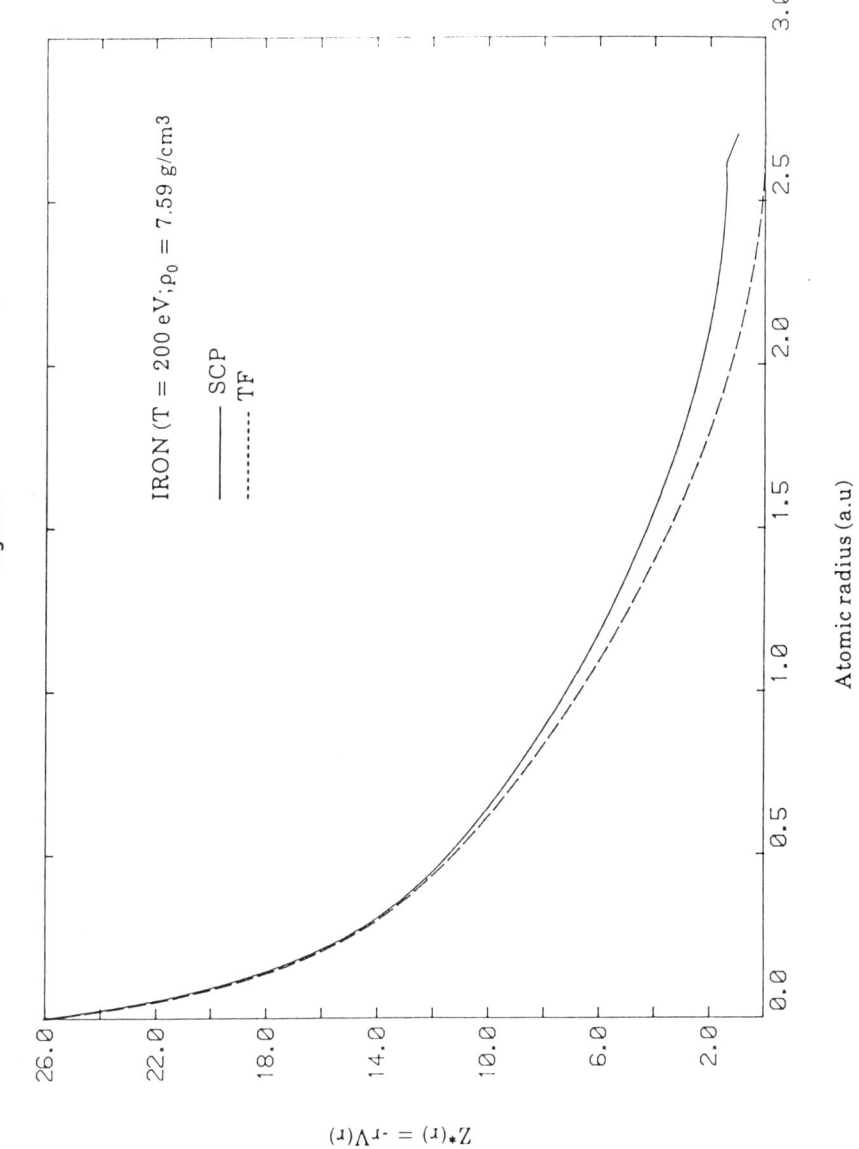

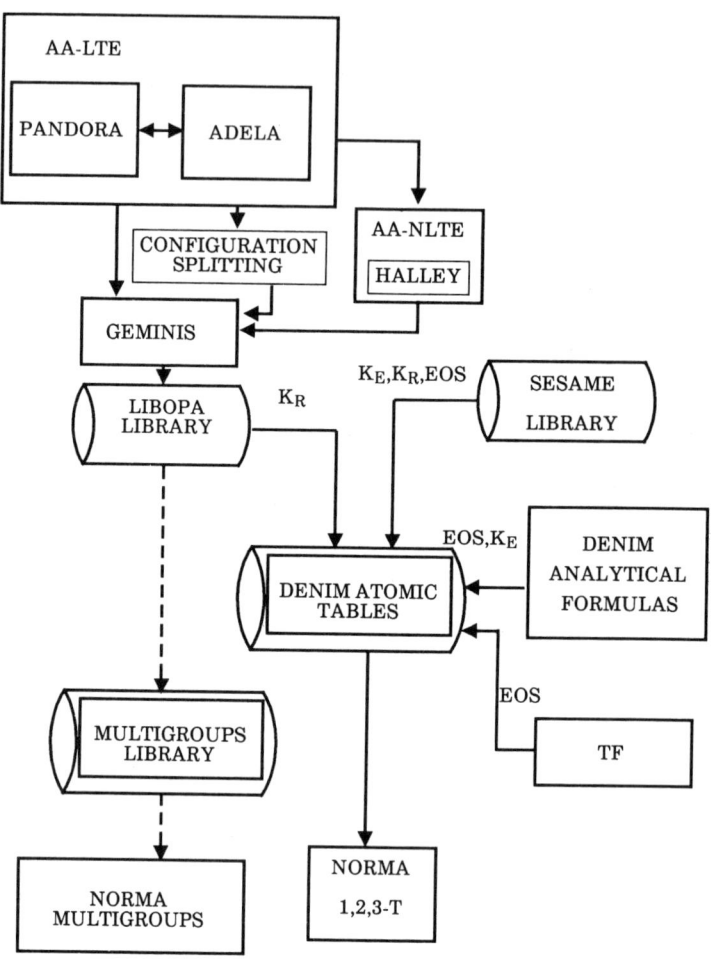

Figure 10

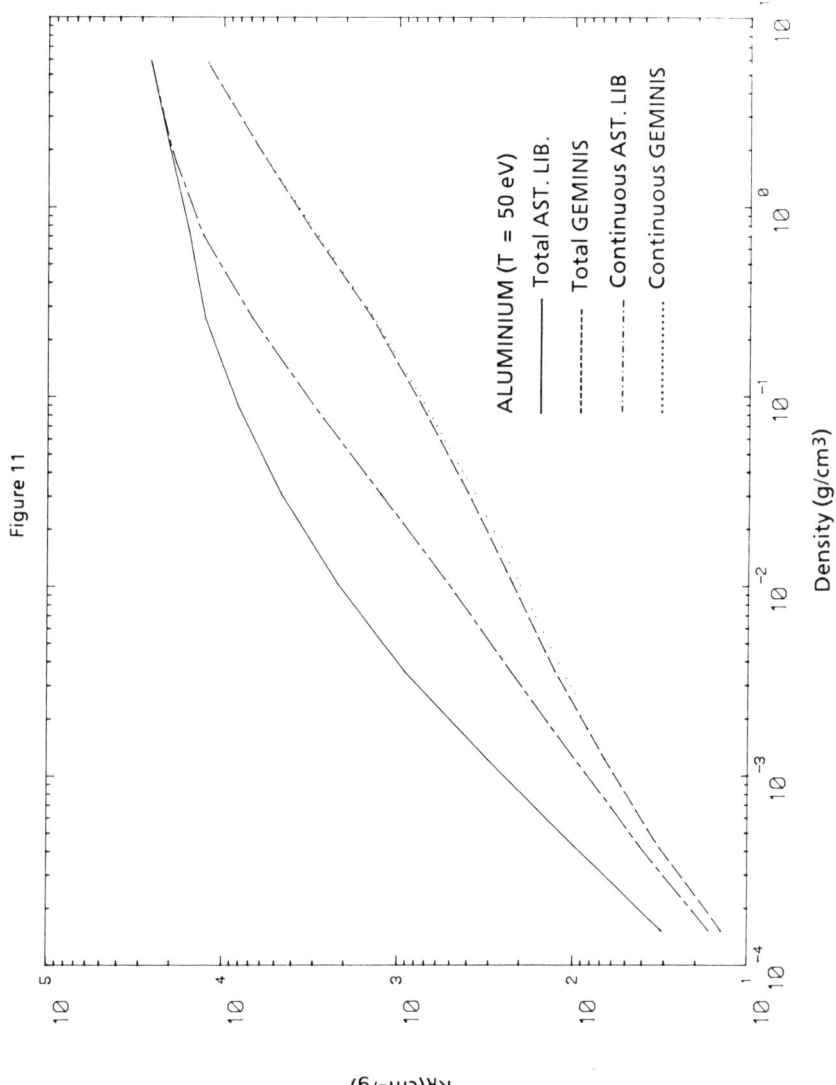

Figure 11

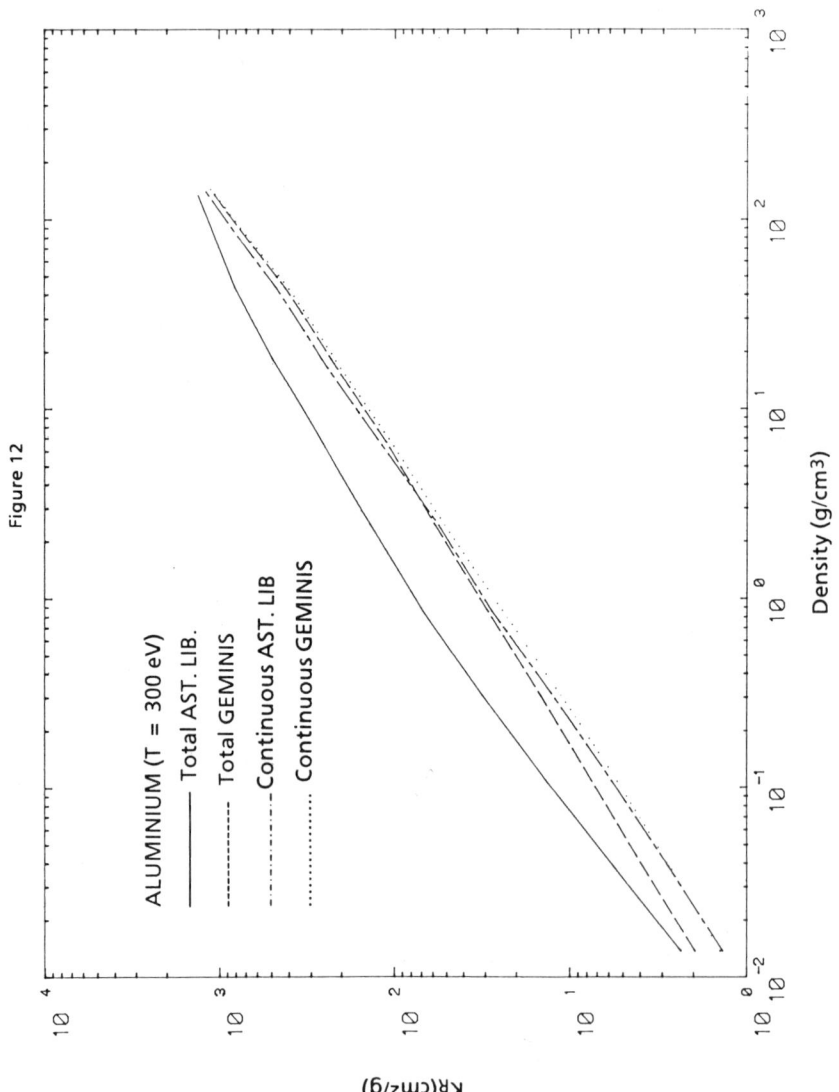

Figure 12

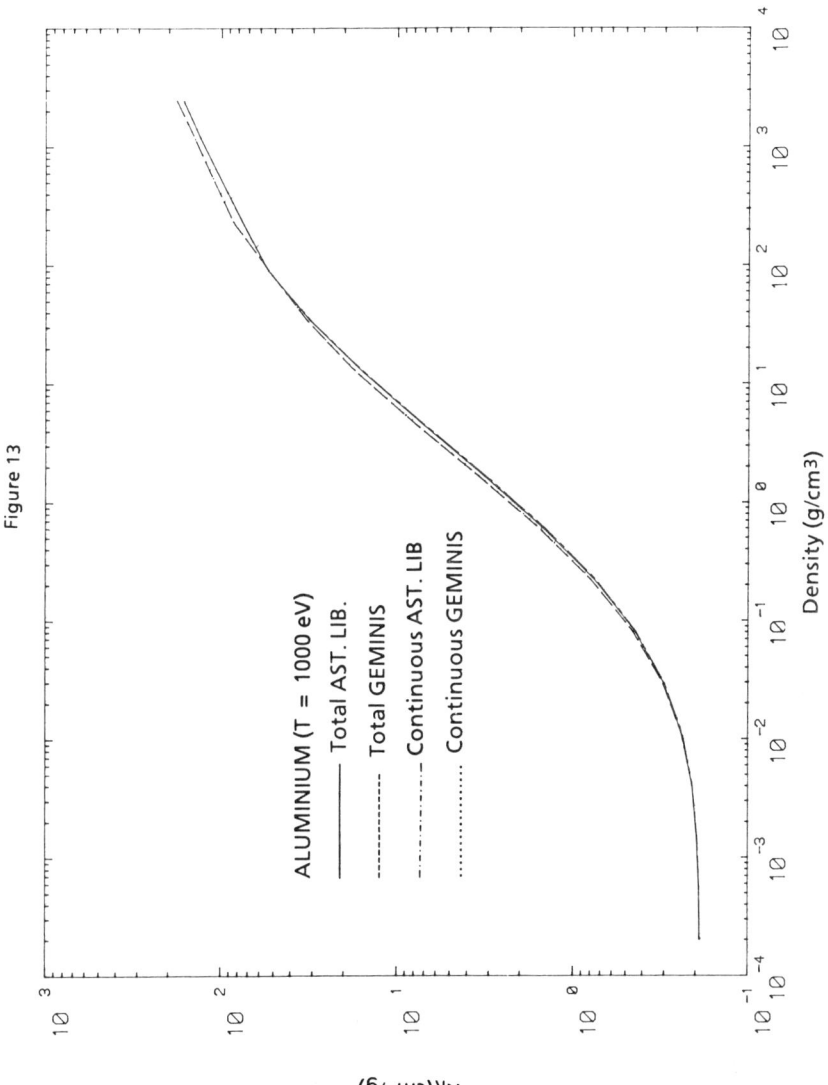

Figure 13

Ion Microfields In Plasmas With Strong Electron-Ion Interactions

- Density-Functional Theory -

M.W.C. Dharma-wardana

National Research Council, Ottawa, Canada K1A OR6

and

François Perrot

Centre d'Etudes de Limeil-Valenton

B.P. 27, 94190 Villeneuve-St-Georges, France

Abstract

We review the existing methods (EM) for the calculation of ion microfields with emphasis on the simplifying physical assumptions inherent in EM. The simple physical picture of a plasma as a one-component classical Debye-like fluid of point ions used in EM needs improvement if strongly coupled plasmas with high charge species, internal structure and strong electron-ion interactions are to be studied. A two-component theory which explicitly treats electron coordinates becomes necessary to these plasmas which are of current interest. We show that density functional theory (DFT) provides a particularly simple self-contained method which can treat such strongly coupled plasmas of arbitrary density and coupling strength.

I. Introduction

The ion microfield distribution $W(\underset{\sim}{E})$ specifies the probability of occurrence of the static (i.e., low frequency) field $\underset{\sim}{E}$ at a test particle placed in a plasma. This has been of great value in describing the line shapes of spectral line-wings in weakly coupled plasmas.[1] With the advent of experiments which probe strongly coupled plasmas, high-Z ions and partially degenerate electrons, the question of extending the domain of validity of the existing methods (EM) for the calculation of microfields becomes important. In fact, line broadening models which distinguish between "radiators" and "perturbers", or even the applicability of the very concept of microfields, have to be examined thoroughly. In this paper we limit ourselves to a re-examination of existing methods (EM) to show that the density functional theory (DFT) of plasmas[2,3] can be very useful in removing the limitations of EM.

We use the Baranger-Mozer (B-M) cluster series for the probability distribution of the microfield, $W(\underset{\sim}{E})$ as the theoretical framework for discussion.[4] In B-M theory, and in the theories of Hooper,[5] Iglesias et al.[6] the Fourier transform $W(k)$ of $W(\underset{\sim}{E})$ is evaluated. In B-M theory, for a plasma of ion density $\bar{\rho}$

$$W(k) = \exp S(k)$$

$$S(k) = \bar{\rho}\, W_1(k) + \frac{1}{2}\bar{\rho}^2\, W_2(k) + \frac{1}{6}\bar{\rho}^3\, W_3(k) + \ldots \qquad (1)$$

The 2nd order term is of the form

$$W_2(k) = \int \phi_1(k)\, \phi_2(k)\, C_2(\underset{\sim}{R}_1, \underset{\sim}{R}_2)\, d\underset{\sim}{R}_1\, dR_2$$

$$C_2(\underset{\sim}{R}_1, \underset{\sim}{R}_2) = g(\underset{\sim}{R}_1, \underset{\sim}{R}_2) - g(\underset{\sim}{R}_1)g(\underset{\sim}{R}_2)$$

$$\phi_1(k) = \exp(i\underset{\sim}{k}\cdot\underset{\sim}{\varepsilon}_1) - 1.$$

The cluster function C_2 involves the pair distribution function $g(\underset{\sim}{R}_1, \underset{\sim}{R}_2)$ and the one-body function $g(R_1)$ and $g(R_2)$, with the test particle at the origin. The higher order cluster terms, e.g., $W_3(k)$ will require a knowledge of the triplet (and higher) correlation functions as well. The ϕ_i functions require an evaluation of the electric field E_i due to the ion at R_i, at the test particle which is at the origin, in the presence of the inhomogeneous distribution of interacting electrons. Thus the calculation of the microfield involves:

(i) A method of calculating ion-ion correlation functions $g(r)$ for a system of interacting ions and electrons, given by the ion distribution $\rho(r)$ and the electron distribution $n(r)$ around the radiator determined self-consistently.

(ii) A method of evaluating the electric field at the test charge, given the ions, their electronic structure, and the distribution of electrons, viz. $n(r)$.

(iii) Evaluating the series to finite order. Any explicit evaluation beyond 2nd order is impracticable in most cases. This leds to the 2nd order B-M microfield.

(iv) Evaluating the effect of the full B-M expansion by some resummation technique. The determination of this "all order" result is the "resummation problem".

Existing methods have treated (i) and (ii) by simplifying the description of the plasma. The plasma is treated as a <u>one-component classical fluid</u> of <u>point ions</u> interacting via a Debye-like effective pair potential. Even the electric field at the test particle is calculated from the derivative $-\nabla_0 U_{ii}(|R-R_0|)$ where $U_{ii}(|\underset{\sim}{R}-\underset{\sim}{R}_0|)$ is the pair potential between an ion at the origin, $\underset{\sim}{R}_0$, and a field ion at $\underset{\sim}{R}$. Thus the electron coordinates are completely removed from the discussion. Within this simple picture, the only remaining problem is essentially the resummation.

Considerable progress has been made during the last two decades regarding the resummation problem, with the simplified physical picture of a classical fluid of point particles tacitly subsumed in the theory without much discussion. An objective of the present study is to take the discussion away from the resummation manipulations and focus on improving the underlying physics. It turns out that the density functional theory (DFT) of plasmas can contribute significantly since DFT enables us to calculate ion-pair distribution functions, electron charge distributions, internal structure of the ions etc., without making the assumption that the electron-ion interaction $V_{ei}(r)$ is weak. DFT treats the plasma as a two-component system, viz., electrons which are quantum mechanical, and ions which are classical, and does not rely on an effective one-component fluid model. The atomic physics, ionization balance, particle distribution functions etc. are calculated within a single self-consistent formalism. Results of DFT calculations for static ion microfields have also yielded the interesting conclusion that in most cases the resummation is perhaps unnecessary. That is, if the pair distributions and the electric fields are correctly calculated, 2nd order B-M is often sufficient (resummations become necessary if electron-ion coupling is weak or zero, as in the OCP limit). Thus some of the need for resummations may have been an artifact of the simplified physical picture. Of course, even within a more accurate physical picture (based on strong electron-ion interactions etc.), resummations may be necessary, depending on plasma conditions. As the existing resummation methods are too strongly wedded to the simpler physical picture, we have introduced[6] a new resummation method, based on the B-M series, and capable of using the DFT model of the plasma. However, here we shall not discuss the resummation problem in detail.

The type of simplifying assumptions made in existing microfield theories become valid if the electron screening is sufficiently strong and the ions are not too close. For degenerate electrons, this is true if the effective electron sphere radius $r_s = (1/(4\pi\bar{n}))^{1/3}$ is less than unity. Since line broadening problems deal with ions with structure, low energy electrons near the radiator are in any case quasi-degenerate and it is possible that an upper limit of validity may be, say $r_s < 2$. In any case, the reduction of the electron-ion fluid to an effective one-component fluid of classical ions has a number of shortcomings. These are avoided in the density functional model which will be discussed in the next section.

II. DFT Calculation of Microfields

In calculating an ion microfield using any of the existing methods (EM)[4-6] we proceed as follows (i) the ion charge $\bar{Z}$ and the free electron density $\bar{n}$ or the ion density $\bar{\rho}$ are assumed given (ii) electric fields are calculated from the gradient of the effective ion-ion pair potential $U_{ii}(r)$ (iii) the pair potential is calculated in a Debye-like linear response form, even for large values of $\bar{Z}$, without using a pseudopotential formulation - i.e., the bare Coulomb interaction is used in linear response theory (iv) the ions are assumed to be point-like, i.e., internal structure is neglected (v) the Debye-like electric field and the Debye-like $U_{ii}(r)$ are used in an all-order ion-ion correlation calculations in evaluating the microfields.

In the DFT calculation[8] the simplifying assumption in each of the above steps is avoided. Thus (i) the ionization balance and the distribution functions are obtained by self-consistently solving two coupled equations, viz., (a) a Kohn-Sham like equation for the electrons moving in the external field of the central ion and the field-ion distribution $\rho(r)$ (b) a screened hyper-netted-chain (HNC) equation for the ion distribution $\rho(r)$ in the screening field of the electron distribution $n(r)$ given by the Kohn-Sham equation. (ii) The electric fields are calculated directly from the screening charge around a field ion. Thus, if we consider the electric field E, at the ion A placed at the origin, due to a "field ion" B placed at $\underset{\sim}{R}$, we have

$$E = -\frac{Z_B}{R^2} + \frac{4\pi}{R^2} \int_0^R r^2 dr \int \Delta n(q) e^{i\underset{\sim}{q}\cdot\underset{\sim}{r}} \frac{d\underset{\sim}{q}}{(2\pi)^3} \qquad (2.1)$$

or equivalently

$$E = -\frac{Z_B}{R^2} + \frac{1}{R^2} \int_0^R \bar{n}[g_{ei}(r)-1]4\pi r^2 dr \qquad (2.2)$$

where $g_{ei}(r)$ is the electron-ion correlation function around <u>a single</u> field ion B placed in a uniform interacting electron gas of density $\bar{n}$. Equation (2.1) or (2.2) is merely a statement of Poisson's equation and says that the electric field due to B, at A, depends on the charge Z_B

and the screening cloud around B integrated up to A. An approximation is made by assuming that $\Delta n(r) = g_{ei}(r)-1$ is spherical even in the presence of the test charge at the origin. But it can be shown that such non-sphericity corrections appear only in third order in a pseudopotential formulation[9] of the calculation of $\Delta n(r)$. Note also that (2.1) or (2.2) involves only one charge pile up, viz., that around the field ion B. This is again exactly consistent with 2nd order pseudopotential theory. In order to clarify some of this, we formally define the pseudopotential by writing

$$\Delta n(q) = V_{ei}^P(q)\chi(q) \tag{2.3}$$

where $V_{ei}^P(q)$ is the pseudopotential describing the electron-ion interaction. $\chi(q)$ is the exact electron gas response function. $\Delta n(q)$ is the full non-linear electron charge pile up around the ion B placed in the electron gas calculated using the Kohn-Sham Schrodinger code. By construction this $\Delta n(q)$, non-linear with respect to the bare Coulomb potential, becomes linear with respect to the pseudopotential $V_{ei}^P(q)$. In such a theory the ion-ion pair potential is given by the 2nd order expression

$$U_{ii}(q) = \frac{4\pi Z_n Z_B}{q^2} + V_A^P(q)\chi(q)V_B^P(q) \tag{2.4}$$

The last term is $V_A^P(q)\Delta n_B(q)$ or, equivalently, $\Delta n_A(q)V_B^P(q)$ and involves only <u>one</u> screening charge. The screening charge is spherical. It is known that such a model is applicable even in very strongly coupled plasmas like liquid metals.[10] Hence we believe that this 2nd order pseudopotential theory is quite adequate for the plasmas of interest to us.

Note that the DFT calculation directly provides $\Delta n(r)$ and hence there is no need to explicitly construct $V^P(q)$ or $\chi(q)$.

If the electron-ion interaction is weak, and if the electron gas is in the Debye regime we can replace $\chi(q)$ by the Debye-Hukel form and $V^P(q)$ by the bare Coulomb form $-4\pi Z_B/q^2$. Then $\Delta n(q) = Z\lambda^2/(q^2+\lambda^2)$ where λ is the Debye screening constant. If this is introduced in (2.1) we get the electric field used in Baranger-Mozer theory.

The pair potential (2.4) contains a purely electrostatic contribution and energy terms arising from exchange and correlation effects. Thus

$$U_{ii}(r) = U(r)_{el} + U(r)_{X-C}$$

Hence the derivative of $U_{ii}(r)$ would contain non-electrostatic terms, except in the classical limit where $U(r)_{X-C} \to 0$. Thus the procedure used by Hooper et al.,[5] Iglesias et al.[6] is in principle correct only for fully classical plasmas. Also, if the ions in a dense plasma have structure (as would be the case in line broadening problems) nearest neighbour distances of approach would involve overlap of bound electron states. Here again $U_{ii}(r)$ would contain non-negligible exchange-correlation corrections and it is probably unwise to use $-\nabla U_{ii}(r)$ for evaluating the electric field. These points are illustrated in fig. 1 where we compare the electric field calculated using eqn. (2.2) and the DFT results for $\Delta n(r)$, with the Debye-Hukel and $-\nabla U_{ii}$ calculations. The system considered is an aluminum plasma, with the coupling parameter $\Gamma = 3.47$, electron density $\bar{n} \simeq 5.9 \times 10^{21}$ electrons/cm^3. The temperature is about eight times the Fermi energy and hence the electrons are nearly classical. Nevertheless, when a low energy electron approaches an ion the potential energy wins out over the electron kinetic energy and the electron moves into a more degenerate behaviour.

It should be noted that the ion-ion pair potential $U_{ii}(r)$ used in fig. 1 to derive $\nabla U_{ii}(r)$ is obtained by inverting[3] the DFT-calculated ion-ion pair distribution function $g_{ii}(r) = \exp(-\beta U_{ii} + N + B)$ where N and B are the nodal and bridge contributions to the potential of mean force. Since $\Gamma \simeq 3$, we take $B = 0$ and evaluate N using the Orstein-Zernike relation. Of course, in Hooper's theory U_{ii} will be simply given by the Debye form $\frac{Z^2}{r} e^{-\lambda r}$ and hence its gradient will agree with the Debye fields. On the other hand, Hooper's pair distribution $g_{ii}(r)$ which is implicit in his cluster-virial calculation would be different from the DFT $g_{ii}(r)$.

We may also ask the question, does Hooper's theory agree with the DFT-microfield in the classical limit, for small microfield parameters r_{mf} ($\simeq r_s/\lambda$)? We have checked by explicit calculation that this is so, as it should be. Another question that may be raised is, if we choose plasma conditions so that the electric field from DFT is closely approximated by a Debye field, would Hooper's theory agree with the DFT-microfield? It can be shown by explicit calculation that a Debye electric field is a good approximation to the DFT electric field at a neutral point in a strongly coupled Al-plasma such that $r_{mf} = 0.8$, $\Gamma = 7.21$ and $r_s = 1$. The temperature of this strongly coupled plasma is only 2.5 times the Fermi energy and yet the calculated electric field is well approximated by the Debye field for distances outside the Wigner-Seitz radius. But the ion-ion interaction for nearest neighbour distances is <u>NOT</u> too well approximated by the Debye model. The microfield at a neutral point calculated for this plasma using DFT, and Hooper's microfield at $r_{mf} = 0.8$ are shown in fig. 2. Thus, even when the electric fields used in the two theories agree, the result for a strongly coupled plasma of Al-ions (with structure) do not agree. It should of course be remarked that Hooper's theory was not designed for such systems although it has been used in a variety of situations, due to the lack of a better theory. The new adjustable parameter exponential approximation (APEX) model is expected to do better as far as strong coupling aspects are concerned, but cannot treat ions of finite size, having internal structures.

This brings us to item (iv) of existing methods, viz., the assumption that ions are point like. In low density plasmas where ions are well separated, the point-ion model is generally satisfactory. In a dense plasma, viz. the Al-plasma of figure 1, (r_s = 3, Γ = 3.47, $\bar{Z}$ = 5.18) the Wigner-Seitz radius r_{WS} = 5.2 a.u. In table 1 we give the electronic structure of the Al-ion and the extension of the various orbitals. It is in fact clear that there is considerable exchange (or tunnelling) of bound electrons between neighbouring Wigner-Seitz cells. The cell model is only appropriate at zero temperature. In the DFT calculation the self-consistently determined ion distribution has been used in calculating the electronic structure. (In Liberman's calculations[11] the ion profile is approximated by a cavity around the central ion, and a uniform background outside the cavity.)

In fig. 3 we show the DFT-microfield distributions calculated using 2nd order B-M theory, at a neutral test point and at a charged test point (Z=1), placed in the Al-plasma with Γ = 3.47, $\bar{n}$ = 5.9 × 10^{21} electrons/cc. This corresponds to a microfield parameter $r_{mf} \simeq r_s/\lambda$ = 0.8. (This parameter is called "a" in Hooper's papers.) We also show the results obtained using the "adjustable parameter exponential approximation" (APEX) method of Iglesias et al.[6] In this method the electric field distribution is fitted, using an adjustable parameter, to approximately satisfy the 2nd moment sum rule for the field distribution. This adjustment procedure probably corrects for some of the errors that may be contained in the underlying physical picture.

As seen from fig. 3, the "all-order" APEX microfield for a neutral point in the Al-plasma is in good agreement with the 2nd order B-M result of DFT. The charged test point results from DFT and APEX do not agree so well. Here we believe that DFT does a better job of treating the test point-field ion pair correlations than does APEX. However, more detailed studies are needed at this stage and will not be taken up here.

In the example given we see that no all-order summation seems to be required. In the all-order resummation of the B-M series given by Perrot and Dharma-wardana,[7] the higher order cluster functions are expressed in terms of 2nd order distribution functions (e.g., (123) → g(12)g(23)g(31)) using a Kirkwood-like superposition approximation. Only chain-like terms, (as in HNC) in the resulting development is retained. This chain approximation to the full B-M series is then improved by demanding that the chain approximation to the n+1th cluster, when integrated over the coordinates of the n+1th atom, reduce correctly to the nth cluster function. This <u>weighted chain sum</u> (WCS) can be easily and efficiently computed as it involves only pair distribution functions etc., of the sort already needed in evaluating the 2nd order B-M result. The WCS approximation has been tested against Monte-Carlo results for the OCP (for $\Gamma < 10$) and found to be quite satisfactory. Hence we can say that the DFT calculation of the 2nd order B-M microfield for a plasma consisting of ions with internal structure, strong V_{ei} interaction etc., can, if necessary, be extended to an all-order evaluation quite efficiently. The theory does not use any adjustable parameters at any stage.

Conclusion

We have argued that the DFT calculations for plasmas can be exploited for going beyond the Debye-fluid type physical model assumed by existing microfield theories. The existing theories may be quite adequate for low density plasmas and even some high density situations. But the DFT model is particularly useful if the electron-ion interaction is considered to be strong, and if internal structure of the ions have to be taken into account.

References

1. See H.R. Griem, Spectral line broadening by plasmas (Academic, New York, 1974)

2. M.W.C. Dharma-wardana and F. Perrot, Phys. Rev. A26, 2096 (1982); F. Perrot and M.W.C. Dharma-wardana, Phys. Rev. A29, 1378 (1984)

3. M.W.C. Dharma-wardana, F. Perrot and G.C. Aers, Phys. Rev. A28, 344 (1983)

4. M. Baranger and B. Mozer, Phys. Rev. 115, 521 (1959); 118, 626 (1960). See H. Pfennig and E. Trefftz, Z. Naturforsch 21A, 697 (1966)

5. C.F. Hooper, Jr., Phys. Rev. 149, 177 (1966); C.A. Iglesias and C.F. Hooper, Jr., Phys. Rev. A25, 1049 (1982)

6. C.A. Iglesias, J.L. Lebowitz and D. McGowan, Phys. Rev. A28, 1667 (1983); C.A. Iglesias and J.L. Lebowitz, Phys. Rev. A30, 2001 (1984)

7. F. Perrot and M.W.C. Dharma-wardana, Physica A (Stat. Phys.) 134, 231 (1985)

8. M.W.C. Dharma-wardana and F. Perrot, Phys. Rev. A xxx, xxx (1986)

9. See W.A. Harrison, Pseudopotentials in the theory of metals (Benjamin, New York 1966).

10. M.W.C. Dharma-wardana and G.C. Aers, Phys. Rev. B28, 1701 (1983)

11. D.A. Liberman, J. Quant. Spectrosc. Radiat. Transfer 27, 335 (1982)

Figure Captions

Fig. 1 Comparison of the electric field at an Al-ion in an Al-plasma ($\bar{n}$ = 5.9 × 10^{21} electrons/cc, Γ = 3.47, r_{mf} = 0.8) calculated with (i) Debye screening (short dashes), (ii) equation (2.2) of DFT (solid line), and (iii) the gradient of the pair potential/Z (long dashes). The right hand panel shows the electric field weighted by r^2, i.e. $r^2 E(r)$.

Fig. 2 The microfields at a neutral point in an Al-plasma with r_{mf} = 0.8, electron sphere radius r_s = 2 calculated using DFT and Hooper's theory. For these plasma conditions, the electric field calculated using eqn. (2.1) IS closely approximated by a Debye model.

Fig. 3 Comparison of DFT and APEX microfields at a neutral point and at a charged point in an Al-plasma with r_s = 3, r_{mf} = 0.8.

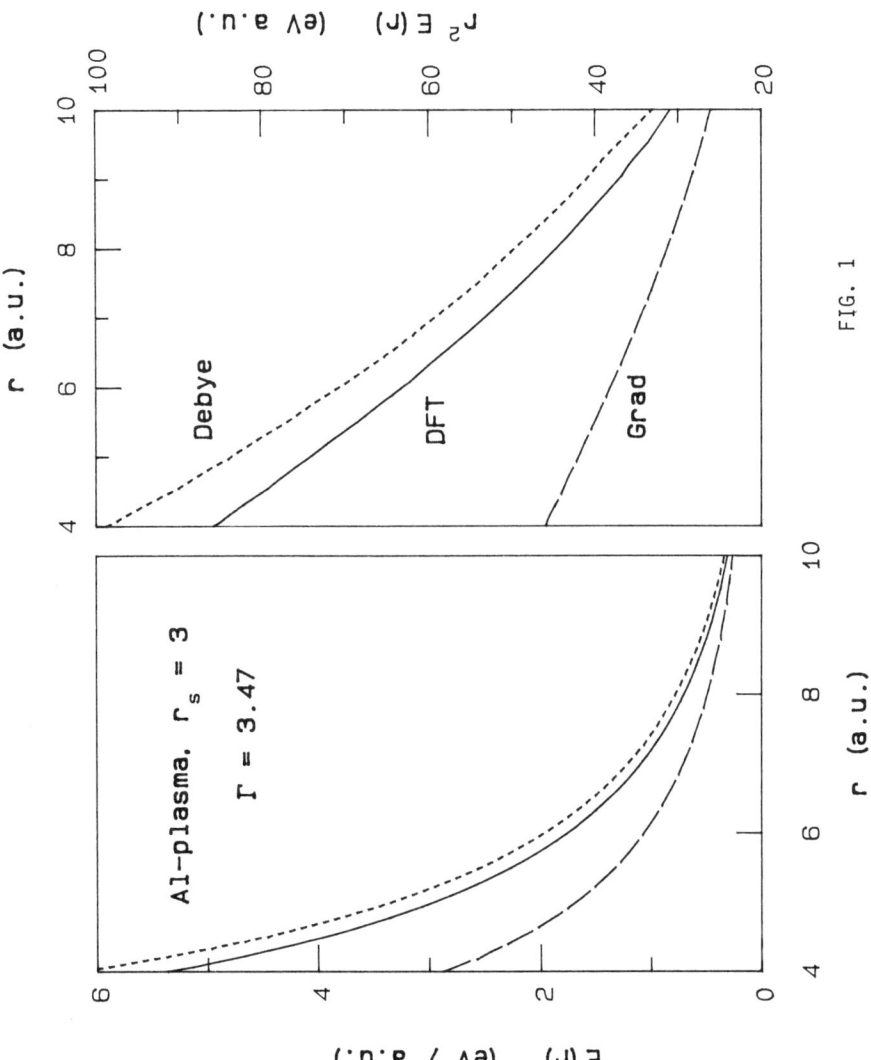

FIG. 1

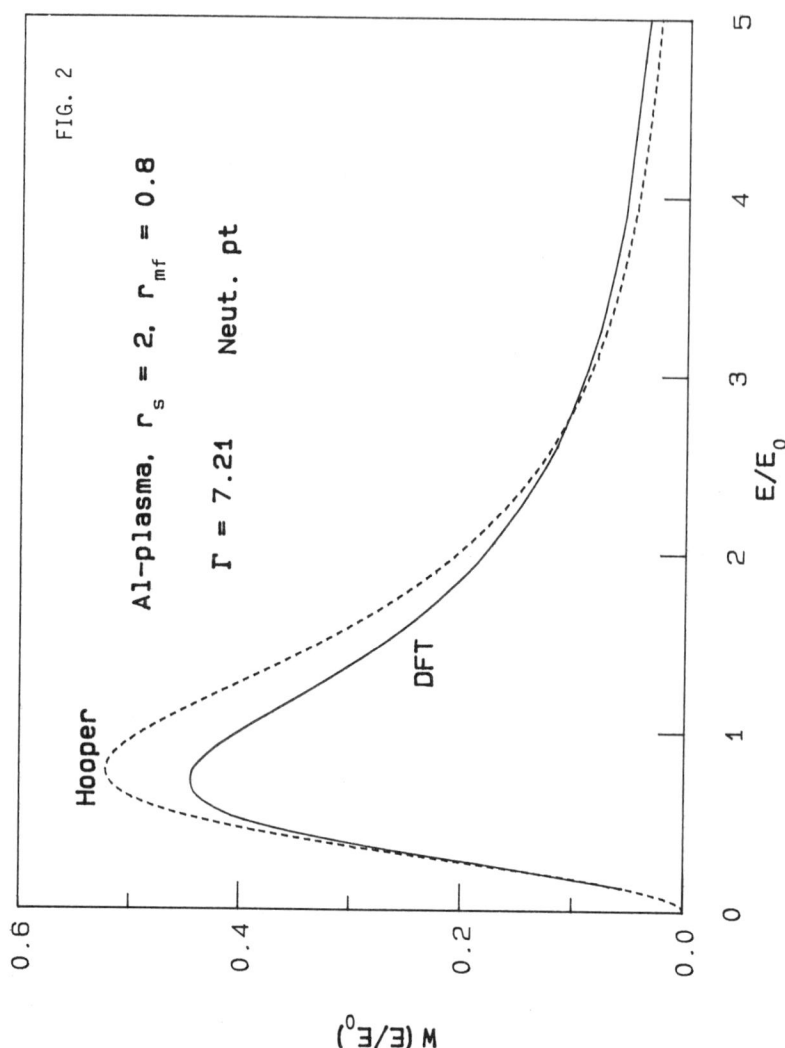

FIG. 2

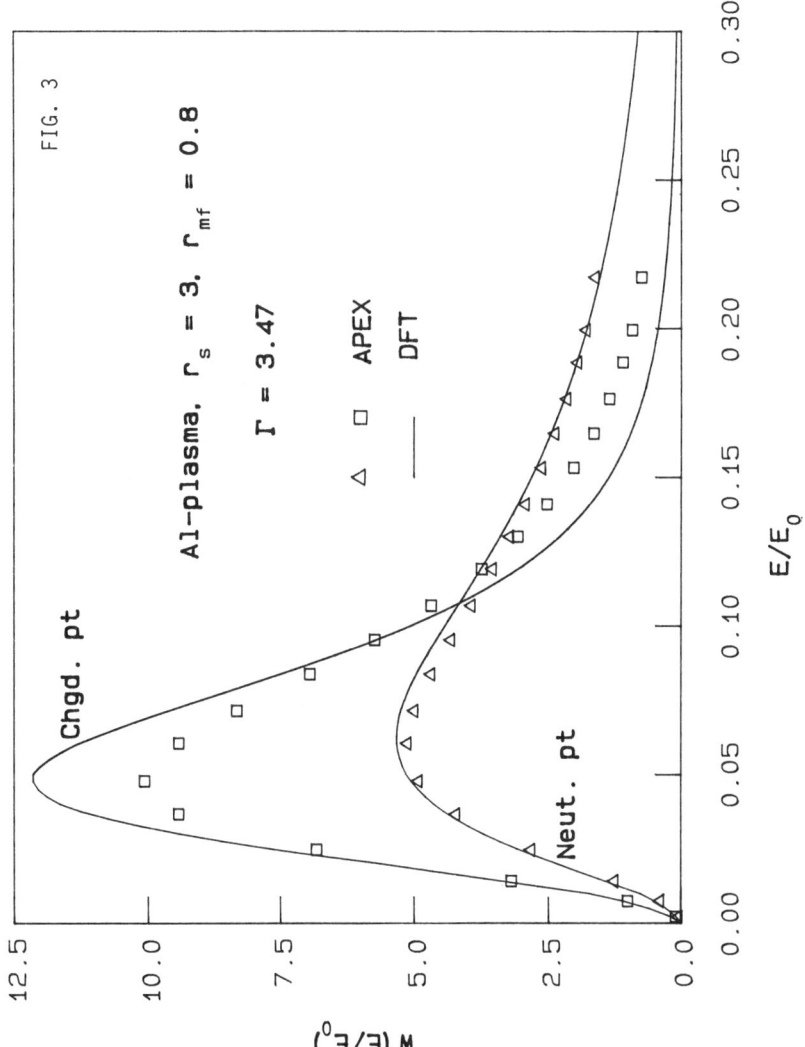
FIG. 3

Table 1. Energy level structure and orbital extension of an $Al^{+\bar{z}}$ ion in (a) H^+-plasma at $r_s = 2.0$, $T/E_F = 5.08$, (b) Al-plasma at $r_s = 3.0$, $T/E_F = 7.624$. The effective charge of Al in (a) and (b) is 5.540 and 5.178 respectively. The Wigner-Seitz radius in (a) and (b) is 3.0 and 5.19 a.u. respectively.

level	energy (a.u.)		occupation		radius (a.u.)	
	(a)	(b)	(a)	(b)	(a)	(b)
1s	−60.249	−59.418	1.000	1.000	0.120	0.121
2s	−7.2160	−6.8330	0.595	0.743	0.563	0.574
2p	−6.0515	−5.6100	0.472 × 3	0.569 × 3	0.505	0.519
3s	−1.1095	−1.0655	0.097	0.067	1.610	1.637
3p	−0.7970	−1.2505	0.86 × 3	0.055 × 3	1.669	1.704
3d	−0.3570	−0.3205	0.073 × 5	0.043 × 5	1.697	1.738
4s	−0.0605	−0.0075	0.065	0.035	4.686	6.014
4p	−0.0028	−	0.063 × 3	−	8.306	−

SOLID DENSITY, LOW TEMPERATURE PLASMA FORMATION IN A CAPILLARY DISCHARGE

D. R. Kania, L. A. Jones
and M. D. Maestas
Physics Division
Los Alamos National Laboratory
Los Alamos, NM 87545

R. L. Shepherd
Department of Nuclear Engineering
The University of Michigan
Ann Arbor, MI 48109

ABSTRACT

We have been able to produce solid density, low temperature plasmas in polyurethane capillary discharges. The initial capillary diameter is 20 µm. The plasma is produced by discharging a one Ohm parallel plate waterline and Marx generator system through the capillary. A peak current of 340 kA in 300 ns heats the inner wall of the capillary, and the plasma expands into the surrounding material. We have studied the evolution of the discharge using current and voltage probes, axial and radial streak photography, axial x-ray diode array and schlieren photography. We have estimated the peak temperature of the discharge to be approximately 10 eV and the density to be near $10^{23} cm^{-3}$. This indicates that the plasma may approach the strongly coupled regime. We will discuss our interpretation of the data and compare our results with theoretical models of the plasma dynamics.

INTRODUCTION

Capillary discharges with diameters in excess of 100 μm have been used to produce soft x-rays (1) and as wall stabilized z-pinches for fusion applications (2). By reducing the initial diameter of the capillary to 20 μm in polyurethane and discharging 340 kA of current in 300 ns through it, a low temperature (10 eV), solid density plasma has been produced. With this source we can measure the resistivity of polyurethane in this temperature and density range.

The dynamics of a capillary discharge can be modeled using a two zone model: a hot inner core and cold capillary material. The column of heated material is constantly expanding by acreating the cold material adjacent to it. Very quickly the size of the column exceeds the initial diameter and a near solid density (the initial void is negligible) plasma continues to expand. In the present discussion a preliminary analysis of the first 100 ns of the discharge will be presented.

EXPERIMENT

The capillaries are 2 cm long and 20 μm in diameter in polyurethane as shown in Figure 1. The capillary is attached to a Marx bank and water line pulse generator which delivers 340 kA in 300 ns to the capillary. The cathode is pointed to assist in guiding the discharge down the capillary which is evacuated from the anode side where the pressure is 10^{-5} torr. The evolution of the discharge was monitored with several diagnostics: current and voltage probes, axial and radial visible streak cameras, laser schlieren, and a four channel axial x-ray diode (XRD) array. The physical orientation of the diagnostics is shown in Figure 2.

The current and voltage are measured with a Rogowski coil and a capacitive voltage probe respectively and are displayed in Figure 3. This data can be used to determine the resistivity of the plasma as a function of time as discussed later in this paper. The radius versus time history of the plasma may be ascertained from the expansion of the luminous front measured with axial and radial streak photography (1 ns time resolution). This correlates well with the expansion velocity measured using laser schlieren photography. Two different laser pulses were used in the schlieren system: 1) a 20 ns ruby laser pulse and 2) a 100 ps frequency doubled Nd glass laser pulse. The best fit to all of the data in the first 100 ns is a constant average expansion velocity of 5.8×10^5 cm/s. Both the luminous expansion front velocity and the schlieren radius (density gradient dependent) agree within experimental errors. The temperature of the plasma is inferred from an axial measurement of the x-ray emission using a 4 channel XRD array. The ratio technique (4) was used to determine an effective blackbody temperature for the three pairs of channels. The fourth channel was used to monitor high energy radiation and was not used in the temperature measurement.

The three temperature versus time plots are shown in Figure 4a. A 50 percent variation in the measured temperatures is observed. The bulk of this variation may be due to uncertainties in the filter transmissions and photocathode quantum efficiency which are not well known in this photon energy range. We will use the average of these three measurements which varies between 7 and 10 eV for the first 100 ns as the temperature (see Figure 4b).

Neglecting end losses and using particle conservation, we can estimate the density of the plasma. After approximately 10 ns the initial void is filled with wall material and can be neglected, the particle density equals solid density, approximately 10^{23} cm^{-3}. An estimation of the average ionization of the plasma can be made using a Saha model. No density effects are included in the calculation, this results in an underestimate of the average ionization of the plasma. The calculated average ionization is 0.5 for the first 100 ns. This yields an electron density and ion density of 5×10^{22} cm^{-3}. Using the experimental values, it is interesting to note that this material is near the strongly coupled plasma regime (4).

MODEL

A simple model of the evolution of the discharge can be formulated if we assume that the plasma is 1) uniform in temperature and density, 2) Ohmically heated and 3) confined within a boundary defined by the measured radius. In addition we assume that all of the energy dissipated in the plasma remains in the plasma. Under these restrictions the temperature at a given time can be determined by equating the enthalpy to the Ohmic dissipation,

$$H(T) = \int_{T_0}^{T} CdT = \int \frac{dt\ I^2(t)\ R(T)}{V(t)} \qquad (1)$$

where $H(T)$ is the enthalpy at a temperature T, T_0 is the initial temperature, $I(t)$ is a specified driving current, $R(T)$ is the temperature dependent plasma resistance, $V(t)$ is the time dependent volume of the plasma column or $\pi r^2 \ell$, and ℓ is the length of the plasma column. Power laws are used to describe the time and temperature dependences of the relevant quantities. The radial expansion has the form

$$r(t) = at^M,$$

where a and M are constants and if M = 1, a is the expansion velocity. Spitzer resistivity is assumed and has the form

$$\eta = bT^{-3/2}$$

where b is a constant. We neglect the effects of changes in the average ionization and the Coulomb logarithm. Finally, the specific heat, which can be integrated to yield the enthalpy, is of the form

$$C = dT^P$$

where d and P are constants. For P=0.5 we have a high temperature plasma model (5). For a sine wave current pulse, equation one can be separated and integrated. A formal solution of the time dependence of the temperature is

$$T(t) = \left\{ \left(\frac{3}{2A} + 1 \right)_t -\gamma \int_0^t dt'\, t'^{\gamma-4M} \sin^2(\omega t') \right\}^{\left(\frac{1}{A + 3/2} \right)}$$

The constants are defined as

$$\delta = \frac{\pi^2 d\, a^4}{A\, I_o^2\, b},$$

$$A = P + 1$$

$$\gamma = \left(\frac{3}{2A} + 1 \right) \left(2M - 1 \right)$$

We note that this result diverges at t=0, but we do not expect the model to be valid at very early times because the initial void is neglected and the physics of the initiation of the channel is not included. The model excludes all loss mechanisms and does not include the detailed physics of the accretion process which is lumped into r(t).

We can use the experimentally determined radius vs. time information, i.e. constant velocity expansion, Spitzer resistivity and a high temperature plasma model for the enthalpy and calculate the time dependent temperature. The use of the above models for the resistivity and specific heat are not completely accurate for the present conditions but the model provides a useful guide for experimental interpretation. Using the above parameters the model underestimates the temperature of the plasma when compared to experiment, this is the standard model that appears in Figure 5. In an attempt to improve the agreement between the model and experimental temperature we raised the resistivity by a factor of 80 to force agreement with the data at 60 ns. Other parameters could have been modified e.g. the specific heat, but the use of Spitzer resistivity is probably the poorest assumption in the model. With this ad hoc correction we see excellent agreement between the data and calculation for late times (see Figure 5).

RESISTIVITY CALCULATION

The measurements can be used to calculate the resistivity of the plasma as a function of time. Following Manheimer (6), the resistivity of the plasma can be calculated from measurements of the voltage, first derivative of the current, I, the radius of the plasma and the radius of the return conductor, r_o. The measured resistivity is

$$\eta_M = \left[\frac{1}{I^2 \ell}\right] \left[(IV) - \frac{2I\dot{I}\ell}{c^2} \left(\ln\left(\frac{r_o}{r}\right)\right) \pi r^2\right]$$

where c is the speed of light. This technique ignores the effects of velocity dependent voltages. A plot of resistivity versus time is shown in Figure 6. We note that the calculated resistivity is in qualitative agreement with the resistivity used in the modified model from the previous section. It is not unreasonable to assume that the resistivity of the plasma under these conditions should exceed the values calculated from Spitzer (7). At this time we wish to draw no firm conclusions from this correlation.

CONCLUSIONS

A near solid density, low temperature plasma can be generated in a capillary discharge of small initial diameter. By further theoretical and experimental study we may be able to infer the resistivity of such a plasma. This will require more detailed measurements and improved modeling, especially of the transport coefficients and average ionization of the plasma.

ACKNOWLEDGEMENTS

This work has been performed under the auspices of the USDOE. The authors would like to thank Physics Division of the Los Alamos National Laboratory for supporting this work.

REFERENCES

1. R. A. McCorkle, Il Nuovo Cimento, 77B, 31 (1983).

2. J. D. Sethian, K. A. Gerber, A. E. Robson and A. W. Desilva, "NRL Dense Z-Pinch Experiments," Proceeding of the Dense Z-Pinches for Fusion Meeting, ed. J. D. Sethian, Washington, D. C., 1984.

3. F. C. Jahoda, E. M. Little, W. E. Quinn, G. A. Sawyer, and T. F. Stratton, Phys. Rev. 119, 843 (1960).

4. S. Ichimaru, Rev. Mod. Phys. 54, 1017 (1982).

5. Ya. B. Zel'Dovich and Yu. P. Raizer, "Physics of Shock Waves and High Temperature Hydrodynamic Phenomena," Academic Press, New York, pp. 201-209, 1966.

6. W. M. Manheimer, Phys. Fluids 17, 1767 (1974).

7. R. Cauble and W. Roz mus, submitted to Journ. Plasma Phys. (1986).

FIGURES

Figure 1:   A drawing of the polyurethane capillary and its position in the experiment. The cathode is pointed to guide the discharge down the capillary. The system is evacuated from the anode side to a pressure of 10^{-5} Torr. The volume around the polyurethane is filled with SF_6 to improve the voltage hold off characteristics. Flat windows are machined into the outside of the capillary block to simplify the radial optical diagnostics.

Figure 2:   A schematic drawing of the diagnostics used in the experiment is shown. The diagnostics include a four channel x-ray diode array, schlieren system and visible streak cameras.

Figure 3:   A plot of current through capillary and voltage across the load versus time for a single discharge is shown.

Figure 4a: Three plots of blackbody temperature versus time derived from the ratio technique for three pairs of x-ray filters. The two filters used are 7500 A Al, 60 $\mu g/cm^2$ of formvar. The third channel has no filter (open).
Figure 4b is the average temperature versus time for the results presented in 4a.

Figure 5:   Three plots of temperature versus time for a capillary discharge are presented: 1) the standard model, 2) the modified model and the average temperature from figure 4b. See the text for details.

Figure 6:   A plot of the measured resistivity (———) versus time is shown. The resistivity was determined from the measured current, voltage and radius according to reference 6. Also included in the plot are Spitzer resistivity (-·-·-·) for these conditions and the resitivity used in the modified model, 80 times the Spitzer resistivity (--------).

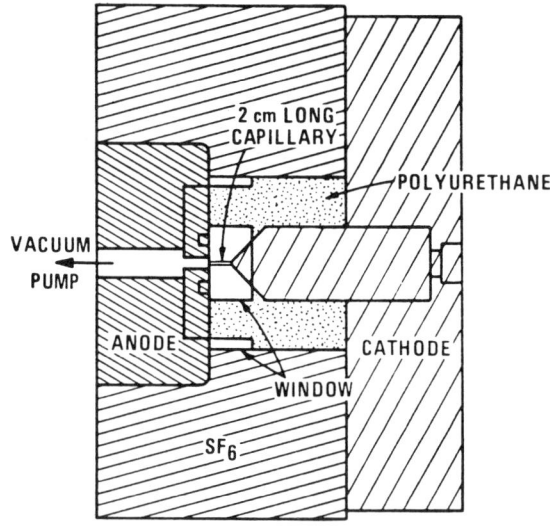

FIGURE 1

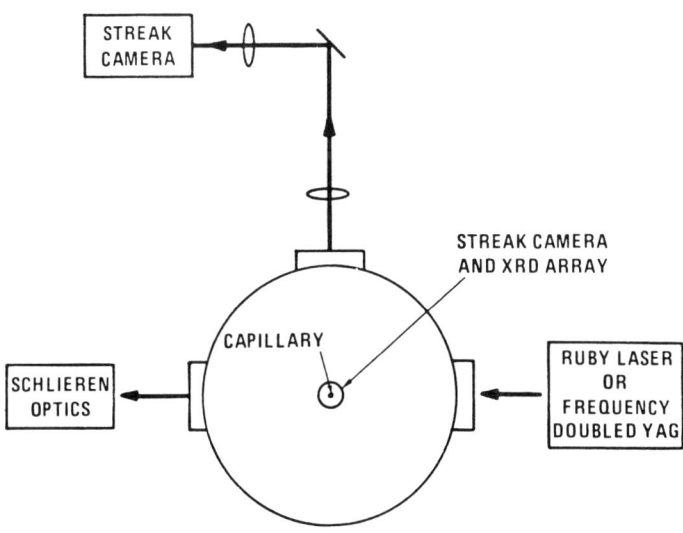

FIGURE 2

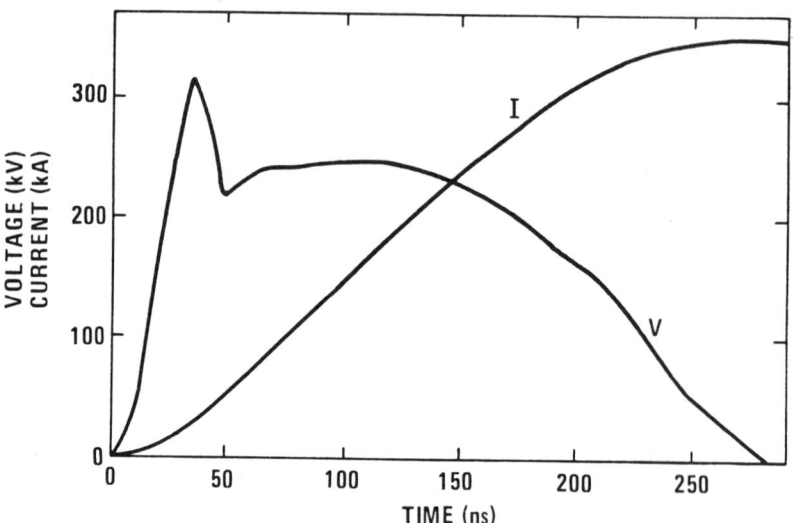

FIGURE 3

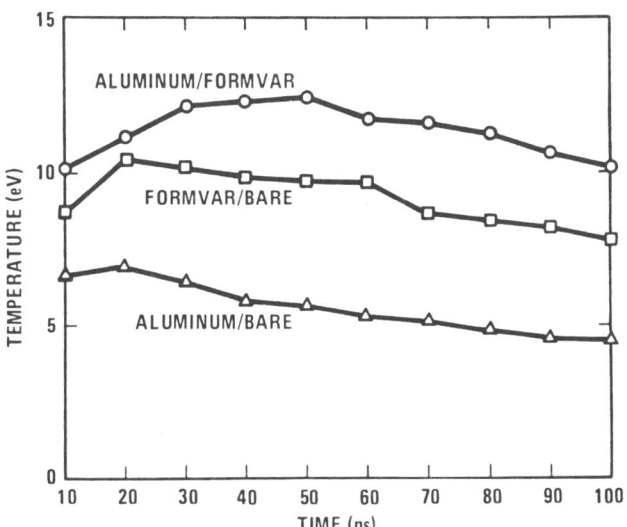

FIGURE 4A

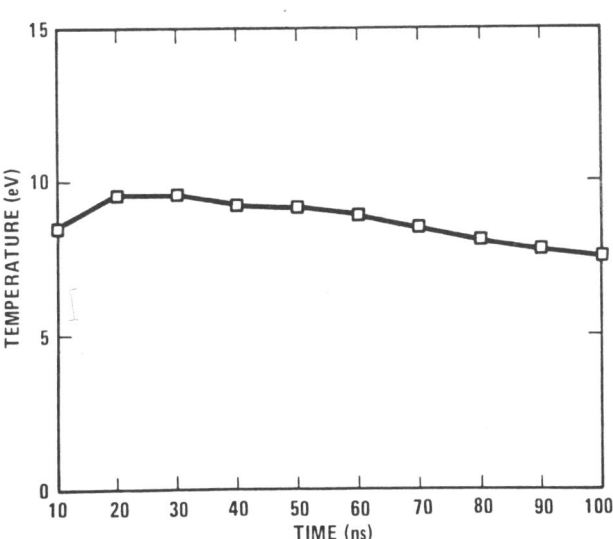

FIGURE 4B

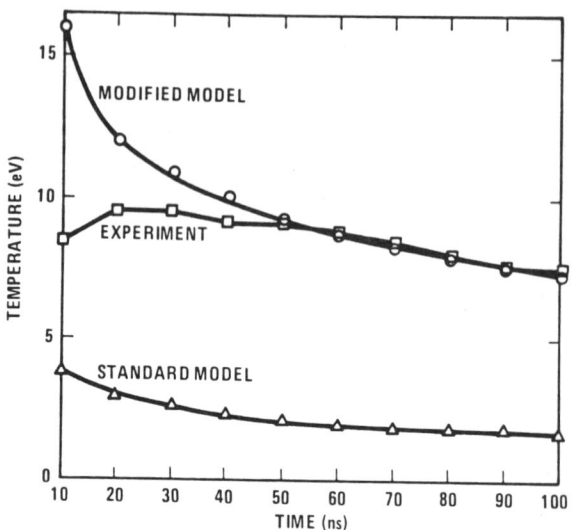

FIGURE 5

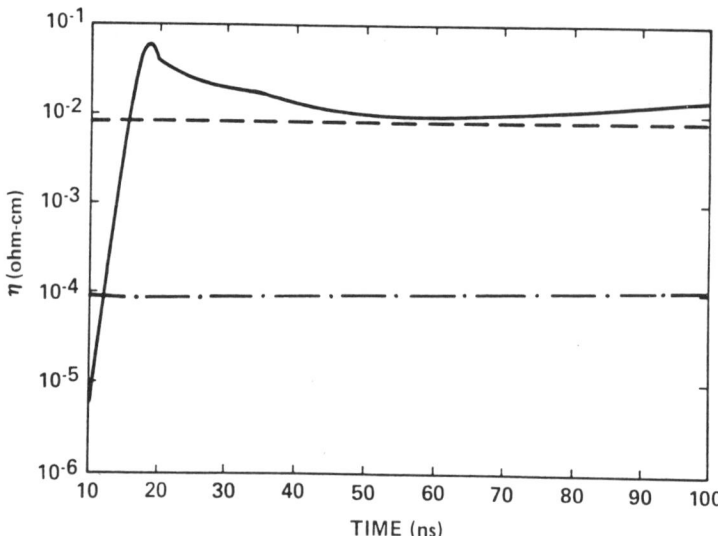

FIGURE 6

A Time-Dependent Ionization Balance
Model for Non-LTE Plasma*

Y.T. Lee, G. B. Zimmerman, D. S. Bailey,
D. Dickson, and D. Kim

University of California
Lawrence Livermore National Laboratory
P. O. Box 808, Livermore, CA 94550

ABSTRACT

We developed a detailed configuration accounting kinetic model which is useful for calculating time-dependent ionization balance and ion level populations in non-LTE plasmas. The model is also self-consistently coupled to the hydrodynamic code LASNEX to design laboratory x-ray laser experiments. Level populations estimated using the model are used to compute spectral line intensities, line ratios, and synethic spectra to fit experimental measurement.

---

* Work performed under the auspices of the U. S. Department of Energy by the Lawrence Livermore National Laboratory under contract number W-7405-ENG-48.

I. INTRODUCTION

Many laboratories are actively investigating radiation spectra emitted from plasmas produced using high-power laser or gas-puff z-pinch. Spectral-line intensities from these plasmas are useful temperature and density diagnostics. In a well-diagnosed plasma, information on the emission spectra can also be used to test the accuracy of theoretical rate coefficients. To analyze the emission spectra, one needs to know the charge-state distribution and ion level populations in a plasma as a function of electron temperature and density.

Since the plasmas of interest are usually produced with electron density $10^{17} - 10^{21} cm^{-3}$, we expect that neither the coronal equilibrium nor Saha-Boltzmann equilibrium can be applied. The coronal model which calculates ionization balance by assuming that the ions are in their ground states can be used only for plasmas with electron density less than $10^{16} cm^{-3}$. In this paper we discuss a detailed configuration accounting kinetic model which is useful for calculating time-dependent ionization balance and ion level populations of non-LTE plasmas produced using high-power laser or gas puff Z pinch.

The model has two options for calculating ion level populations. The default option treats all the ionization stages in a plasma using hydrogenic data. The other option treats certain ionization stages using detailed non-hydrogenic energy levels and rate coefficients.

In the default option, we assume that each energy level couples with all the levels of the same ionization stage, but only to the ground state of the next ionization stage through ionization and recombination processes. This approximation allowed us to develop a particularly efficient method of solving for the populations. We generate the ground state and excited state energy levels for each ionization stage by using screening constants. There is one energy level per principal quantum number. The following atomic processes are included: (1) electron collisional excitation and de-excitation, (2) radiative emission and absorption, (3) electron collisional ionization and photoionization, (4) radiative recombination, (5) three-body recombination, and (6) dielectronic recombination. All the rate coefficients are scaled from the results for hydrogenic ions. Using the default option, our model is useful for calculating charge-state distribution and radiative power of arbitrary non-LTE plasmas.

In the other option, the code obtains the energy levels and collisional and radiative data for certain ionization stages from a data file. The data file contains energy levels, oscillator strengths, collisional excitation and ionization rate coefficients, photo-ionization cross sections, and auto-ionization rates. This option makes it possible to apply the model to analyze emission spectra of non-LTE plasmas and to simulate laboratory x-ray laser experiments.

Our model is also self-consistently coupled to the hydrodynamic code LASNEX.[1] We use the electron temperature, electron density, and photon density from LASNEX to compute ion level populations. The average ionization state, emissivity, photon absorption cross section, and binding energy are calculated using the new ion level populations. LASNEX uses these quantities in the equations of radiation transport, hydrodynamics, and electron energy and number balance to update the electron temperature, electron density, and photon density for the next time step.

In the next section, we discuss the hydrogenic option of our model. Both time-dependent ionization balance and steady-state ionization balance are discussed in detail. We also compare our results for average ionization state and charge-state abundance to an average atom model and recent experimental measurement. In Section III, we discuss the non-hydrogenic option of our model. As an application, we use our model to validate a method proposed to estimate ionization balance data from experimentally measured emission spectra.

II. HYDROGENIC APPROXIMATION

A. Time-Dependent Ionization Balance

In the hydrogenic approximation, the populations for the ground state N_i^1 and excited state $N_i^\ell (\ell>1)$ of the ion at its i-th ionization stage are described by the following equations:

$$\frac{dN_i^\ell}{dt} = -R_i(\ell)N_i^\ell + \sum_k t_i^{\ell k} N_i^k + S_{i+1}(\ell) N_{i+1}^1 + \sum_k \left[R_{i-1}(k) N_{i-1}^k - S_i(k) N_i^1 \right] \delta_{\ell,1} \quad , \quad (1)$$

where $t_i^{\ell k} = (n_e C_{k\ell}^e + B_{k\ell} U_{h\nu}) \quad \ell > k$,

$t_i^{\ell k} = (n_e C_{k\ell}^d + A_{k\ell} + B_{k\ell} U_{h\nu}) \quad \ell < k$,

$t_i^{\ell\ell} = -\sum_{k<\ell}(n_e C_{\ell k}^d + A_{\ell k} + B_{\ell k} U_{h\nu}) - \sum_{k>\ell}(n_e C_{\ell k}^e + B_{\ell k} U_{h\nu})$,

and n_e = electron density. The symbol $A_{k\ell}$ is the Einstein coefficient for spontaneous emission from state k to state ℓ. The symbol $B_{k\ell}$ is the Einstein stimulated emission coefficient for a transition induced by radiation of energy density $U_{h\nu} d(h\nu)$ in the energy range $h\nu$ to $h\nu + d(h\nu)$.

The matrix t_i couples all the energy levels within the i-th ionization stage by collisional or radiative transition. $C_{\ell k}^e$ and $C_{\ell k}^d$ are respectively the collisional excitation and de-excitation rate coefficients for a transition between level ℓ and level k. The term $R_i(\ell)$ represents the total ionization rate from level ℓ including electron collisional ionization and photoionization. The total recombination rate $S_{i+1}(\ell)$ includes radiative recombination, three-body recombination, and dielectronic recombination. The term $\sum_k R_{i-1}(k) N_{i-1}^k$ represents the increase of the ground state population due to ionization. The term $\sum_k S_i(k) N_i^1$ represents the decrease of the ground state population due to recombination. These two terms exactly cancel each other when the plasma reaches steady-state equilibrium. At steady-state the total recombination rate of a given ionization stage is equal to the total ionization rate of the lower ionization stage. Although in this model each excited level is coupled to the ground state of the next ionization stage only, the coupling to the excited states of the next ionization stage is, in general, negligible for electron densities less than 10^{22} cm^{-3}.

The equation for excited state level populations can be rewritten as:

$$\frac{dN_i^\ell}{dt} = -R_i(\ell)N_i^\ell + \sum_{k>1} \tilde{t}_i^{\ell k} N_i^k + S_{i+1}(\ell) N_{i+1}^1 \qquad (2)$$

$$+ (n_e C_{1\ell}^e + B_{1\ell} U_{h\nu}) N_i^1 \qquad \ell > 1$$

where the matrix $\tilde{t}_i$ couples the excited states together. The coefficients $C_{1\ell}^e$ and $B_{1\ell}$ are the electron collisional excitation rate and the radiative absorption cross section, respectively, for the transition from ground state to level ℓ.

By solving Eq. (2), we obtain the following difference relation for the excited state level population

$$(N_i^\ell)^{t+\Delta t} = (c_i^\ell)^t + \Delta t [a_i^\ell (N_i^1)^{t+\Delta t} + b_i^\ell (N_{i+1}^1)^{t+\Delta t}] \qquad (3)$$

where Δt = time step for the kinetic calculation. The term c_i^ℓ is given in terms of the level populations. Both the coefficients a_i^ℓ and b_i^ℓ are expressed in terms of the matrix $\tilde{t}_i$ and the ionization and recombination rate coefficients.

Substituting Eq. (3) into Eq. (1) gives

$$(N_i^1)^{t+\Delta t} = \left[\Delta t\, F_i^0 + F_i^1 (N_{i-1}^1)^{t+\Delta t} + F_i^3 (N_{i+1}^1)^{t+\Delta t} \right] \qquad (4)$$

$$/ (1 - \Delta t\, F_i^2)$$

where the coefficients F_i^0, F_i^1, F_i^2, and F_i^3 depend on a_i^ℓ, b_i^ℓ, c_i^ℓ, and the ionization and recombination coefficients. Equation (4) for the ground state populations can be easily solved using any method for a tridiagonal matrix.

B. Steady-State Ionization Balance

In steady-state equilibrium, the level populations are determined by the solution of Eq. (1) when dN_i^ℓ/dt is set to zero for every level. Eq. (1) then reduces to the following form:

$$\sum_k (T_i^{\ell k}) N_i^k = S_{i+1}(\ell) N_{i+1}^1 \quad , \tag{5}$$

where

$$T_i^{\ell k} = R_i(\ell) \delta_{\ell k} - t_i^{\ell k} \quad .$$

By solving the level populations in terms of the inverse of the matrix T_i, we have

$$N_i^k = \sum_\ell (T_i^{-1})^{k\ell} S_{i+1}(\ell) N_{i+1}^1 \quad . \tag{6}$$

Using these results for the level populations, we write the total ionization rate of the i-th ionization stage as

$$\alpha_i N_{i+1}^1 = \sum_{k,\ell} R_i(k)(T_i^{-1})^{k\ell} S_{i+1}(\ell) N_{i+1}^1 \tag{7}$$

and the total recombination rate for the i+1-th ionization stage as

$$\beta_{i+1} N_{i+2}^1 = \left[\sum_k S_{i+1}(k) \right] \left[\sum_\ell (T_{i+1}^{-1})^{\ell\ell} S_{i+2}(\ell) \right] N_{i+2}^1 \quad . \tag{8}$$

At steady-state equilibrium, the total ionization rate of the i-th ionization stage is exactly equal to the total recombination rate of the i+1-th ionization stage. The results are

$$\alpha_i N_{i+1}^1 = \beta_{i+1} N_{i+2}^1 \quad 1 \leq i \leq Z_n - 1 \quad , \tag{9}$$

where Z_n = the nuclear charge of the ion.

The excited state populations are determined from solutions of Eqs. (6) and (9). The results are

$$N_i^\ell = \left[\sum_k (T_i^{-1})^{\ell k} S_{i+1}(k)\right](\beta_{i+1}/\alpha_i)(\beta_{i+2}/\alpha_{i+1})\cdots(\beta_{Z_n}/\alpha_{Z_n-1}) N_{Z_n+1}, \quad (10)$$

$$N_{Z_n+1} = \frac{N_t}{1 + \sum_{i\ell k}(T_i^{-1})^{\ell k} S_{i+1}(k)(\beta_{i+1}/\alpha_i)(\beta_{i+2}/\alpha_{i+1})\cdots(\beta_{Z_n}/\alpha_{Z_n-1})},$$

where N_t = total ion density and $1 \leq i \leq Z_n$

If we are only interested in the total population of an ionization stage, we can use Eqs. (6) through (8) to write the total ionization rate as

$$\phi_i N_i = \frac{\alpha_i}{\sum_{\ell k}(T_i^{-1})^{\ell k} S_{i+1}(k)} N_i \quad , \qquad (11)$$

and the total recombination rate as

$$\psi_{i+1} N_{i+1} = \frac{\beta_{i+1}}{\sum_{\ell k}(T_{i+1}^{-1})^{\ell k} S_{i+2}(k)} N_{i+1} \quad , \qquad (12)$$

where $N_i = \sum_\ell N_i^\ell$. Both ϕ_i and ψ_i represent the total ionization rate and recombination rate, respectively. These rates depend on the nuclear charge, ionization state, electron temperature, and electron density.

The populations of each ionization stage can be obtained from the following relations:

$$\phi_i N_i = \psi_{i+1} N_{i+1} \quad , \quad 1 \leq i \leq Z_n \quad . \qquad (13)$$

The solution for this set of equations is written as

$$N_i = (\psi_{i+1}/\phi_i)(\psi_{i+2}/\phi_{i+1})\cdots(\psi_{Z_n+1}/\phi_{Z_n}) N_{Z_n+1} \quad , \qquad (14)$$

where $N_{Z_n+1} = N_t / \left[1 + \sum_{i=1}^{Z_n}(\psi_{i+1}/\phi_i)(\psi_{i+2}/\phi_{i+1})\cdots(\psi_{Z_n+1}/\phi_{Z_n})\right]$.

At very high election density, where electron collisional de-excitation rate dominates over the radiative spontaneous emission rate, Eq. (13) reduces to Saha-Boltzmann equilibrium of the form

$$N_i/N_{i+1} = (G_i n_e \lambda_e^3 / 2G_{i+1}) \, e^{I_i/kT_e} \quad , \tag{15}$$

where $G_i = \sum g_{i,\ell} \, e^{-E_i^\ell/kT_e}$,

$g_{i,\ell}$ = statistical weight for level ℓ of i-th ionization stage,
E_i^ℓ = excitation energy for level ℓ of i-th ionization stage,
I_i = ionization potential of i-th ionization stage,
T_e = electron temperature,
λ_e = free electron thermal de-Broglie wavelength.

At very low electron density, where radiative spontaneous emission rate dominates over electron collisional de-excitation rate, Eq. (13) reduces to coronal equilibrium of the form,

$$N_i/N_{i+1} = \sum_\ell S_{i+1}(\ell)/R_i(1) \tag{16}$$

where $S_{i+1}(\ell)$ is the total recombination rate including radiative recombination and dielectronic recombination.

C. Atomic Model

Each level of an ion is described by a set of occupation numbers $\{P_m\}$. The symbol P_m which equals to $0, 1, 2, \ldots 2m^2$ gives the number of electrons in the m-th shell of the ion. As an example, the ground level of a Ne-like ion is represented by $P_1=2$, $P_2=8$, $P_m=0$, $m \geq 3$, and the first excited state is represented by $P_1=2$, $P_2=7$, $P_3=1$, $P_m=0$, $m \geq 4$. The principal quantum number of a level is determined by its valance electron. This model neglects all the $\Delta n=0$ transitions and assumes the sublevels are in statistical equilibrium.

We calculated energy levels for each ionization stage using hydrogenic screening constants. The ionization potential for a level ℓ of i-th ionization stage is taken as

$$I_i^\ell = E[\{P_m\}_i] - E[\{\tilde{P}_m\}_{i+1}] \tag{17}$$

where

$$E[\{P_m\}_i] = \sum_m P_m \varepsilon(Q_m, m)$$

$$Q_m = Z_n - \sum_{n<m} \sigma(m,n) P_n - \frac{1}{2} \sigma(m,m)(P_m - 1)$$

$$\varepsilon(Q_m, m) = \frac{1}{2} m_e c^2 \left(\frac{\alpha Q_m}{m}\right)^2 \left[1 + \left(\frac{2m}{m+1} - \frac{3}{4}\right)\left(\frac{\alpha Q_m}{m}\right)^2 + \frac{1}{6}\left(\frac{\alpha Q_m}{m}\right)^4\right]$$

The $\tilde{P}_m$ are equal to P_m except for the shell in which an electron is removed by ionization processes. The symbols m_e, c, and α are the mass of the electron, speed of light, and fine structure constant, respectively. The term Q_m represents the effective charge of the m-th shell. And this effective charge includes screening from electrons in inner shells only. The binding energy for the m-shell $\varepsilon(Q_m, m)$ is calculated using the Pauli approximation to the Dirac equation. The $\sigma(m,n)$ are hydrogenic screening constants.

Several sets of screening constants have been published in the literature. A set which was derived by Mayer[3] using perturbation theory has been widely used. The ionization potentials computed using the Mayer screening constants are reasonably accurate for highly-stripped ions but are grossly over-estimated for nearly neutral ions.[4] Recently, More obtained a set of screening constants by least-squares fitting to a data base containing 800 ionization potentials.[2] This new set of screening constants gives satisfactory results for both highly-stripped ions and ions closed to neutral. We adapt this set in our calculation of ground state ionization potential.

The electron collisional excitation and ionization rate coefficients are scaled from the results of Golden and Sampson[5] for hydrogenic ions. The inverse rates are calculated using the detailed balance relation. We use the Burgess and Merts[6,7] formula to calculate dielectronic recombination rates for both the $\Delta n = 1$ and $\Delta n = 0$ transitions. In calculating the $\Delta n = 0$ transitions, we used Post's[8] results for both the average excitation energy and oscillator strength. The radiative recombination rate is computed using Kramers' classical formula[9] together with a bound-free Gaunt factor. The radiative spontaneous emission rates are calculated using hydrogenic oscillator strengths. All the rate coefficients are represented by simple analytic formulas.

D. Results and Comparison

Since non-LTE calculations based on an average atom model have been employed in many laser-produced plasma simulations,[10] it is of interest to compare the average ionization states given by our model and an average atom model.[11] In Fig. 1a-1b, we show a contour plot for the ratio for the average ionization state of our model to an average atom model Z_{DCA}/Z_{AA} in the (T_e, ρ) plane for cesium. One can see that for $T_e > 20$ eV, the agreement is ± 20%. The largest discrepancies occur at high densities and low temperatures, where one expects pressure ionization and multiple excited states to be important. The hydrogenic model which includes only singly excited electron states is not expected to be applicable at these temperatures and densities.

In the following figures (2-8) we presented results for a steady-state non-LTE plasma. In Fig. 2, we plot the total ionization rate and total recombination rate for SeXXIIV and SeXXV ionization stages as a function of electron temperature at an electron density 5×10^{20} cm^{-3}. These calculations employ the dielectronic recombination rate coefficient given by the Burgess-Merts formula. The total ionization rate for Ne-like ions becomes equal to its total recombination rate at the electron temperature approximately 1580.0 eV. At this temperature, the Ne-like ion population is approximately equal to the F-like ion population. The Ne-like ion population is also expected to be greater than the F-like ion population below the temperature 1580.0 eV and become less than the F-like ion population above this temperature.

In Fig. 3, we compare the total ionization rate to the ground state ionization rate for SeXXIIV and SeXXV ionization stages as a function of electron temperature at an electron density 5×10^{20} cm^{-3}. The comparison shows that the total ionization rate is greater than the ground state ionization rate by large factors 5 to 10. The enhanced ionization is due to contribution from the excited levels which are populated by collisional excitation from the ground state. Both the collisional excitation rate and excited state ionization rate at these temperatures and densities are large compared to the ionization rate of the ground state. This will result in a charge state abundance curve which peaks at a lower temperature than a coronal calculation.

In Fig. 4, we plotted the average ionization state of a Se plasma as a function of electron temperature at an electron density 5×10^{20} cm^{-3}. Also plotted in the figure are the average ionization states given by the LTE model and coronal equilibrium model. The LTE ionization states are calculated using Saha-Boltzmann equation. The coronal equilibrium model computes ionization balance in a plasma by assuming the ions are in their ground states. The figure shows that the average ionization state given by our model is intermediate between coronal equilibrium and LTE. The LTE model significantly over-estimates the average ionization state at all temperatures because these plasmas are not dense enough. The coronal model underestimates the ionization state at low temperature due to the neglect of excited state ionization. However, the coronal model and our model agree at high electron temperature where step-wise ionization contribution is negligible.

In Figs. 5a-5b, the charge-state abundances of a hot Se plasma are plotted as a function of electron temperature at electron densities 5×10^{18} cm^{-3} and 5×10^{20} cm^{-3}. Also plotted in the figure are charge-state abundance given by the coronal equilibrium model. We see that our model agrees with the coronal model at low density 5×10^{18} cm^{-3}, but disagree by more than one charge state at higher density 5×10^{20} cm^{-3}. The Ne-like ions (Se XXV) dominate the charge-state distribution over a wide range of temperatures because of their closed atomic shell structure.

In Figs. 6a-6b, we plot the charge-state abundance of a hot Se plasma as a function of electron density at electron temperatures of 500 eV and 1000 eV. We see that excited state ionization starts to affect the charge-state abundance at electron density 5×10^{18} cm^{-3} at the temperature 500 eV and does not contribute to the charge-state abundance until the electron density is 10^{20} cm^{-3} at the temperature 1000 eV. This is due to the strong temperature dependence of the ground state ionization rate coefficients.

In all of the calculations discussed previously, we estimated the dielectronic recombination rate coefficient using the Burgess-Merts formula. To test these rate coefficients, we compared them to a recent detailed calculation[12] which computed the Auger and radiative rates of individual doubly excited states using Multi-Configuration Dirac-Fock method.[13] The comparison is shown in Fig. 7. We see that our dielectronic recombination rate coefficient agrees with the quantum calculation to better than 20% over a wide range of temperatures. The effects of dielectronic recombination on the ionization balance of a hot Se plasma is shown in Fig. 8.

Figure 9 compares the charge-state distribution of a hot steady-state krypton plasma to experimental measurement.[14] Both the electron temperature and density were estimated using the K-shell spectroscopy of phosphorus impurities seeded in the plasmas. The charge-state distribution was inferred from the L- and M-shell emission spectra. Also plotted in the figure are results of a calculation using the code XRASER.[15] Our results agree reasonably well both with the experimental measurement and the other theoretical calculation.

Since one of the motivations of our model was to design x-ray laser experiments, we show the results of a prototype Nova exploding foil target.[16] This design uses a 500 Å Se foil illuminated by a 0.5 μm laser of 400 psec FWHM duration at an intensity of 5×10^{13} W/cm^2. In Fig. 10a are shown the resulting LASNEX (T_e, n_e) time histories, which are approximately (900 eV, 5×10^{20} cm^{-3}) at the peak of the laser pulse. In Fig. 10b, we show the resulting charge state distributions, which has reached steady state by the peak of the laser. Finally, we compare in Fig. 10c the output x-ray energy of the average atom result to that of our model. Although the integrated outputs agree to 10%, there is obviously much more details in our results which are important in interpreting experimental results.

In Fig. 11, we plot the radiative cooling rates of an optically thin Se plasma as a function of electron temperature at an electron density 5×10^{20} cm^{-3}. The contributions due to bound-bound (lines), bound-free (recombination), and free-free (bremsstrahlung) are plotted separately. At low electron temperature, the line emission is totally dominant, but bremsstrahlung becomes dominant at a temperature of about 20 keV. Radiative recombination does not contribute significantly to the total radiative cooling rate at these electron densities. Our calculation neglects the line emission from Δn = 0 transitions because their contribution to the total radiative cooling rate is in general negligible at the densities 10^{20} cm^{-3}.

III. NON-HYDROGENIC APPROXIMATION

A. Basic Equations

Our model obtains the level populations for the ionization stages which are described using non-hydrogenic atomic data from the solution of the following rate equations

$$\frac{dN_\ell}{dt} + \sum_k (W_{k\ell} N_\ell - W_{\ell k} N_k) = S_\ell \tag{18}$$

where $W_{\ell k}$ are the collisional plus radiative rate for the transition between levels ℓ and k. The term S_ℓ gives the contribution to the population of level ℓ due to ionization or recombination from these ionization stages. The summation sums over all the levels in the non-hydrogenic ionization stages. The level populations for the rest of the ionization stages are calculated using Eqs. (3) and (4) in the hydrogenic approximation.

The model reads an atomic data file to obtain the energy levels, statistical weights, oscillator strengths, electron collisional excitation and ionization rate coefficients, photoionization cross sections, and autoionization rates. The rate coefficients and cross sections are fit to simple analytic expressions and the fitting coefficients are stored in the data file.

These data files which are also read by other x-ray laser design code such as XRASER are generated by a computer package, called ADAM.[17] Briefly, ADAM receives the content of the model to be made from a generator deck and then creates a model from an atomic physics data base or by using of simple analytic formulas. The data base contains primary atomic data which could be obtained from experimental measurement or more elaborated atomic code calculation.

This option makes it possible to apply the model to analyze non-LTE emission spectra and design laboratory x-ray laser experiments.

B. Results and Discussion

Figure 12 shows the charge-state distribution of a Se plasma at the electron density 5×10^{20} cm^{-3} and electron temperature 1000 eV. The solid curve represents the results of a calculation using the default hydrogenic option of our model. The dashed curve represents the results of a calculation using an atomic model which treats the Ne-like ionization stage in non-hydrogenic approximation. The dotted curve represents the results of a calculation using an atomic model which treats both Ne- and F-like ionization stages in non-hydrogenic approximation. The atomic model has 58 energy levels for Ne-like ion and 113 energy levels for F-like ion.

The energy levels for the n = 2 and n = 3 states of Ne- and F-like ions are computed using the multi-configuration Dirac-Fock code YODA.[18] The oscillator strengths for the transitions between these states are also computed using YODA.

Figure 13 shows the average ionization state of a Se plasma at the electron density 5×10^{20} cm^{-3} as a function of electron temperature. These comparisons in Figs. 12 and 13 show that the calculation using a hydrogenic model produces results which agree reasonably well with calculations using a more detailed atomic model.

C. Ionization Balance Derived from Emission Spectra

In this section we apply our model to validate a method proposed to measure ionic charge-state abundance in non-LTE plasmas. This method obtains the ratio of populations for two ionization stages from the relative intensities of the integrated x-ray flux in the spectral lines between two configurations seen in each of these ions produced in a plasma. For example, the ratio of populations for Ne-like ions to F-like ions is measured by comparing the relative spectral line intensities arising from the 2p - 3d transitions of Ne-like and F-like ions and scaling the relative intensities to the number of 2p electrons in each ion. This method assumes the 3d level populations are dominantly produced by electron collisional excitation from the ground state. We have estimated the cascade contribution to the 3d level populations in both F-like and Ne-like ions to be less than 20% for most laboratory plasmas produced using a z-pinch or a high-power laser.

Stewart[14] has applied this method together with a satellite line model for Ne-like ions to measure the krypton charge-state distribution in plasmas produced using a z-pinch. Results of his measurement are plotted in Fig. (9) together with our calculation. Recently, Bailey[19] used the same method to measure ionic charge-state abundance of Ni-like, Co-like, and Fe-like ions in europium plasmas produced with laser-irradiated micro-dot targets.

To test this method for measuring ionization balance in non-LTE plasmas, we calculated all the spectral line intensities arising from the 2p - 3d transitions in Ne-like and F-like selenium ions. We then used these relative intensities to compute the ratio of populations for Ne-like ions to F-like ions. Results are plotted in Fig. (14) together with the ratio of populations predicated by our model. We see that the population ratios obtained from the relative intensities of 2p - 3d transition agree within 20% with the results given by the ionization balance model over a wide range of electron temperatures and densities. This suggests that ionization balance can be measured by comparing relative intensities for a set of transitions seen in each ionization stages.

IV. SUMMARY

We developed a detailed configuration accounting kinetic model for calculating time-dependent ionization balance and ion level populations in non-LTE plasmas. The model has two options for calculating level populations. The default option treats all the ionization stages in hydrogenic approximation. Rate coefficients for both collisional and radiative transitions are calculated using analytic formulas. The other option treats certain ionization stages in non-hydrogenic approximation. The model obtains all of the atomic data for these ionization stages from a datafile. The option makes it possible to apply the model to simulate x-ray laser and plasma spectroscopy experiments.

We have also coupled the model self-consistently to the hydrodynamics code LASNEX. The model uses the electron density and temperature and photon intensity from LASNEX to compute the time-dependent level populations. The average ionization state, emissivity and photon cross section, and binding energy are computed using these new level populations. LASNEX uses these quantities in the equation of hydrodynamics, photon transport, and conservation of energy.

References

1. G. B. Zimmerman, Lawrence Livermore National Laboratory Report No. UCRL-74811 (1973). Also Laser Program Annual Report 1984, Lawrence Livermore National Laboratory Report No. UCRL-50021-84 (1984).

2. R. M. More, "Atomic Physics in Inertial Confinement Fusion," Lawrence Livermore National Laboratory, Report No. UCRL-84991 (1982).

3. H. Mayer, Los Alamos Scientific Laboratory, Report No. LA-647 (1947).

4. R. M. More, H-Division Quarterly Report, Lawrence Livermore National Laboratory, UCRL-50028-79-2 (1979).

5. L. Golden and D. Sampson, Astrophys. J. 170, 181 (1971).

6. A. Burgess, Astrophys, J. 141, 1588 (1965).

7. A. L. Merts, R. D. Cowan, and N. H. Magee, Jr., Los Alamos National Laboratory, Report No. LA-6220-MS (1976).

8. D. E. Post, R. V. Jenen, C. B. Tarter, W. H. Grasberger, and W. A. Lokke, At. Data Nucl. Table 20 239 (1977).

9. H. A. Kramer, Phil. Mag. 271, 836 (1923).

10. M. Rosen, D. Phillion, V. Rupert, W. Mead, W. Kruer, J. Thompson, H. Kornblum, V. Slivinsky, G. Caporaso, M. Boyle, and K. Tirsel, Phys. of Fluids 22 2020 (1979). T. Yabe and C. Yamanaka, Comments Plasma Physics, Controlled Fusion, 1 No. 4, 169, (1985).

11 W. A. Lokke and W. H. Grasberger, Lawrence Livermore National Laboratory, Report No. UCRL-52276 (1977).

12. M. Chen, Lawrence Livermore National Laboratory, Livermore, California, private communication (1985).

13. I. Grant, J. Phys. B 9, 2777 (1976).

14. R. E. Stewart, Ph.D. dissertation, University of California, Davis (1985) (unpublished).

15. P. L. Hagelstein, Lawrence Livermore National Laboratory, Report No. UCRL-53100 (1981).

16. M. Rosen, Lawrence Livermore National Laboratory, Livermore, California, private communication (1985).

17. J. K. Nash, W. L. Morgan, R. K. Jung, and D. R. Kim, (1986) (to be published in the Proceedings of Third International Conference on Radiative Properties of Hot Dense Matter, Williamsburg, Va., October 14-18, 1985).

18. P. L. Hagelstein and R. K. Jung. (1985) (submitted to J.Q.R.S.T.).

19. J. Bailey, Sandia National Laboratories, Albuquerque, NM (private communication).

Figures

Figure 1a   Contour plot of Z_{DCA}/Z_{AA} the ratio for the average ionization of our model to an average atom model for a LTE cesium plasma in (T_e,ρ) space.

Figure 1b   Contour plot of Z_{DCA}/Z_{AA} the ratio for the average ionization of our model to an average atom model for a steady state optically thin cesium plasma in (T_e,ρ) space.

Figure 2   Total ionization rate and recombination rate for a selenium plasma vs. electron temperature at electron density of 5×10^{20} cm^{-3}. The solid curves represent the total ionization rate ϕ_i and the dashed curves represent the total recombination rate ϕ_{i+1}.

Figure 3   Comparison of the total ionization rate to the ground state ionization rate for a selenium plasma at electron density 5×10^{20} cm^{-3}.

Figure 4   Average ionization state vs. electron temperature at electron density of 5×10^{20} cm^{-3}. Comparison of our model to the LTE model and the coronal equilibrium model.

Figure 5a   Charge-state abundance vs. electron temperature at electron density of 5×10^{18} cm^{-3}. Comparison of our model to the coronal equilibrium model.

Figure 5b   Charge-state abundance vs. electron temperature at electron density of 5×10^{20} cm^{-3}. Comparison of our model to the coronal equilibrium model.

Figure 6a   Charge-state abundance vs. electron density at electron temperature of 500 eV.

Figure 6b   Charge-state abundance vs. electron density at electron temperature of 1000 eV.

Figure 7   Comparison of the dielectronic recombination rate coefficient calculated using the Burgess-Merts formula to a recent calculation using Multi-Configuration Dirac-Fock Method (M.C.D.F.).

Figure 8   Dependence of the charge-state abundance of a selenium plasma on the dielectronic recombination rate coefficients. The symbol γ is the dielectronic recombination multiplier. $N_e = 5 \times 10^{20}$ cm^{-3} and $T_e = 1000$ eV.

Figure 9   Charge-state distribution for a krypton plasma at an electron temperature of 800 eV and an electron density of 3×10^{20} cm^{-3}.

Figure 10 (a) Time history of (T_e, N_e) for two zones of an x-ray laser exploding foil target design. (b) Time dependence of charge states SeXXIII to SeXXVII for an x-ray laser foil target design. (c) Comparison of our model and an average atom model output spectra for an x-ray laser foil target design.

Figure 11 Radiative cooling rate for an optically thin selenium plasma vs. electron temperature at electron density 5×10^{20} cm^{-3}. The contributions due to bound-bound (lines), bound-free (recombination) and free-free (bremsstrahlung) are plotted separately.

Figure 12 Charge-state distributions of a selenium plasma calculated using different atomic models at an electron density of 5×10^{20} cm^3. (———) hydrogenic model, (-----) non-hydrogenic approximation for SeXXV and SeXXVI, (·····) non-hydrogenic approximation for SeXXV.

Figure 13 Average ionization states vs. electron temperature of selenium plasmas calculated with different atomic models at an electron density 5×10^{20} cm^{-3}. (———) hydrogenic model, (o) non-hydrogenic approximation for SeXXV and SeXXVI.

Figure 14 Ratio of populations for Ne-like selenium ions to F-like selenium ions vs. electron temperature at electron densities 10^{20} cm^{-3} and 10^{21} cm^{-3}. The solid curves represent results given by the ionization balance model. The dashed curves represent results obtained by comparing relative intensities of 2p - 3d transitions seen in each ionization stages.

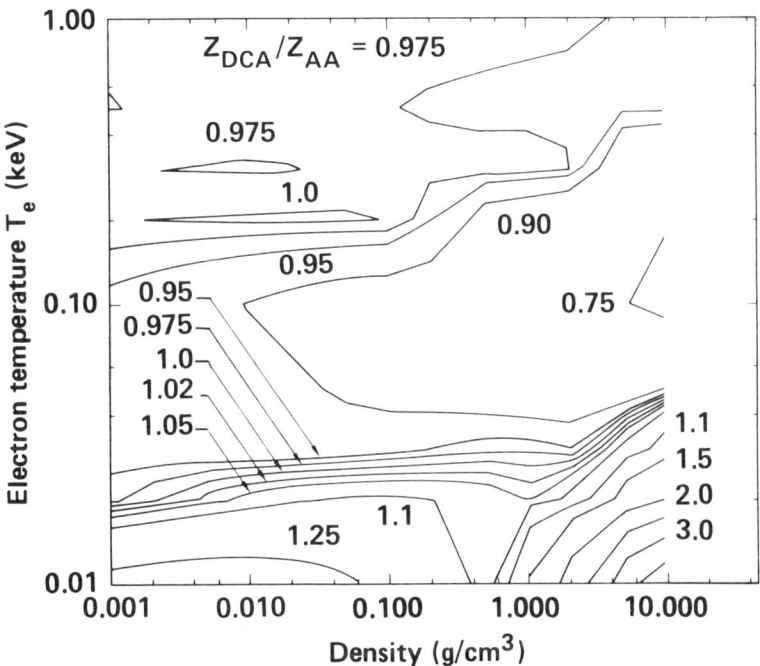

Figure 1a

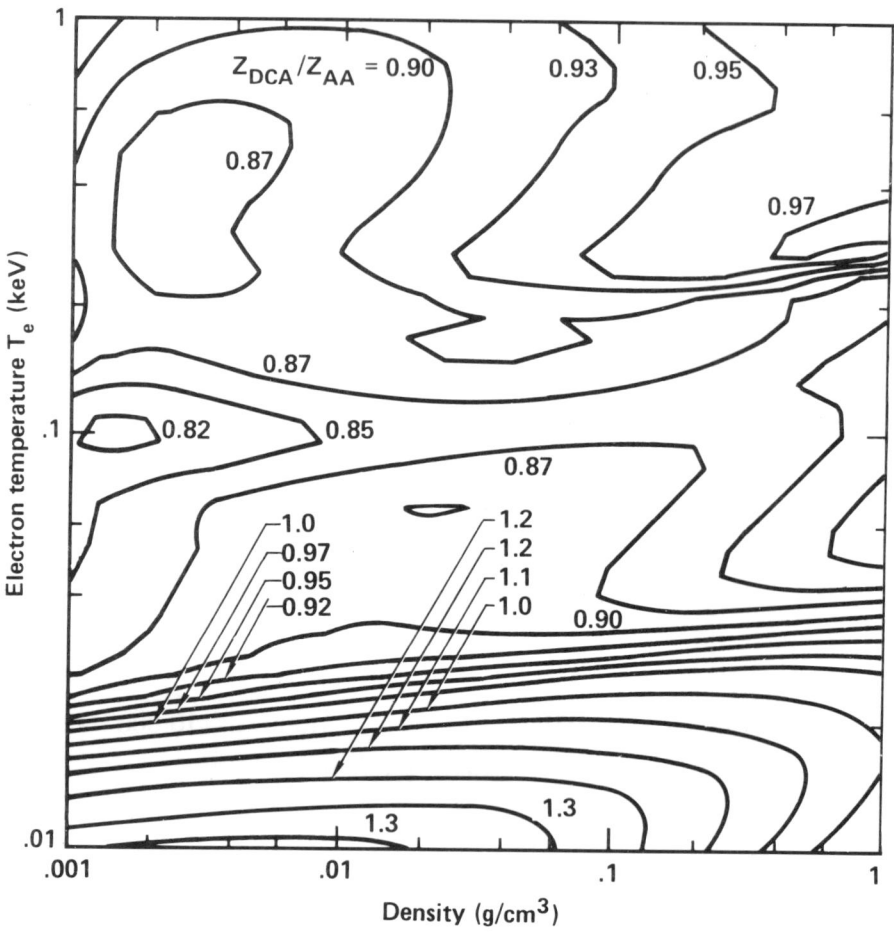

Figure 1b

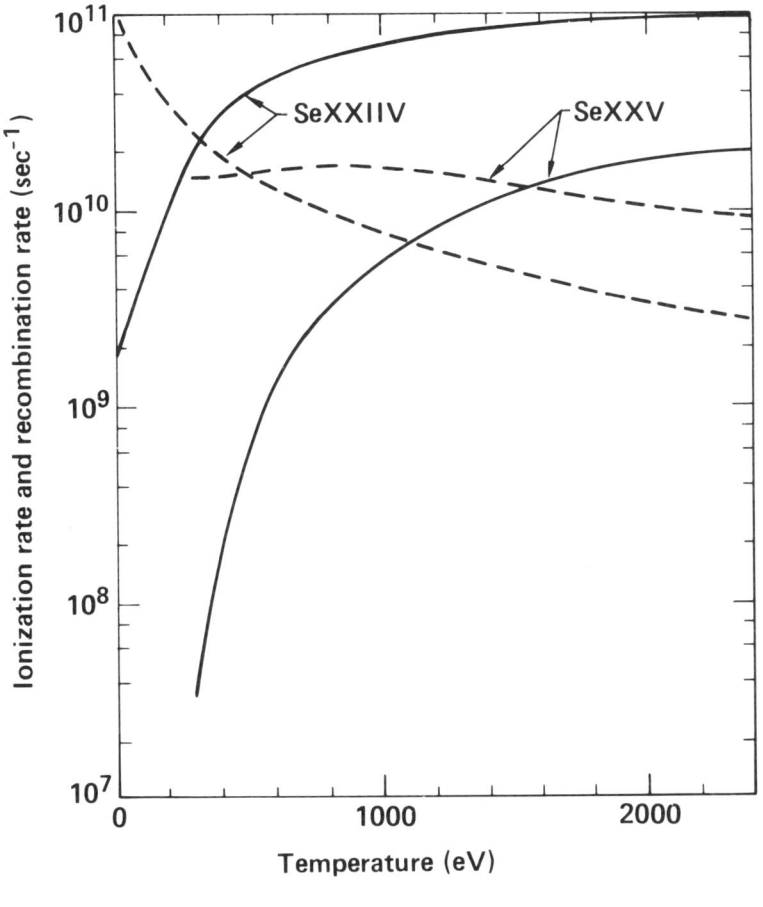

Figure 2

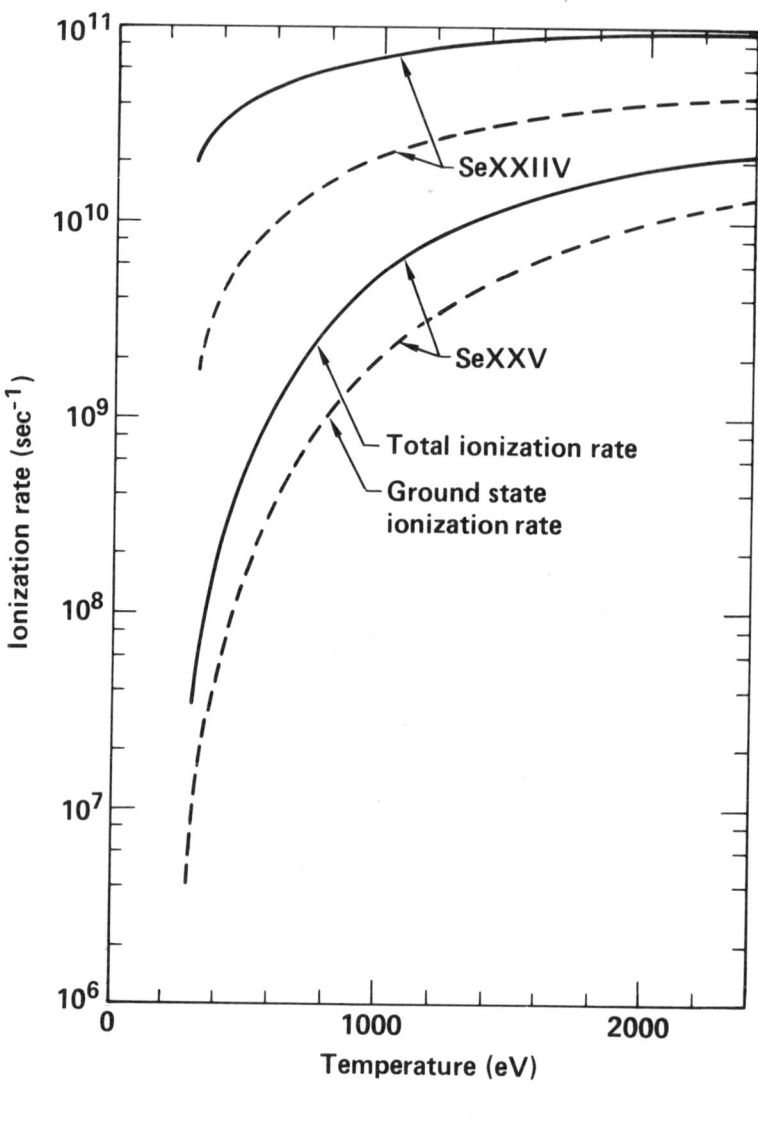

Figure 3

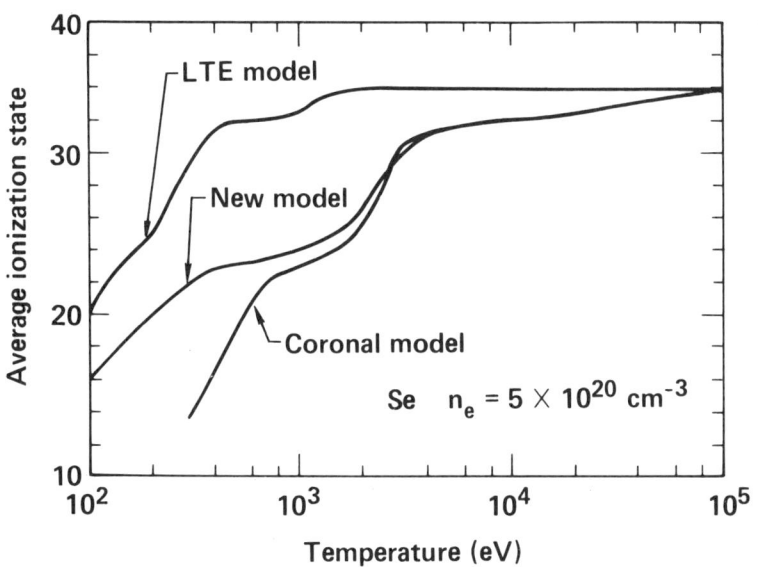

Figure 4

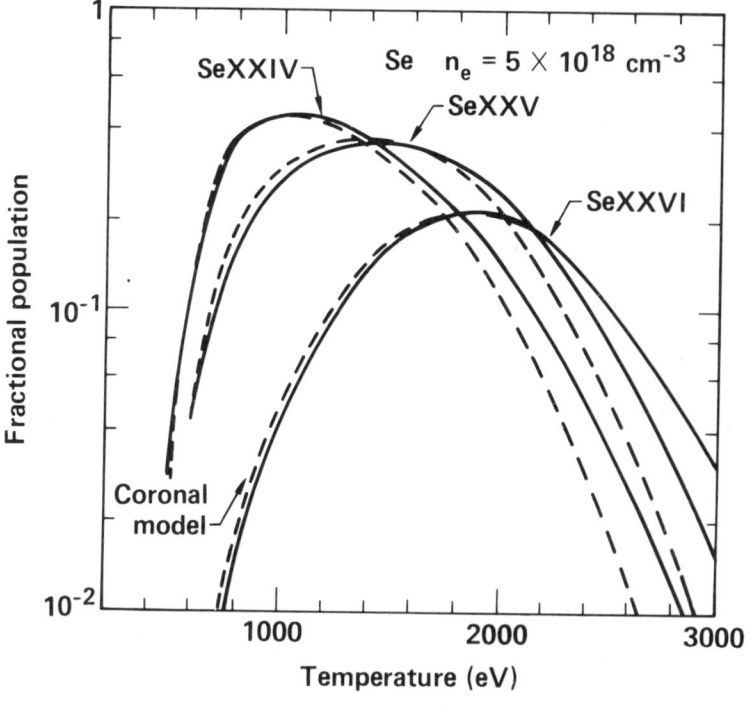

Figure 5a

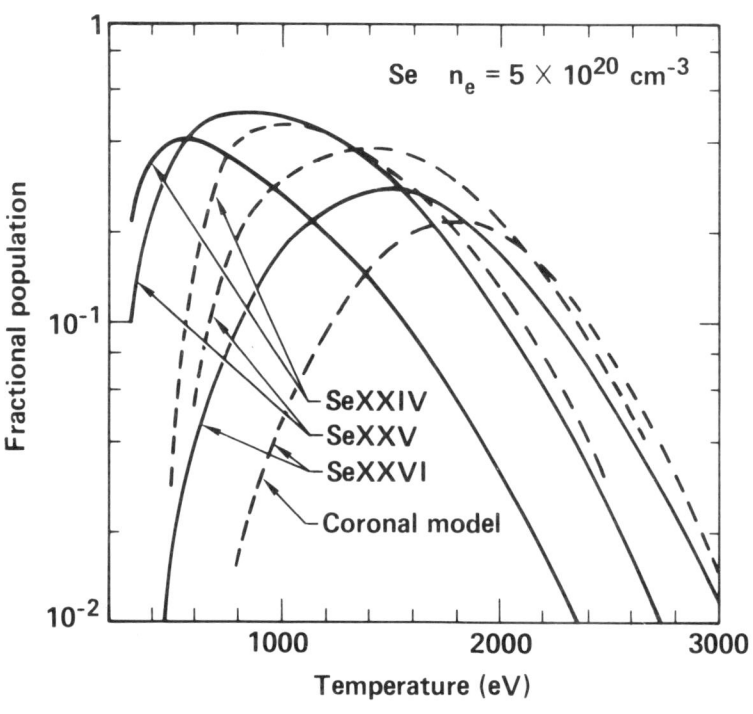

Figure 5b

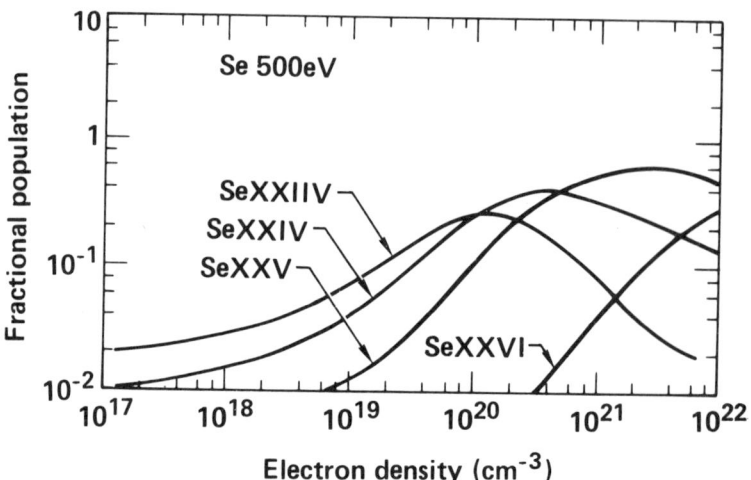

Figure 6a

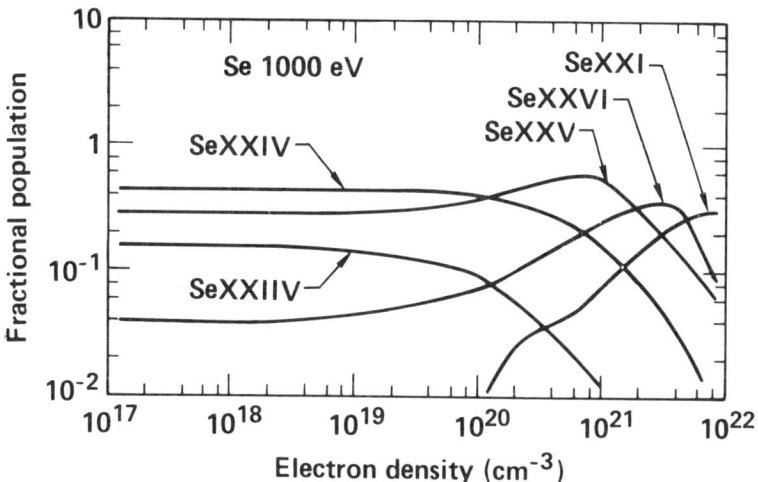

Figure 6b

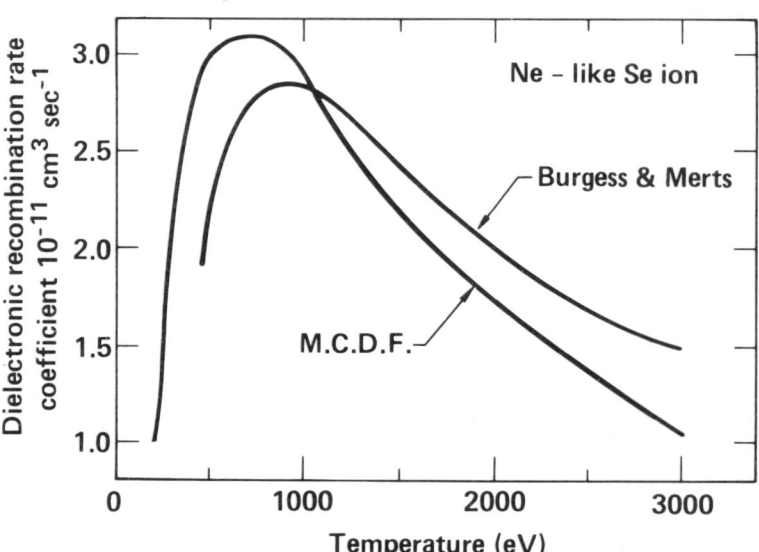

Figure 7

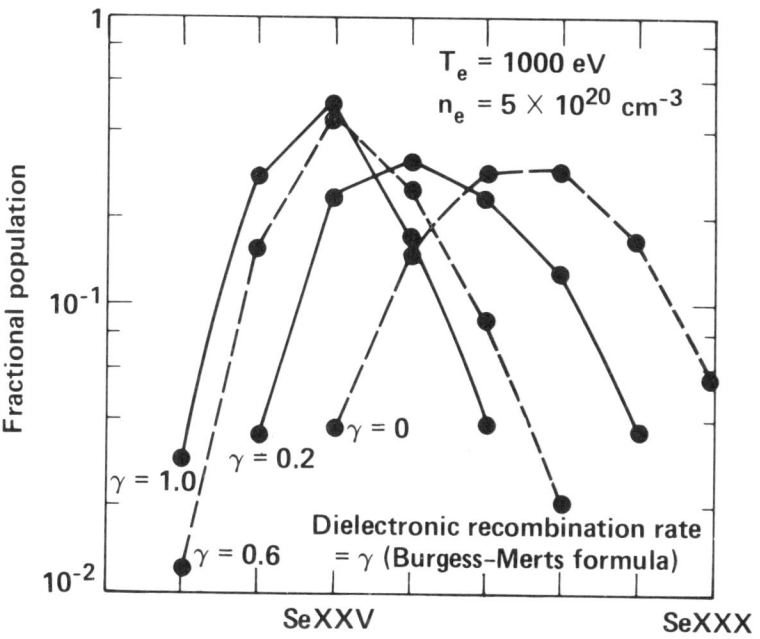

Figure 8

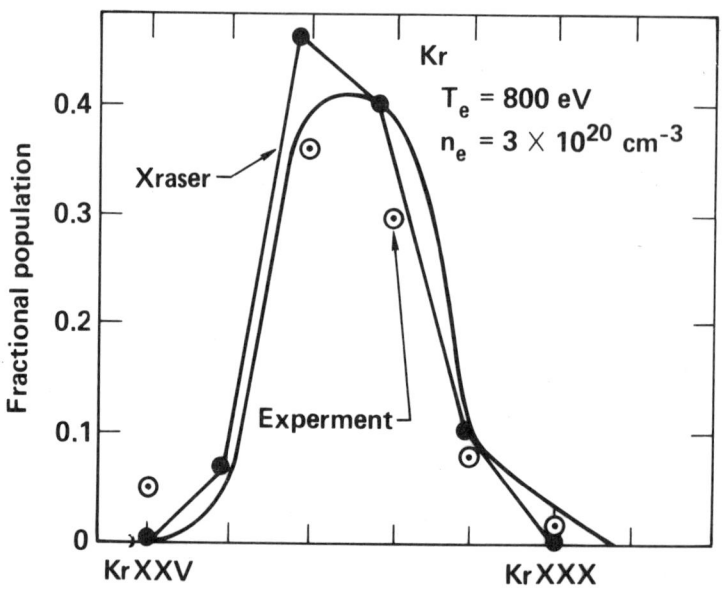

Figure 9

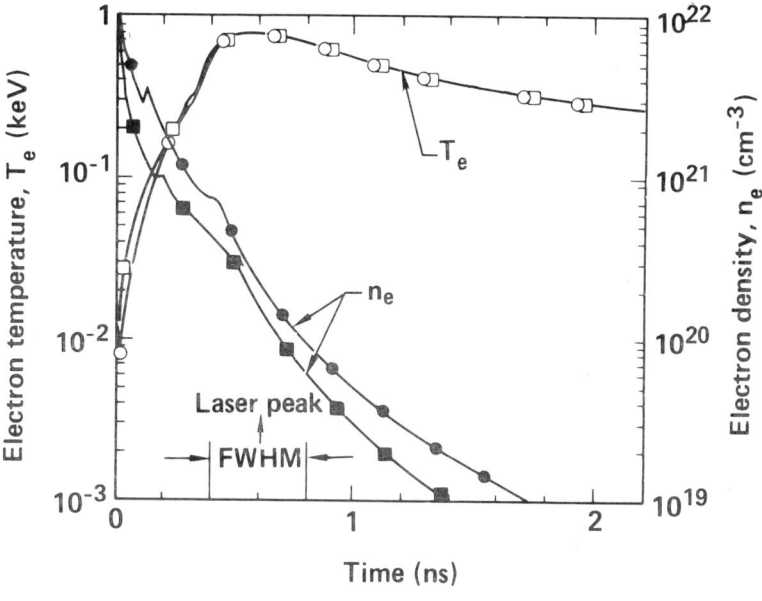

Figure 10a

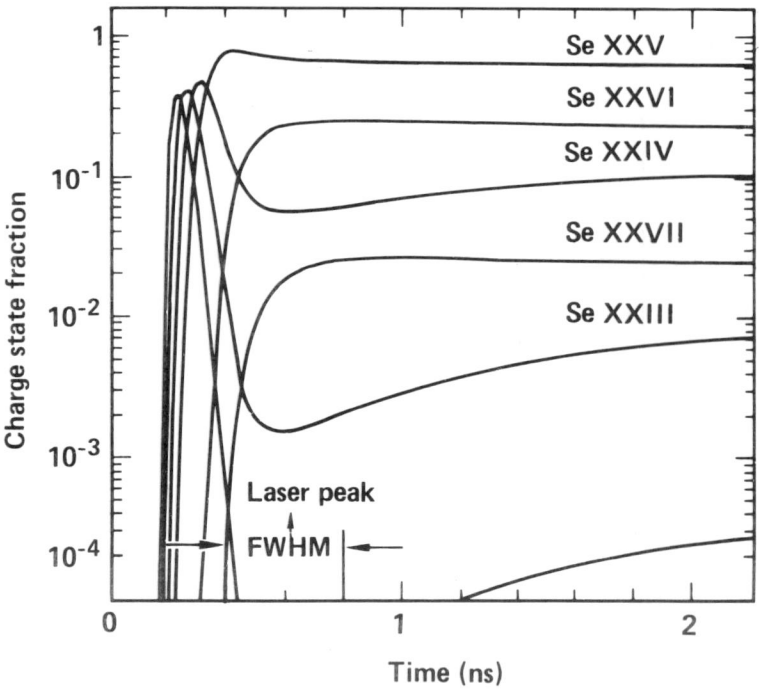

Figure 10b

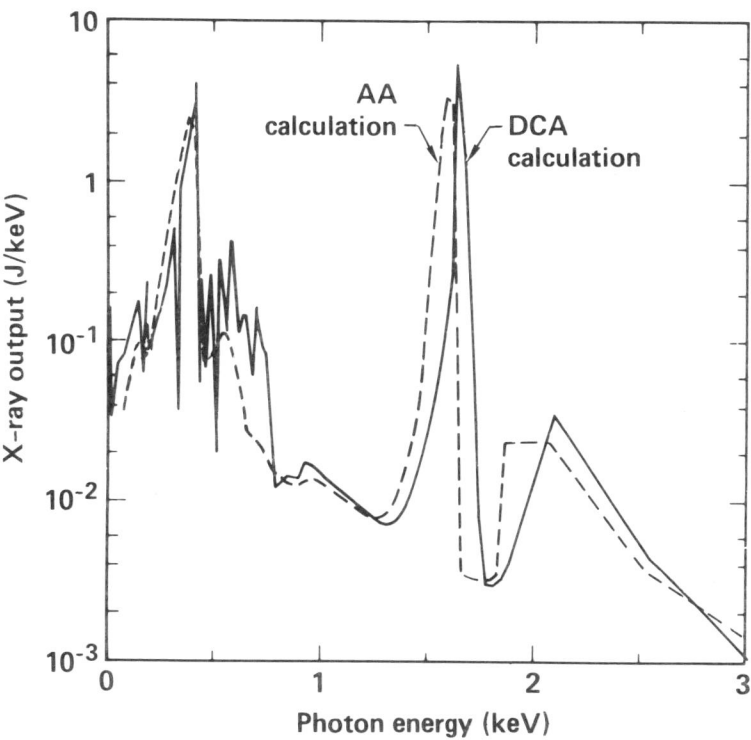

Figure 10c

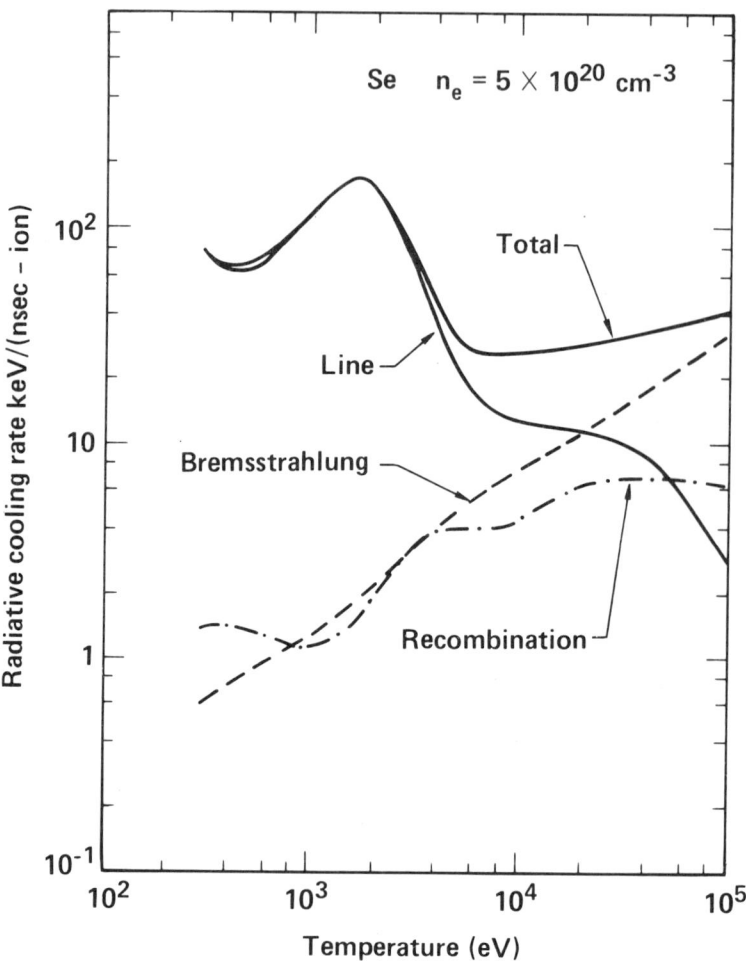

Figure 11

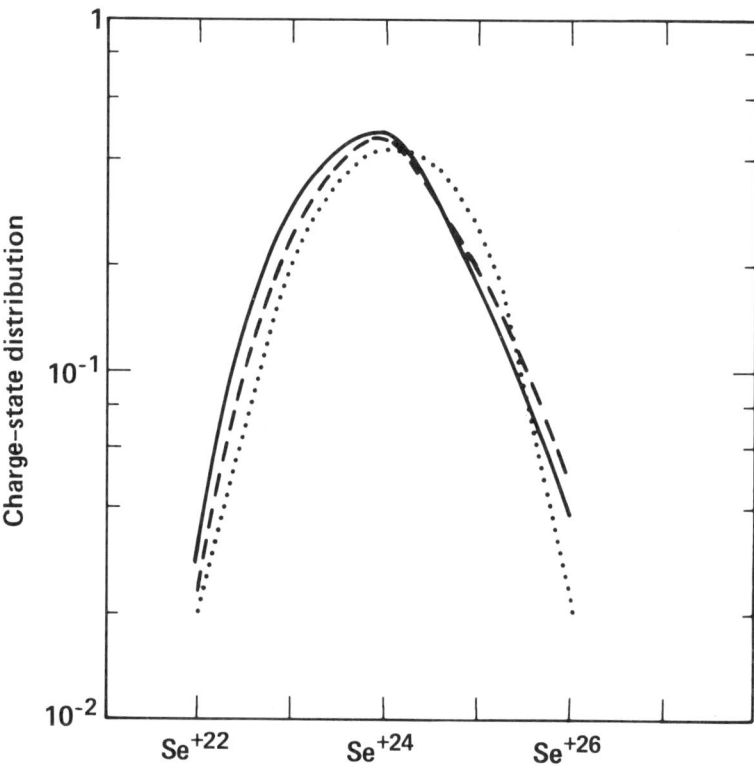

Figure 12

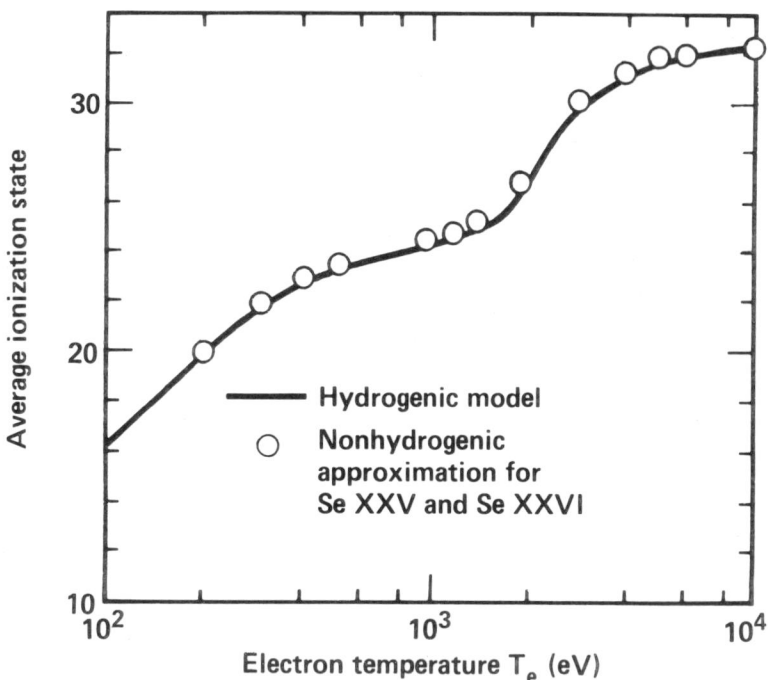

Figure 13

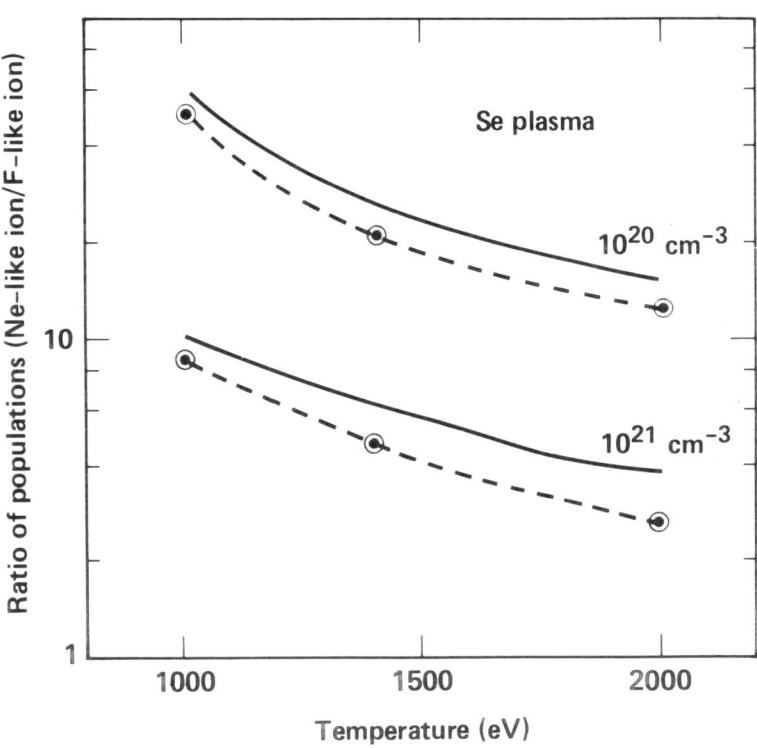

Figure 14

The Ponderomotive force:

Derivation of the Stress-Tensor for High-Frequency Radiation in Isotropic Dispersive Matter by Generalisation of the Helmholtz Method.

B J B Crowley

Atomic Weapons Research Establishment, Aldermaston, Reading, U.K.

Abstract

A formal proof is presented of the generalised Helmholtz stress tensor for high frequency transverse electromagnetic radiation in an isotropic translucent medium with an arbitrary frequency-dispersive refractive index $n(\omega;\rho,T)$. The refractive index is taken to be dependent on the local density ρ and temperature T which themselves may be slowly-varying functions of position and time. Using approximations valid in the geometric-optics limit, the basic Helmholtz method, as hitherto used to determine the stress-tensor in magnetostatics and electrostatics (and also for very low frequency electromagnetic waves), is generalised to high frequencies. This method identifies the stress-tensor from the virtual work term produced by a local gauge transformation representing an arbitrary infinitesimal deformation of fields and material. The result supports earlier conjectures concerning the general validity of the Helmholtz expression (when expressed in terms of the radiation free energy).

The result is used to obtain an expression for the ponderomotive force due to unpolarised radiation with an arbitrary spectral distribution represented by amplitudes (proportional to the rms fields) that are slowly-varying functions of both position and time. Particular care is taken over the momentum-of-light problem and its ramifications regarding the manner of any decomposition into fields/radiation and matter. Consideration is given to the consequences of non-uniform motion of the material medium.

A number of other useful results are produced in passing: identification of several propagation and adiabatic invariants of the radiation field; and the provision of the complete thermodynamic relations for arbitrary infinitesimal quasistatic processes in an open system containing both matter and radiation.

Keywords
Ponderomotive force, radiation pressure, Helmholtz stress-tensor, geometric optics, Helmholtz method, thermodynamics, adiabatic invariants, gauge transformation, electrodynamics, dispersion, refraction, electrostriction, electrocaloric effect, momentum-of-light, Abraham force.

Ⓒ Controller HMSO, London, 1985.

Introduction

The ponderomotive force [1,49] in optically dense matter due to high frequency electromagnetic radiation, such as laser light [2-6], high-temperature thermal radiation [7,34,35] etc., has been attracting a considerable interest over the last decade or so for a variety of reasons. In particular, its effect on the macroscopic interaction of high-power laser light with matter is now a feasible subject of experimental study [3] within the wider context of ICF-motivated studies of laser-target interactions. Much importance is attached to non-linear and pseudo non-linear effects generated by the coupling of a locally linear interaction to the necessarily non-linear dynamical response of the material fluid or plasma system. In the presence of fairly long wavelength coherent laser light not too far above the plasma frequency, the material may be able to exhibit a response on the length scale of the wavelength (the Raman regime) giving rise to nonlinear effects such as filamentation and self-focussing [8-13,50], and non-linear wave interactions [14-23,50] such as stimulated Brillouin and Raman scattering. At shorter wavelengths, or if the radiation is insufficiently coherent, the response may be charactised by much longer length scales and will be locally linear with respect to the rms fields. Of course, the complete system of equations remains nonlinear with the result that (pseudo-) nonlinear effects may be induced on the length- and time-scales characterising, for example, the laser pulse. The close connection with the local effective field is another important aspect of this problem [40,42, 44-45] which makes it relevant to microcsopic dynamics as well as macroscopic dynamics.

The interest in such effects is compounded by the lack of a general theory which hitherto rigorously extends only to quasistatic fields [1] and low frequency radiation characterised by frequencies well below the lowest resonance [24] or by wavelengths long on other relevant scales [25]. Several theoretical formulations of the ponderomotive force have been developed using kinetic models with specific application to plasmas [26-31]. Such models inevitably incorporate simplifications or approximations regarding the treatment of the material (eg. collisionless free-electron plasma). Such theories are often directed at studying the first order corrections (involving spatial and temporal derivatives of the field amplitudes) to the original ponderomotive force expression of Boot et al [26] and to determining additional effects such as arising in the presence of self-generated or induced quasi-static magnetic fields and longitudinal-mode plasma oscillations [25,30-31]. Other derivations of the high-frequency form, such as [32-33], take a more general view. However these are usually found to contain an unproven assumption or conjecture at an early stage of the proof. Also some of the methods and concepts employed are unconvincing when applied to very high frequency freely-propagating radiation. Often the momentum-of-light-problem is not properly resolved, and is more often than not completely overlooked, leaving the theory in an unsatisfactory state for reliable determination of higher order corrections. Indeed alternative proposals for such correction terms (arising from slow time-variation of the intensity) may be merely consequences of various improper treatments of this particular aspect.

This paper considers carefully the electrodynamical interaction of high frequency radiation (not near resonance where there is anomalous dispersion) with a locally isotropic material continuum whose electromagnetic response is linear and characterised by a frequency-dispersive refractive index, $n(\omega)$. By making use of approximations valid in geometrical optics [36], a first-time complete formal proof is given of the leading terms in the electromagnetic stress-tensor for high-frequency radiation. The approximations made are valid, not only for a transparent medium, but also for one in which there is weak absorption. In this manner, a fully self-consistent treatment of (weak) dispersion is provided. The rigorous treatment of strong (ie anomalous) dispersion, which is characterised by a failure of geometrical optics and the lack of a sensible group velocity, however remains an open problem.

The approximation of local isotropy is principally one of convenience so as to keep the theory as simple as possible by avoiding complications due to spatial dispersion. We are aware that local isotropy in the absence of radiation does not strictly imply the same when polarised radiation is present due to the symmetry being broken by the introduction of a plane of polarisation [48]. However the assumption is perfectly valid when the local field is an unpolarised superposition of different polarisation states (eg thermal radiation). Although the following is presented as being applicable to a single wave mode, the assumed linearity of the electromagnetic interactions and the absence of any dependence on phase means that generalisation to a superposition of wave states (as is done at equations 73ff) is simple and straightforward. A representation of the radiation field in terms of independent *photons* is perfectly sensible for this purpose.

For the purpose of this derivation however, we use Maxwellian electrodynamics to represent the electromagnetic radiation field in terms of a set of "external" fields, **E, D, H, B** which satisfy the source-free Maxwell's equations:

$$\text{div } \mathbf{D} = 0 \tag{1}$$

$$\text{div } \mathbf{B} = 0 \tag{2}$$

$$\text{curl } \mathbf{E} + \frac{\partial \mathbf{B}}{\partial t} = 0 \tag{3}$$

$$\text{curl } \mathbf{H} - \frac{\partial \mathbf{D}}{\partial t} = 0 \tag{4}$$

where the influence of the medium is introduced via the permittivity and permeability *operators* $\boldsymbol{\epsilon}$ and $\boldsymbol{\mu}$ such that:

$$\mathbf{D} = \epsilon_0 \boldsymbol{\epsilon} \mathbf{E} \; , \quad \mathbf{B} = \mu_0 \boldsymbol{\mu} \mathbf{H} \; , \tag{5}$$

and which possess analytic eigenvalues $\epsilon(\omega)$ and $\mu(\omega)$ in the frequency domain according to:

$$\boldsymbol{\epsilon} e^{-i\omega t} = \epsilon(\omega) e^{-i\omega t} \; , \quad \boldsymbol{\mu} e^{-i\omega t} = \mu(\omega) e^{-i\omega t} \; . \tag{6}$$

These eigenvalues may be complex and thus permit a consistent treatment of absorption. Induced currents in the medium are also included via the imaginary part of $\epsilon(\omega)$ in terms of which, for example, the (real part of) AC conductivity $\sigma(\omega)$ is given by:

$$\mathrm{Re}\,\sigma = \omega\epsilon_0 \,\mathrm{Im}\,\epsilon \qquad (7)$$

Note that this description avoids any need to distinguish between real and polarisation currents.

It is permissable, in the first instance, for the eigenvalues ϵ and μ in (6) to retain a slow dependence on time which expresses the effect of any time dependence of the properties of the medium, such as temperature T, density ρ, and any slowly-varying fields. Pitaevskii [25] demonstrates that such a time dependence due to the changing properties of the medium leads to an additional contribution to the imaginary parts of ϵ and μ as given by, for example: $\epsilon(\omega,t) = \epsilon_s(\omega,t) + \tfrac{1}{2}i\,\partial^2 \epsilon_s(\omega,t)/\partial\omega\partial t$, where $\epsilon_s(\omega,t) = \epsilon_s(\omega;\rho(t),T(t)...)$, $\epsilon_s(\omega;\rho,T...)$ being the permittivity in an otherwise similar medium whose properties are static. This would result in κ appearing in equations (19) below being replaced by
$\kappa + (\omega/nc)(n\,\partial^2 n/\partial\omega\partial t + (\partial n/\partial t)(\partial n/\partial\omega))$. However we find that this result is not sufficiently general to apply to freely propagating radiation due to neglect of frequency modulation induced by a time-dependent medium. In the optical regime, with which this paper is particularly concerned, such additional contributions to the absorption appear likely to be negligible except perhaps when they give rise to *instabilities* or interesting *non-linear effects*. For simplicity of presentation of the essential ideas, the formulation which follows is initially, and for the most part, given for a stationary medium whose properties do not vary with time. The generalisation to media whose properties do vary in time and which may be in a state of non-uniform motion is treated subsequently towards the end of this paper.

We consider the field equations for a quasimonochromatic wave whose dominant frequency ω lies within a *transparency range* [1] when the imaginary parts of $\epsilon(\omega)$ and $\mu(\omega)$ are small to first order leading to first-order small phase differences between the **E, D, H, B** fields. If moreover this frequency is sufficiently high and the corresponding wavelength is sufficiently short, then we can justify a *local* description of the wave field in terms of the rms fields $\bar{E}$, $\bar{D}$, $\bar{H}$, $\bar{B}$ whose directions are given conventionally in terms of the polarisation, aside from an irrelevant sign ambiguity, and whose derivatives are small on both frequency and wavelength time and length scales. This is the basis of the geometric optics approximation. An additional consequence of the above assumptions which is incorporated into the geometric optics approximation is that the mean energy propagation velocity (associated with the Poynting vector) is the *group velocity*, $v = \partial\omega/\partial\mathbf{k}$, where $\mathbf{k}$ is the real part of the (complex) wavevector,

$$\mathit{k} = \mathbf{k} + \tfrac{1}{2}i\kappa \qquad (8)$$

the imaginary part of which yields the attenuation coefficient, κ. Note that the translucency conditions imply that κ/k is a first-order small quantity ($\kappa/k \ll 1$). The dispersion relations for the local electromagnetic field are:

$$k^2 c^2 = \epsilon \mu \omega^2 \tag{9}$$

$$kc = n\omega \tag{10}$$

where $c = 1/\sqrt{(\epsilon_0 \mu_0)}$ is the velocity of light and n is the refractive index, which is real and positive. Following straightforwardly from the above are:

$$n = \text{Re}((\epsilon\mu)^{1/2}), \qquad \text{Re } n > 0, \tag{11}$$

$$\kappa = \frac{\omega}{nc} \text{Im}(\epsilon\mu) . \tag{12}$$

In order to treat dispersion properly we must give careful consideration to the nature of operators such as ϵ and μ defined above. Let us consider a general operator α defined in terms of its spectrum $\alpha(\omega)$. We shall apply this operator to quasimonochromatic fields such as $A(t) = \sqrt{2} \bar{A}(t)\cos(\omega t + \delta)$ and $B(t) = \sqrt{2} \bar{B}(t)\cos(\omega t)$. Provided that the Fourier transform of $B(t)$ vanishes on the real frequency axis outside the circle of convergence centred at the frequency ω and constrained by the singularities of the operator kernel $\alpha(z)$, then the result of applying α to B is given by the following (which results from Taylor-series expansion of the kernel about $z = \omega$):

$$\alpha B = \text{Re}\left[\alpha(\omega) B + \sum_{n=1}^{\infty} \frac{1}{n!} \frac{\partial^n \alpha}{\partial \omega^n} \left(i\frac{\partial}{\partial t} - \omega\right)^n B\right]$$

where $B(t) = \sqrt{2} \bar{B}(t) \exp(-i\omega t)$. The derivation of this equation shows that it continues to apply if α and even ω retain an explicit residual time-dependence — except that in the latter case we must continue to treat ω and $\partial/\partial t$ as commuting operators. [Here, and throughout what follows, $\partial/\partial t$, when applied to a *function* of frequency (such as $n, \epsilon, \mu, \alpha, \nu$ etc), denotes the time derivative at (*inter alia*) constant frequency. However, when applied to fields etc which are not functions of ω, then $\partial/\partial t$ denotes the usual partial time derivative at a fixed point in space. (Thus we could quite sensibly write: $\partial(U\nu)/\partial t = \nu \partial U/\partial t + U \partial \nu/\partial t + U(\partial \nu/\partial \omega)(\partial \omega/\partial t)$.)] We now assume that $\bar{B}$ is sufficiently slowly varying to permit neglect of its second and higher order derivatives. Hence, if $\langle\rangle$ denotes the time average on a scale long compared with $1/\omega$ so that

$$\langle AB \rangle = \bar{A}\bar{B} \cos(\delta),$$

then:

$$\langle A\alpha B \rangle = \langle AB \rangle \left[\text{Re}\alpha - \text{Im}\left(\frac{\partial \alpha}{\partial \omega}\right) \frac{1}{\bar{B}} \frac{\partial \bar{B}}{\partial t} - \tan(\delta)\left[\text{Im}\alpha + \text{Re}\left(\frac{\partial \alpha}{\partial \omega}\right) \frac{1}{\bar{B}} \frac{\partial \bar{B}}{\partial t}\right]\right]$$

(13)

We distinguish two principal classes of operator: those, such as ϵ and μ, which introduce only small phase-shifts, ie $\text{Im}\alpha \ll \text{Re}\alpha \Rightarrow \arg\alpha \ll 1$; and those such as $\partial/\partial t$ which introduce phase shifts $\sim \pi/2$. Then, taking α to be in the first class, and stipulating that δ is also first-order small,

$$\langle B\alpha B \rangle \simeq \langle BB \rangle \text{Re}\alpha \simeq \overline{B}\overline{B}\,\text{Re}\alpha$$

$$\simeq \overline{B}\,\overline{\alpha B}$$

$$\Rightarrow \overline{\alpha B} = \overline{B}\,\text{Re}\alpha \; ; \qquad (14)$$

and

$$\langle A\,\frac{\partial B}{\partial t}\rangle = \langle AB \rangle \left[\frac{1}{\overline{B}}\,\frac{\partial \overline{B}}{\partial t} + \omega\tan(\delta)\right] \qquad (15)$$

Now, substituting $B = \alpha A$ in (15) and making use of (14),

$$\langle A\,\frac{\partial}{\partial t}\,\alpha A\rangle = \langle A\alpha A \rangle\left[\frac{1}{\overline{\alpha A}}\,\frac{\partial \overline{\alpha A}}{\partial t} + \omega\,\frac{\text{Im}\alpha}{\text{Re}\alpha}\right]$$

$$= \overline{A}\left[\frac{\partial \overline{\alpha A}}{\partial t} + \omega\overline{A}\,\text{Im}\alpha\right] \qquad (16)$$

Alternatively, we can calculate the same quantity in a different way as follows:

$$\langle A\,\frac{\partial}{\partial t}\,\alpha A\rangle = \langle A\left[\frac{\partial}{\partial t},\alpha\right]A\rangle + \langle A\alpha\,\frac{\partial}{\partial t}\,A\rangle$$

where, by direct application of (13), and allowing for the fact that the frequency may be changing in time due to interaction with the medium,

$$\langle A\left[\frac{\partial}{\partial t},\alpha\right]A\rangle = \overline{A}^2\left[\text{Re}\,\dot{\alpha} - \text{Im}\,\frac{\partial\dot{\alpha}}{\partial\omega}\,\frac{1}{\overline{A}}\,\frac{\partial\overline{A}}{\partial t}\right] \simeq \overline{A}^2\,\text{Re}\,\dot{\alpha},$$

$$\langle A\alpha\,\frac{\partial}{\partial t}\,A\rangle = \overline{A}^2\left[\text{Re}(-i\alpha\omega) - \text{Im}\,\frac{\partial(-i\alpha\omega)}{\partial\omega}\,\frac{1}{\overline{A}}\,\frac{\partial\overline{A}}{\partial t}\right]$$

$$= \overline{A}^2\left[\omega\text{Im}\alpha + \text{Re}\,\frac{\partial(\alpha\omega)}{\partial\omega}\,\frac{1}{\overline{A}}\,\frac{\partial\overline{A}}{\partial t}\right] .$$

where $\dot{\alpha} = \frac{\partial\alpha}{\partial t} + \frac{\partial\alpha}{\partial\omega}\,\dot{\omega}$, where, for freely propagating radiation,

$$\dot{\omega} = \frac{\partial\omega}{\partial t} = -\frac{\omega\nu}{c}\,\frac{\partial n}{\partial t}$$

(equation 85). Hence

$$\langle A\,\frac{\partial}{\partial t}\,\alpha A\rangle = \overline{A}\left[\overline{A}\,\text{Re}\,\dot{\alpha} + \omega\overline{A}\,\text{Im}\alpha + \text{Re}\,\frac{\partial(\alpha\omega)}{\partial\omega}\,\frac{1}{\overline{A}}\,\frac{\partial\overline{A}}{\partial t}\right] ,$$

whereupon, by comparison with (16),

$$\frac{\partial\overline{\alpha A}}{\partial t} = \overline{A}\,\text{Re}\,\frac{\partial\alpha}{\partial t} - \overline{A}\,\frac{\omega\nu}{c}\,\frac{\partial n}{\partial t}\,\text{Re}\,\frac{\partial\alpha}{\partial\omega} + \text{Re}\,\frac{\partial(\alpha\omega)}{\partial\omega}\,\frac{\partial\overline{A}}{\partial t} . \qquad (17)$$

Note that we could not have obtained (17) simply by differentiating (14). Equation (14) can nevertheless be differentiated with respect to position coordinates and other time-independent variables in the usual manner. However, had we been considering a situation involving *spatial dispersion*, then relations similar to (15-17) above would apply in respect of the spatial coordinates and wavevector components. Note well that the terms frequency (temporal) dispersion and spatial dispersion refer to properties of the operators ϵ and μ without reference to the dispersion relation.

Equation (17) is the completely general form of this result which takes full account of any time-dependence of the medium properties.

With the aforementioned approximations in mind and if we neglect, for the present, any time variation in the properties of the medium ($\partial \epsilon / \partial t = \partial \mu / \partial t = 0$), we find that the rms electromagnetic fields satisfy the following general first-order field equations:

$$\text{div } \bar{D} + \frac{1}{2} \kappa \cdot \bar{D} = 0 \qquad (18a)$$

$$\text{div } \bar{B} + \frac{1}{2} \kappa \cdot \bar{B} = 0 \qquad (18b)$$

$$\text{curl } \bar{E} + \frac{\partial \bar{B}}{\partial t} + \frac{1}{2}(\kappa \times \bar{E}) = 0 \qquad (19a)$$

$$\text{curl } \bar{H} - \frac{\partial \bar{D}}{\partial t} + \frac{1}{2}(\kappa \times \bar{H}) = 0 \qquad (19b)$$

$$\bar{E} \cdot \bar{D} = \bar{B} \cdot \bar{H} = k \cdot (\bar{E} \times \bar{H})/\omega . \qquad (20)$$

In a stationary isotropic medium, when the vectors E and H (and hence $\bar{E}$ and $\bar{H}$) are respectively parallel to D and B ($\bar{D}$ and $\bar{B}$) while the wavevector k is parallel to κ, the solutions of these equations are *transverse waves* such that

$$k \times \bar{E} = \omega \bar{B} , \quad k \times \bar{H} = -\omega \bar{D} , \quad \bar{E} \cdot \bar{H} = 0. \qquad (21)$$

$$\text{div } \bar{D} = \text{div } \bar{B} = 0 . \qquad (22)$$

Therefore, making use of (10),

$$\bar{E} \times \bar{H} = \overline{EH} \, \hat{k} = \frac{c}{n} \bar{E} \cdot \bar{D} \, \hat{k} = \frac{c}{n} \overline{ED} \, \hat{k}$$

$$= \frac{c}{n} \bar{H} \cdot \bar{B} \, \hat{k} = \frac{c}{n} \overline{HB} \, \hat{k} ,$$

and hence,

$$\bar{D}/\bar{H} = \bar{B}/\bar{E} = n/c . \qquad (23)$$

We draw attention to the fact that, in arriving at the above conclusions, it was not necessary to make the common assumption that the operators ϵ and μ are equivalent to multiplication by ordinary numbers. The present formulation, which includes a first-order treatment of absorption, permits these operators to give rise to first-order small phase-shifts and neglects only second (and higher) order terms.

Now, introducing the (time-averaged) *energy density* U, and equating the energy flux, $U\nu$, to the time-averaged Poynting vector, $\langle E \times H \rangle = \bar{E} \times \bar{H}$, yields, with the aid of (10 & 20-21):

$$\bar{E} \times \bar{H} = \frac{c}{2n}(\bar{E}\cdot\bar{D} + \bar{H}\cdot\bar{B})\hat{k} = U\nu \tag{24}$$

which gives the mean energy density according to:

$$U = \frac{c}{2n\nu}(\bar{E}\cdot\bar{D} + \bar{H}\cdot\bar{B}) \tag{25}$$

which, upon noting that (for $n^2 = \epsilon\mu$):

$$\frac{c}{\nu} = \frac{\partial}{\partial\omega}(n\omega) = \frac{n}{2}\left[\frac{1}{\epsilon}\frac{\partial}{\partial\omega}(\epsilon\omega) + \frac{1}{\mu}\frac{\partial}{\partial\omega}(\mu\omega)\right]$$

is seen to be in agreement with the result given by Landau and Lifshitz [1,49] for the situation of an electromagnetic wave in a transparent weakly-dispersive medium.

Now, using (21) and (25), we can write the average *Maxwell stress tensor* M in the following way:

$$M \equiv \bar{E}\bar{D} + \bar{H}\bar{B} - \frac{1}{2}(\bar{E}\cdot\bar{D} + \bar{H}\cdot\bar{B})\mathbf{1}$$

$$= -\frac{1}{2}(\bar{E}\cdot\bar{D} + \bar{H}\cdot\bar{B})(\hat{k}\hat{k}) = -\frac{\nu n}{c} U (\hat{k}\hat{k}) \tag{26}$$

The contribution of a stress-tensor to a body force is obtained by taking its divergence (in general, by contraction on the second index). Taking the divergence of M in this manner, and making use of (11,14 & 18-19), yields:

$$\nabla\cdot{}^T M = \bar{D} \times \frac{\partial\bar{B}}{\partial t} + \frac{\partial\bar{D}}{\partial t} \times \bar{B} + \frac{1}{2n}\left[\bar{E}\cdot\bar{D} + \bar{B}\cdot\bar{D}\right](n\mathbf{k} - \nabla n) \tag{27}$$

Now, calculating the time-derivatives according to (17) using (21) yields:

$$\frac{\partial\bar{D}}{\partial t} = -\frac{\partial}{\partial t}\left[\frac{k}{\omega} \times \bar{H}\right] = -\frac{\partial k}{\partial\omega} \times \frac{\partial\bar{H}}{\partial t} = -\frac{1}{\nu}\hat{k} \times \frac{\partial\bar{H}}{\partial t},$$

and similarly:

$$\frac{\partial\bar{B}}{\partial t} = \frac{1}{\nu}\hat{k} \times \frac{\partial\bar{E}}{\partial t}.$$

Hence, upon substituting accordingly into (27) while making further use of (21), we obtain:

$$\nabla\cdot{}^T M = \frac{n}{\nu c}\frac{\partial}{\partial t}(\bar{E} \times \bar{H}) + \frac{1}{2n}(\bar{E}\cdot\bar{D} + \bar{B}\cdot\bar{D})(n\mathbf{k} - \nabla n)$$

$$= \frac{\partial}{\partial t}\left[\frac{U}{\omega}k\right] + \frac{U\nu}{c}(n\mathbf{k} - \nabla n), \tag{28}$$

in which $(U/\omega)k$ is recognisable as the Minkowski momentum density

[38] (which is the density of *pseudomomentum* $\hbar k$ [40]), which, together with M, Uv, and U comprise the components of the Minkowski energy-momentum tensor. Unlike the Abraham tensor [39] which describes the superimposed "external" electromagnetic field as if it were a decoupled entity, this tensor is not in general symmetric in an arbitrary Lorentz frame; it is guaranteed symmetric only in the local co-moving frame, and then only if the medium is isotropic. The nature of the incompleteness of the Minkowski tensor has been pointed out by Peierls [40] (among others) who showed that some momentum that must be associated with the radiation is actually carried in the medium in (mechanical) kinetic form. A detailed discussion of this, the so-called *momentum-of-light* problem [38-45], is given elsewhere [37] by this author where it is shown that the canonical decomposition of the total energy-momentum tensor rests upon the practical requirement that the material (thermokinetic) part of the stress 3-tensor be Galilean invariant to $O(u/c)$. This results in a decomposition of the rotationally-invariant total energy-momentum tensor into components (one of which is the Minkowski tensor) each containing a (non-rotationally invariant) pseudotensor part. An alternative decomposition which separates out the Abraham tensor yields rotationally invariant tensors but leaves some of the momentum of the radiation associated with the material-related component of the tensor which is then not Galilean invariant to $O(u/c)$. Accepting the interpretation of the Abraham tensor as being the part which relates to the pseudo-decoupled electromagnetic-field subsystem, it is demonstrated below that the momentum-of-light dichotomy manifests itself as an additional contribution to the ponderomotive force. Another point to beware of is that any *ad hoc* imposition of rotational invariance on a partial tensor, such as the Minkowski tensor, is usually completely unjustified.

The above formulation in terms of first-order-approximate field equations provides a *local* description of the radiation field. Because of the approximations, it is not adequate for the purpose of describing changes in the field during propagation over large time and distance scales. For this *global* description of the field, we use methods based on *geometric optics*.

Thermodynamic relations

Before giving an account of the geometric optics description of the field, we use the above to derive the usual thermodynamic relations in terms of the local field variables.

The Helmholtz free energy of the field is obtained by considering a general variation of the macroscopic fields subject to an isothermal, isochoric variation of the medium. For such a process, the change in the free energy of the field is precisely given by the net flow of electromagnetic energy into a stationary homogeneous element of the material medium. Therefore, letting φ be the instantaneous free energy density of the field, we have:

$$\delta \int \varphi dV = -\delta t \int (E \times H) \cdot dS$$

$$\equiv -\delta t \int \nabla \cdot (E \times H) dV$$

$$\equiv -\delta t \int (\mathbf{H} \cdot \text{curl } \mathbf{E} - \mathbf{E} \cdot \text{curl } \mathbf{H}) dV$$

whence, using Maxwell's equations (3-4),

$$\delta \int \varphi dV = \delta t \int (\mathbf{H} \cdot \frac{\partial \mathbf{B}}{\partial t} + \mathbf{E} \cdot \frac{\partial \mathbf{D}}{\partial t}) dV \ .$$

Since these integrals are taken over an arbitrary fixed volume of material that remains at rest, we must have:

$$\delta \varphi = \mathbf{H} \cdot \delta \mathbf{B} + \mathbf{E} \cdot \delta \mathbf{D} \tag{29}$$

in which **H** and **E** appear as generalised forces, and **B** and **D** respectively as the corresponding displacements. (In the same sense that the corresponding thermodynamic relations for homogeneous material systems are independent of the equations-of-state, we can reasonably regard the above as being a general thermodynamic relation (valid for constant ρ and T) that is essentially independent of the constitutive field equations.) The function $\varphi(\mathbf{B},\mathbf{D})$ is the generating function for the fields, **H**, **E** , in the usual manner.

In the (general) case of a *linear* medium, equation (29) can be integrated to yield the well-known result,

$$\varphi(\rho,t,\mathbf{D},\mathbf{B}) = \frac{1}{2} (\mathbf{B} \cdot \mathbf{H} + \mathbf{D} \cdot \mathbf{E}) \ . \tag{30}$$

(In the case of an *intrinsically* non-linear medium – eg. strong-field situations – equation (29) remains valid, but not (30).)

We now construct the time-average of (30) for quasimonochromatic fields. However, it is important to realise that, in so doing, one introduces an additional variable, namely the wave frequency ω, into the description of the field. Thus, taking the time average of (30), making use of (14), yields straightforwardly that

$$\bar{\varphi}(\rho,T,\bar{\mathbf{D}},\bar{\mathbf{B}},\omega) = \frac{1}{2} (\bar{\mathbf{B}} \cdot \bar{\mathbf{H}} + \bar{\mathbf{D}} \cdot \bar{\mathbf{E}}), \tag{31}$$

while the general variation of $\bar{\varphi}$ with the medium properties held fixed becomes

$$\delta \bar{\varphi} = \frac{\partial \bar{\varphi}}{\partial \bar{\mathbf{B}}} \cdot \delta \bar{\mathbf{B}} + \frac{\partial \bar{\varphi}}{\partial \bar{\mathbf{D}}} \cdot \delta \bar{\mathbf{D}} + \frac{\partial \bar{\varphi}}{\partial \omega} \delta \omega \tag{32}$$

in which the first two terms continue to represent the generalised work δW. The third term is associated with a generalised field-entropy ζ and field-temperature θ (which are unrelated to material entropy and temperature except by analogy) as can be seen by recalling the earlier expression for the field energy U , equation (25), which differs from the free-energy (31) due to the presence of a non-zero "TS" term. Thus, with the aid of (25), we find

$$U - \bar{\varphi} = \theta \zeta = - \left[\frac{n v}{c} - 1 \right] U \tag{33}$$

while, referring back to (32) we have

$$\zeta \delta\theta = -\frac{\partial \bar{\varphi}}{\partial \omega} \delta\omega . \qquad (34)$$

Hence, upon eliminating ζ from between (33) and (34),

$$\left[\frac{n\nu}{c} - 1\right] U \frac{\delta\theta}{\theta} = \frac{\partial \bar{\varphi}}{\partial \omega} \delta\omega . \qquad (35)$$

But, upon differentiating (30) directly while making use of (14), (20) and (11), together with

$$\delta\bar{E} = \delta\left[\frac{1}{\epsilon_0 \text{Re}\epsilon}\right]\bar{D} + \text{Re}\frac{\partial}{\partial \omega}\left[\frac{\omega}{\epsilon_0 \epsilon}\right]\delta\bar{D}$$

which follows from (17) and from which the the partial derivatives of $\bar{E}$ for constant $\bar{D}$ follow directly, we also find

$$\frac{\partial \bar{\varphi}}{\partial \omega} = \frac{1}{2}\left[\bar{B}\cdot\bar{B}\frac{\partial}{\partial \omega}\left[\frac{1}{\mu_0 \text{Re}\mu}\right] + \bar{D}\cdot\bar{D}\frac{\partial}{\partial \omega}\left[\frac{1}{\epsilon_0 \text{Re}\epsilon}\right]\right]$$

$$= -\frac{1}{2}\left[\bar{B}\cdot\bar{H}\frac{1}{\text{Re}\mu}\frac{\partial}{\partial \omega}\text{Re}\mu + \bar{D}\cdot\bar{E}\frac{1}{\text{Re}\epsilon}\frac{\partial}{\partial \omega}\text{Re}\epsilon\right]$$

$$= -\frac{1}{2}\left[\bar{B}\cdot\bar{H} + \bar{D}\cdot\bar{E}\right]\frac{1}{n}\frac{\partial n}{\partial \omega}$$

$$= \left[\frac{n\nu}{c} - 1\right]\frac{U}{\omega} , \qquad (36)$$

with equation (11) having been replaced by the now-appropriate approximation, $n^2 = \text{Re}(\epsilon)\text{Re}(\mu)$. Equations (35-36) now yield that

$$\frac{\delta\theta}{\theta} = \frac{\delta\omega}{\omega} \quad \Rightarrow \quad \theta \propto \omega .$$

Therefore we can define, for a quasimonochromatic wave of frequency ω, a generalised field-temperature $\theta = \hbar\omega/K$ (where K is Boltzmann's constant) and a corresponding entropy $\zeta = -(K/\hbar)\partial\bar{\varphi}/\partial\omega$.

[In the first instance, the introduction of Planck's constant $\hbar$ here is quite arbitrary. However it does have a natural significance when one considers the *entropy per photon* of the quantised (boson) field. The photon density is $U/\hbar\omega$. The entropy per photon is therefore

$$\left[\frac{\zeta}{U/\hbar\omega}\right] = \left[1 - \frac{n\nu}{c}\right]K$$

use having been made of (36). Then recalling Boltzmann's famous equation, $S = K\ln\Omega$, we find that the thermodynamic "probability" $1/\Omega$

is given by

$$\frac{1}{\Omega} = \exp\left[\frac{n\nu}{c} - 1\right]$$

which is in (0,1) if and only if $n\nu/c < 1$. Therefore $1/\Omega$ thus defined is a sensible probability only if in cases of normal dispersion to which the validity of the present formalism has already been restricted.]

It may appear that absorption has been neglected in the derivation of (29-30). That this is not so can be confirmed by returning to equations (19) from which one readily obtains

$$-\kappa \cdot (\bar{E} \times \bar{H}) \equiv -\frac{1}{2}(\kappa \times \bar{E}) \cdot \bar{H} + \frac{1}{2}(\kappa \times \bar{H}) \cdot \bar{E}$$

$$= \bar{H} \cdot \text{curl } \bar{E} - \bar{E} \cdot \text{curl } \bar{H} + \bar{H} \cdot \frac{\partial \bar{B}}{\partial t} + \bar{E} \cdot \frac{\partial \bar{D}}{\partial t}$$

$$= \nabla \cdot (\bar{E} \times \bar{H}) + \frac{\partial \bar{\varphi}}{\partial t}, \qquad (37)$$

which shows that $\bar{\varphi}$ satisfies the appropriate conservation equation containing the absorption term $\kappa \cdot (\bar{E} \times \bar{H})$.

Finally, by considering the establishment of the electromagnetic field in an initially field-free medium without disturbing the material density or temperature, it is easily deduced that the the total Helmholtz free-energy F is given by:

$$F(\rho, T, \bar{B}, \bar{D}, \omega) = F^0(\rho, T) + \bar{\varphi}(\rho, T, \bar{B}, \bar{D}, \omega)V \qquad (38)$$

where F^0 is the free-energy when there is no macroscopic field present.

Geometrical optics description of radiation in an inhomogeneous weakly-absorbing dispersive medium.

We consider the application of geometrical optics to the propagation of a narrow pencil Π of rays, whose local mean direction of propagation is $\hat{k}$, in an inhomogeneous continuous medium in which $\nabla \text{Re}\epsilon$, $\nabla \text{Re}\mu$, $\nabla \text{Im}\epsilon$, $\nabla \text{Im}\mu$ and (hence) ∇n are all mutually parallel in the direction of the unit vector $\hat{n}$. The aim is to determine how the direction and intensity of the beam vary during propagation over large distances/times. The former objective is met by Snell's laws of refraction which, in the case of propagation in a continuous medium, take the form [36]

$$d\hat{k}/ds \equiv \hat{k} \cdot \nabla \hat{k} = -\hat{k} \times (\hat{k} \times \nabla \ln(n)) \qquad (39)$$

which possess the corollary that the areal projection of the base of the wavevector cone onto $\hat{n}$ is a constant of the motion, ie

$$\frac{d}{ds} (k^2 \hat{k} \cdot \hat{n} \, d\Omega) = 0 , \qquad (40)$$

where $d\Omega$ is the solid angle spanning the directions within the pencil of rays. The result (40) is a direct consequence of local translational invariance in the directions normal to $\hat{n}$, and can therefore be assumed to apply even if the medium exhibits time dependence.

In order to determine the amplitude variation, we return to the unapproximated forms of Maxwell's equations (1-4) whence, by elimination of **H** and **E** respectively, we obtain in the Fourier-transform space (r,ω):

$$\nabla^2 \mathbf{E} + k^2 \mathbf{E} + \frac{1}{\mu} \nabla \mu \times \text{curl } \mathbf{E} = 0$$

$$\nabla^2 \mathbf{H} + k^2 \mathbf{H} + \frac{1}{\epsilon} \nabla \epsilon \times \text{curl } \mathbf{H} = 0$$

where k^2 is given by (9). The procedure for tackling these equations is set out in [46], and involves defining local axes with the z-axis in the direction of $\hat{n}$ and the y-axis in the direction normal to $\hat{k}$. The local translational invariance leads to identifying the x-dependence of the solution as being of the form $\exp(-iKx)$ where K is a local constant of the motion ($dK/ds = 0$). For $K \neq 0$, we consider separately the two independent polarisation modes, viz **E** in the y-direction which leads to the E-field equation:

$$\mu \frac{\partial}{\partial z} \frac{1}{\mu} \frac{\partial E}{\partial z} + \left[k^2 - K^2 \right] E = 0$$

and **H** in the y-direction, in which case we have the H-field equation:

$$\epsilon \frac{\partial}{\partial z} \frac{1}{\epsilon} \frac{\partial H}{\partial z} + \left[k^2 - K^2 \right] H = 0$$

By means of the respective replacements, $\Psi = E/(\mu\mu_0)^{1/2}$ and $\Psi = H/(\epsilon\epsilon_0)^{1/2}$, the above pair of equations are reduced to the common form,

$$\frac{\partial^2 \Psi}{\partial z^2} + \left[k^2 - K^2 \right] \Psi = 0, \qquad (41)$$

which is a standard wave equation of the Schrödinger type. In the weak-absorption geometric optics limit, the required solutions of this equation are based upon the WKB approximant:

$$\left[k^2 - \text{Re} K^2 \right]^{-1/4} \exp\left[i \int^z (k^2 - K^2)^{1/2} dz \right]$$

which is the saddle-point approximation to a narrow superposition of solutions representing Π. Exploiting the saddle-point approximation for the $\omega \leftrightarrow t$ transform yields the formal solution of the original equations, in the geometric optics limit, as:

$$\Psi(r,t) \simeq \Psi_0 \frac{(2\omega)^{3/2}}{(k\cdot\hat{n})^{\frac{1}{2}}} \exp\{iS(r,t)\} \qquad (42)$$

from which one obtains the observable fields by taking the real part, and where

$$S = \int^r k\cdot ds - \omega t$$

together with (since, in the weak-absorption limit, the trajectories are in $\mathbb{R}^3$):

$$k = \nabla \mathrm{Re} S \;,\; \kappa = \kappa\, \hat{k} \;,\; k = k + \tfrac{1}{2}i\kappa$$

where κ is given by (12). Hence we have that,

$$\mathrm{curl}\; k = 0. \qquad (43)$$

Geometric optics retains essential information about the pulse shape and beam profile through the amplitude function $\Psi_0(r,t)$ whose space and time derivatives (on the relevant scales of wavelength and frequency) are almost negligible. Within a steady-state beam, Ψ_0 is constant along any given ray. In the lowest order of approximation, derivatives of Ψ_0 would be neglected completely along with those of the attenuation factor. However, we choose to retain the effect of these low-order derivatives in the theory by incorporating them into a single slowly-varying factor $A(r,t) = |\Psi_0|^2 \exp\{-\int \kappa ds\}$ within the intensity. Recalling that, for the two independent polarisation modes, $\mathrm{Re}\Psi$ respectively gives $(\mu_0\mu)^{\frac{1}{2}} \bar{E}$ and $(\epsilon_0\epsilon)^{\frac{1}{2}} \bar{H}$ which, according to (23) and (14), have the same rms magnitudes. Hence we conclude, with the aid of (11), that the general result (which turns out to be independent of the polarisation) for the product of the rms E and H fields is as follows:

$$\bar{E}\bar{H} = \frac{n}{c} \langle [\mathrm{Re}\Psi]^2 \rangle = \frac{n}{c} A(r,t) \frac{\omega^3}{k\cdot\hat{n}} \qquad (44)$$

in which the principal intensity variation is due to the $1/k\cdot\hat{n}$ factor. This factor diverges at *caustics* which are not treated properly by the WKB type of approximations used above. However it turns out that this deficiency of the geometric optics approximation is of no consequence in the present application.

Outline of the Helmholtz gauge method for determining the electromagnetic contribution to the stress tensor

The Helmholtz method for determining the stress tensor in the presence of a steady-state field involves considering a hypothetical process in which a material medium containing the field is subject to a quasistatic deformation which we shall here represent by means of a continuous displacement field $\xi(r)$ in terms of which the *strain tensor* e is

$$e_{ij} = \frac{1}{2}\left[\frac{\partial \xi_i}{\partial x_j} + \frac{\partial \xi_j}{\partial x_i}\right] \tag{45}$$

All components of the strain are assumed to be first-order small ($\ll 1$). The electromagnetic field, here consisting of quasiunidirectional quasimonochromatic radiation, is subject to accompanying adiabatic deformation represented by a *local gauge transformation* with respect to $\xi(r)$. The above formalism is used to establish the *adiabatic invariants* of the field under such a transformation, from which, using the earlier derived thermodynamic relations, an expression representing the work done during the process is derived. The (total) stress tensor (which is symmetric) is then deduced from the form of this expression.

The deformation of the electromagnetic fields is subject to the following *ab initio* requirements:

(i) The radiation field is considered to be an external field in the sense that the intrinsic properties and locations of the sources are not themselves subject to the deformation except as they need to be altered in order to accomplish the deformation.

(ii) The radiation is allowed to interact with the medium throughout the entire process and any variation of the field (after leaving the source) is determined solely by the deformation and is not due to any additional explicit time-dependence.

(iii) At all times throughout the process, the electromagnetic field remains a valid solution of Maxwell's equations here applied in an approximated form appropriate to the geometric optics limit.

(iv) The deformation of the field is represented as a deformation e of the manifold charted by the rays and constant phase surfaces such that, in the presence of an accompanying deformation ξ of the material medium, the system of rays remains invariant with respect to the material points. (ie. A ray that traverses any two material points, continues to traverse the same two points when subject to the deformation).

In the absence of longitudinal shear (see below), we consider the same deformation to be simultaneously applied to the medium and to the field, in which case the system of rays and constant phase surfaces are both invariant with respect to the material points. (Applying the same deformation to field and matter avoids having to make a decomposition of the system, so that, at this stage, we do not have to worry about the "momentum-of-light" problem.)

In general, the strain tensor e of the electromagnetic field is taken to be the 4-component projection of the total strain $\tilde{e}$ applied to the material given by

$$e = \tilde{e} - (\tilde{e}\cdot\hat{k})\hat{k} - \hat{k}(\hat{k}\cdot\tilde{e}) + 2(\tilde{e}:\hat{k}\hat{k})\hat{k}\hat{k} .$$

which retains the invariant property of the rays with respect to the material.

The need to impose this constraint on the field strain tensor e arises because a *transverse* wave can only be subject to 4 independent local deformations, namely compression/extension in the three orthogonal directions along and normal to k, and rotation of the plane of polarisation (as represented by the transverse shear (xy) component of the strain), whereas a general strain comprises 6 independent components. Moroever a longitudinal shear deformation of the constant-phase surfaces produces an accompanying refraction of the rays so as to preserve their orthogonality with the wavefronts, with the result that they can no longer transect the same material points when the latter are subject to the same deformation. Thus we see that any field representing transverse electromagnetic radiation cannot, in the sense of (iii-iv) above, be subject to *longitudinal shear* (which are the xz and yz components of the strain when z is in the local direction of the wavevector). However if the material alone is subject to a pure longitudinal shear, then the ray trajectories, as given in the geometric optics approximation, remain fixed in space. Moreover, as the displacement is everywhere along the ray directions, the rays remain invariant with respect to the material, so that only the phase of the wave changes when viewed at a fixed point in the material. Therefore the requirement that, when the material is subject to a pure longitudinal shear, there is no accompanying deformation of the field is consistent with (iii) and (iv) above. In the case of a general deformation of the medium, the strain tensor e to which the local field is subject is taken to be material strain with the longitudinal shear components projected out. (Refraction, or bending of rays, due to the transformation, is associated with the derivatives of the components of e — rather than with shear components of $\tilde{e}$) In the presence of caustics, the ray manifold may be multi-valued in $\mathbb{R}^3$ in which case we consider a separate calculation to apply to each layer of the ray manifold, and subsequently construct the total stress tensor by superposition (see below).

It is convenient, in the first instance, to confine the discussion to the case $\tilde{e} = e$ when there is no longitudinal shear. The generalisation to situations when the material deformation includes longitudinal shear is made later.

If necessary, in order to permit (iii) and (iv), the deformation may be considered to be applied *locally* within a small element of material of volume V. This of course relies on the assumption that the stress-tensor is dependent only on local properties of the fields.

We define an *adiabatic invariant* of the wave field to be a property of the field within V that changes *negligibly* due to a *finite* quasistatic variation of the type described above. In a transparent medium, the propagation of radiation in the geometric optics limit is governed by the Liouville transport equation [47,33,37], $Df/Dt = 0$, where $f(k,r,t)$ is the phase-space distribution function and $D/Dt = \partial/\partial t + \{f, \omega\}$. This equation states that f, and hence $I^* \equiv I/n^2\omega^3$ where I is the specific intensity, is a conserved property of a wave packet or beam element as it propagates through phase space and is so *independently of any motion of the medium*. We say that f and I^* are *propagation invariants* in a time-dependent medium. Remembering that the transformation leaves the ray paths fixed in relation to the material within V, it follows, given that the related source property is also invariant, that such propagation invariants are also adiabatic invariants. Thus, in a transparent medium, I^* is an adiabatic invariant. In an absorbing medium, I^* is modified by absorption in the surrounding medium which will be affected by the variational process. However we are only interested in the local behaviour within V. Therefore we can choose to compensate for absorption in the surrounding medium by effectively varying the source intensity so as to constrain I^* to be constant at some reference point within V (and fixed relative to the medium). Thus, in a transparent medium, I^* is a natural adiabatic invariant when the constraint is imposed on the source. In an absorbing medium, I^* itself is directly constrained to be an adiabatic invariant. In either case:

(vi) The Lorentz-invariant intensity I^* is constrained to remain constant at a reference point fixed in the medium within V. If the dimensions of V are chosen small enough (compared with $1/\kappa$) then I^* can be considered to be invariant anywhere within V.

Identification of adiabatic-invariants

We now proceed to identify the other adiabatic invariants given in terms of the parameter set $\{\omega, \Delta\omega, \Delta\Omega, \hat{k}, I\}$ which represents Π. In addition to I^*, an additional four such invariants are identifiable from which the complete variation of Π can be determined.

At equation (40) it was noted that $k^2 \hat{k} \cdot \hat{n} \Delta\Omega$ is a propagation invariant in a time-dependent medium. This quantity is clearly unchanged in any region in which $\kappa = 0$, and hence can be regarded as an invariant property of the source. It therefore follows by a previous argument that $k^2 \hat{k} \cdot \hat{n} \Delta\Omega$ is an adiabatic invariant.

A deformation of the wave field in accordance with (iv) above has the property that the integral

$$\int_1^2 \mathbf{k} \cdot d\mathbf{s}$$

is invariant over the deformation for any pair of material points 1 and 2. Moreover the geometric optics property of this integral that, by Fermat's principle, it is a minimum with respect to variations in the ray path, is thus also preserved by the deformation. Therefore the deformed ray paths remain given in accordance with geometric optics showing that (iii) and (iv) above are consistent. Now, since curl $\mathbf{k} = 0$, writing $\mathbf{k} = \nabla\psi$, where $\psi = \text{Re}S$, the variation of

$$\int_1^2 \mathbf{k} \cdot d\mathbf{s} = \psi_2 - \psi_1$$

vanishes for arbitrary choices of 1 and 2 if and only if ψ is an invariant at each material point in the medium (aside from an arbitrary global gauge transformation). Thus ψ is an adiabatic-invariant gauge field.

In particular, if we consider V to be a rectangular element of dimensions Δx, Δy, Δz with Δz measured in the local direction of $\hat{k}$, then the above leads to $\Delta k_x \Delta x$, $\Delta k_y \Delta y$ and $k \Delta z$ being identified as further adiabatic invariants where Δk_x and Δk_y are the wavevector uncertainties in the x and y directions respectively (being the maximum x- and y- components of k on the surface of the wave vector cone). (cf diffraction by rectangular slit). Taking the product of these three invariants yields

$$\Delta k_x \Delta k_y k \Delta x \Delta y \Delta z = k^3 \left[\frac{\Delta k_x \Delta k_y}{k^2} \right] V \propto k^3 V \Delta \Omega$$

from which we deduce that $k^3 V \Delta \Omega$ is an adiabatic invariant of Π within V, a property which we can regard as being suitably independent of the shape of the volume V.

The last adiabatic invariant of Π that we require is identified by recognising the quantised nature of the field and that the deformation process necessarily preserves the separate identities of the individual photons. If we regard each photon as a wave-train of duration τ_c - the *coherence time* which is proportional to $1/\Delta \omega$ - then the number of oscillation periods counted at a quasistatic point during the passage of any single photon is clearly an invariant property of that photon. Therefore $1/\omega \tau_c \propto \Delta \omega / \omega$ is an adiabatic invariant.

Note that we are here not dealing with a closed system. Although V contains a fixed amount of material, radiation is allowed to propagate freely through the boundaries. The number of photons within V at any instant is given by

$$N = f V k^2 \Delta k \Delta \Omega = (\Delta k / k) f k^3 V \Delta \Omega$$

so that, since f and $k^3 V \Delta \Omega$ are adiabatic invariants as already noted, we find that N varies in proportion to $\Delta k / k = c \Delta \omega / n \omega v \propto c / n v$ where v is the group velocity. Hence N is an adiabatic invariant only in a non-dispersive medium. What *is* necessary is that the total number of photons in the universe is unaltered as a direct consequence of adiabatic work. Therefore, if P_r is the radiation pressure defined by $P_r = -\partial (\varphi V)/\partial V$, we require that $P_r dV + N \hbar d\omega = 0$ which states that all work done on the radiation is accounted for by changes in the specific photon energy ($\hbar \omega$), and which implies [52] that it is the Massieu function $(\zeta - U/\omega)V$ that is the conserved thermodynamic quantity for the system. It is easily verified that this is indeed so and moreover we could have invoked this argument as an alternative means of deriving the adiabatic invariance of $\Delta \omega / \omega$. (Also note that, in the present context, it is necessary to distinguish between the three possible definitions of "adiabatic", *viz* "slow in respect of a defined system - not necessarily closed - that is thermally insulated from its surroundings"; "isentropic" and "of a closed system". In the case of radiation, where frequency plays the rôle of temperature, by "thermally insulated" we simply mean that that frequency of radiation within the system is not influenced by the frequency of radiation present in the surroundings. This is necessarily true given that there is no process whereby photons can directly exchange energy. It is however interesting to speculate on the generalised consequences of non-linear photon coupling.)

Finally, upon referring back to (44) and writing the energy-flux or Poynting vector,

$$F = \bar{E} \times \bar{H} = \bar{E}\bar{H}\hat{k} = A \frac{\omega^2}{\hat{k}\cdot\hat{n}} \hat{k} , \qquad (46)$$

in terms of the parameters of Π gives,

$$F = I\hat{k}\Delta\omega\Delta\Omega \equiv \left[\frac{I}{n^2\omega^3}\right]\left[\frac{\Delta\omega}{\omega}\right] \omega^4 n^2 \hat{k}\Delta\Omega$$

$$\equiv \left\{I^* (\Delta\omega/\omega) k^2\hat{k}\cdot\hat{n} \Delta\Omega\right\} \omega^2 c^2 \hat{k}/\hat{k}\cdot\hat{n}$$

whereupon

$$A = c^2 \left\{I^* (\Delta\omega/\omega) k^2\hat{k}\cdot\hat{n} \Delta\Omega\right\} . \qquad (47)$$

which expresses the slowly-varying intensity-normalisation factor $A(r,t)$ in terms of the parameters of Π. The expression in {} brackets in the above is, by reference to the preceding, recognised as an adiabatic invariant and hence so also is $A(r,t)$.

The following expresses the list of adiabatic invariants so far identified as a set of constraints to which the local deformation of the field is considered to be subject:

$$\delta\psi = 0 \qquad (48)$$

$$\delta(k^2\hat{k}\cdot\hat{n} \Delta\Omega) = 0 \qquad (49)$$

$$\delta(k^3 V \Delta\Omega) = 0 \qquad (50)$$

$$\delta(\Delta\omega/\omega) = 0 \qquad (51)$$

$$\delta(I^*) = 0 \qquad (52)$$

$$\delta(A) = 0 \qquad (53)$$

where δ here denotes the variation at a point that is fixed in the material. (The notation ∂ shall be used to denote a variation at a fixed point in space.) Upon eliminating $\Delta\Omega$ from between (49) and (50) yields that

$$\delta(\hat{k}\cdot\hat{n}/kV) = 0 \qquad (54)$$

which demonstrates the important feature of the deformation process that the caustics are fixed relative to the medium and therefore we do not have to worry about the generation of zero-order terms due to infinitesimal displacements of (pseudo-infinite) caustics. Thus, in the present context, the geometric-optics approach is not invalidated by the presence of caustics despite their inadequate representation in the first instance by divergent wavefunctions.

Determination of the stress-tensor for high-frequency radiation in an isotropic material medium

We begin by introducing the field $\mathbf{Q} = \bar{\mathbf{D}} \times \bar{\mathbf{B}}$ so that, making reference to (46),

$$\mathbf{Q} \equiv \bar{\mathbf{D}} \times \bar{\mathbf{B}} = \frac{n^2}{c^2} \bar{\mathbf{E}} \times \bar{\mathbf{H}} = A \frac{k}{\hat{k} \cdot \hat{n}} \mathbf{k} = A \frac{k}{\hat{k} \cdot \hat{n}} \nabla \psi \ . \qquad (55)$$

and proceed by using the first constraint (48) to determine the variation $\delta \mathbf{Q}$ at the point of reference (which is fixed in the medium) in terms of the displacement field $\boldsymbol{\xi}(r)$ and corresponding strain tensor $\mathbf{e}$:

$$\left. \begin{array}{rcl} \delta\psi & = & 0 \\ \delta\psi & \equiv & \partial\psi + \boldsymbol{\xi} \cdot \nabla\psi \end{array} \right\} \Rightarrow \partial\psi = -\boldsymbol{\xi} \cdot \nabla\psi$$

$$\therefore \quad \partial(\nabla\psi) \equiv \nabla \partial\psi = -\nabla(\boldsymbol{\xi} \cdot \nabla\psi)$$

and hence:

$$\delta(\nabla\psi) \equiv \partial(\nabla\psi) + \boldsymbol{\xi} \cdot \nabla(\nabla\psi) = -\nabla(\boldsymbol{\xi} \cdot \nabla\psi) + \boldsymbol{\xi} \cdot \nabla(\nabla\psi)$$

which becomes, upon invoking a standard vector identity,

$$\delta(\nabla\psi) = -(\nabla\psi) \cdot \nabla\boldsymbol{\xi} - \nabla\psi \times \text{curl } \boldsymbol{\xi} \ .$$

Therefore, using (55),

$$\delta \mathbf{Q} \equiv \delta\left[A(k/\hat{k}\cdot\hat{n}) \nabla\psi \right]$$
$$= -\mathbf{Q} \cdot \nabla\boldsymbol{\xi} - \mathbf{Q} \times \text{curl}\boldsymbol{\xi} + \mathbf{Q}\delta\ln(k/\hat{k}\cdot\hat{n}) + \mathbf{Q}\delta\ln(A)$$

But, by (53) and (54) respectively,

$$\delta\ln(A) = 0,$$
$$\delta\ln(k/\hat{k}\cdot\hat{n}) = -\delta\ln V = -\delta V/V,$$

and therefore,

$$\delta \mathbf{Q} = -\mathbf{Q} \cdot \nabla\boldsymbol{\xi} - \mathbf{Q} \times \text{curl}\boldsymbol{\xi} - \mathbf{Q}\delta V/V \ . \qquad (56)$$

Taking the scalar product with $\mathbf{Q}$ yields:

$$\mathbf{Q} \cdot \delta\mathbf{Q} = -(\mathbf{Q}\mathbf{Q}):\nabla\boldsymbol{\xi} - Q^2 \delta V/V$$
$$= -(\mathbf{Q}\mathbf{Q}):\mathbf{e} - Q^2 \delta V/V$$
$$= -Q^2 \left[(\hat{k}\hat{k}):\mathbf{e} + \frac{\delta V}{V} \right] \ .$$

But, from (24) and (55),

$$Q^2 = \frac{n^4 \nu^2}{c^2} U^2.$$

Hence:

$$\mathbf{Q}\cdot\delta\mathbf{Q} = -\frac{n^4\nu^2}{c^2} U^2 \left[(\hat{k}\hat{k}):e + \frac{\delta V}{V} \right] . \tag{57}$$

Calculating this same quantity in another manner starting from,

$$\mathbf{Q}\cdot\mathbf{Q} \equiv (\bar{\mathbf{D}} \times \bar{\mathbf{B}})\cdot(\bar{\mathbf{D}} \times \bar{\mathbf{B}}) = (\bar{\mathbf{D}}\cdot\bar{\mathbf{D}})\cdot(\bar{\mathbf{B}}\cdot\bar{\mathbf{B}})$$

gives, using (14), (23), (20) and finally (25):

$$\mathbf{Q}\cdot\delta\mathbf{Q} \equiv \frac{1}{2}\delta(\mathbf{Q}\cdot\mathbf{Q}) = \bar{\mathbf{B}}^2\bar{\mathbf{D}}\cdot\delta\bar{\mathbf{D}} + \bar{\mathbf{D}}^2\bar{\mathbf{B}}\cdot\delta\bar{\mathbf{B}}$$

$$= \frac{\bar{\mathbf{B}}\bar{\mathbf{D}}}{\bar{\mathbf{E}}\bar{\mathbf{H}}} \left[\bar{\mathbf{B}}\cdot\bar{\mathbf{H}}\ \bar{\mathbf{E}}\cdot\delta\bar{\mathbf{D}} + \bar{\mathbf{D}}\cdot\bar{\mathbf{E}}\ \bar{\mathbf{H}}\cdot\delta\bar{\mathbf{B}} \right]$$

$$= \frac{n^2}{2c^2} \left[\bar{\mathbf{B}}\cdot\bar{\mathbf{H}} + \bar{\mathbf{D}}\cdot\bar{\mathbf{E}} \right] \left[\bar{\mathbf{E}}\cdot\delta\bar{\mathbf{D}} + \bar{\mathbf{H}}\cdot\delta\bar{\mathbf{B}} \right]$$

$$= \frac{n^3\nu}{c^3} U \left[\bar{\mathbf{E}}\cdot\delta\bar{\mathbf{D}} + \bar{\mathbf{H}}\cdot\delta\bar{\mathbf{B}} \right] . \tag{58}$$

Hence, upon equating (57) and (58), while recalling (26) & (29), we find:

$$\bar{\mathbf{E}}\cdot\delta\bar{\mathbf{D}} + \bar{\mathbf{H}}\cdot\delta\bar{\mathbf{B}} = -\frac{n\nu}{c} U \left[(\hat{k}\hat{k}):e + \frac{\delta V}{V} \right]$$

$$= \mathbf{M}:e - \bar{\varphi}\frac{\delta V}{V} , \tag{59}$$

which gives the essential relation between the strain and the resulting change in the electromagnetic field. Equation (59) is a key result of this work.

The final step involves calculating the work done δW during the process. From the first and second Laws of Thermodynamics applied to both material and radiation we have:

$$\delta(E + UV) = T\delta S + \omega\delta(\zeta V) + \delta W$$

where E represents the material contribution to the internal energy, and T and S are the material temperature and entropy respectively. Hence we find that the change in the total free-energy,

$$F = F^0 + \bar{\varphi}V \equiv E + UV - TS - \omega\zeta V, \tag{60}$$

is given by:

$$\delta F = \delta(F^0 + \bar{\varphi}V) = -S\delta T - V\zeta\delta\omega + \delta W. \tag{61}$$

In the absence of any external electromagnetic field, the above relations respectively become:

$$\delta E^0 = T\delta S^0 + \delta W^0 \tag{62}$$

$$F^0 = E^0 - TS^0 \tag{63}$$

$$\delta F^0 = -S^0 \delta T + \delta W^0. \tag{64}$$

We now *define* the *total radiative stress tensor* σ such that

$$\delta W = \delta W^0 + V\sigma{:}e \tag{65}$$

$${}^T\sigma = \sigma \tag{66}$$

where ^{T} denotes transpose (ie. σ is symmetric). Then, combining the above relations so as to eliminate specifically material-related terms,

$$\frac{1}{V} \delta(\bar{\varphi}V) = -\frac{1}{V}(S - S^0)\delta T - \zeta\delta\omega + \sigma{:}e \tag{67}$$

from which, in the case of an isothermal process ($\delta T = 0$), we have

$$\frac{1}{V}\delta(\bar{\varphi}V) = \sigma{:}e - \zeta\delta\omega \tag{68}$$

while, by application of the chain rule, the variation in $\bar{\varphi}(\bar{B}, \bar{D}, \omega, \rho, T)$ for this process is:

$$\delta\bar{\varphi} = \frac{\partial\bar{\varphi}}{\partial \bar{B}} \cdot \delta\bar{B} + \frac{\partial\bar{\varphi}}{\partial \bar{D}} \cdot \delta\bar{D} + \frac{\partial\bar{\varphi}}{\partial\rho}\delta\rho + \frac{\partial\bar{\varphi}}{\partial\omega}\delta\omega$$

$$= \bar{H}\cdot\delta\bar{B} + \bar{E}\cdot\delta\bar{D} + \rho\frac{\partial\bar{\varphi}}{\partial\rho}\frac{\delta\rho}{\rho} - \zeta\delta\omega$$

where we have made use of (29) and (34) (with $\theta = \omega$). Finally, invoking (59), and using that $\delta V/V = -\delta\rho/\rho = \text{trace}(e) \equiv e{:}1$, we have

$$\delta(\bar{\varphi}V) = M{:}e - \rho\left[\frac{\partial\bar{\varphi}}{\partial\rho}\right]1{:}e - \zeta\delta\omega$$

which, by comparison with (68), yields that

$$\sigma{:}e = M{:}e - \rho\left[\frac{\partial\bar{\varphi}}{\partial\rho}\right]1{:}e \ .$$

Now, since this holds for an arbitrary small 4-component deformation e, the non-longitudinal-shear components of the stress tensor are given by:

$$\sigma = M - \rho\left[\frac{\partial\bar{\varphi}}{\partial\rho}\right]1 \tag{69}$$

or, by making reference to (26) and (31),

$$\sigma = -\bar{\varphi}\,\hat{k}\hat{k} - \rho\left[\frac{\partial\bar{\varphi}}{\partial\rho}\right]\mathbf{1} \tag{70}$$

which expresses σ entirely in terms of the free-energy in a suggestively simple form. In precisely the same manner as (36) we have that:

$$\frac{\partial\bar{\varphi}}{\partial\rho} = -\bar{\varphi}\,\frac{1}{n}\,\frac{\partial n}{\partial\rho}\,, \tag{71}$$

and hence, with the aid of (25) and (31),

$$\sigma = M + U\rho\,\frac{v}{c}\,\frac{\partial n}{\partial\rho}\,\mathbf{1} \tag{72}$$

The transverse-wave stress tensor σ is thus defined to have vanishing longitudinal shear components and therefore has the property that $\sigma:(e - \tilde{e}) \equiv 0$, so that no change results from a formal substitution of e by $\tilde{e}$ in equations (56) onwards. As the field is invariant under the strain $\tilde{e} - e_{\perp}$ the stress σ thus correctly gives the work due to a general strain $\tilde{e}$. Then since $\tilde{e}$ (unlike e) is single valued on $\mathbb{R}^3$, we can now calculate the total stress by a linear superposition over all components of the radiation field in the manner indicated below.

The result expressed by equations (69-72) is a straightforward generalisation of the quasistatic Helmholtz stress-tensor [1,24] in accord with earlier conjectures [7,18,32,33,44,43]. We note that it is the form of the expression expressed in terms of the Helmholtz free-energy density that is unchanged by the generalisation, whereas alternative expressions, such as (72), which give σ in terms of the intensity or energy density may involve the group velocity.

The above results so far only apply to quasimonochromatic quasiunidirectional radiation. However, with the already-implicit assumption that M and σ depend *linearly* on the field intensity, they are easily generalised to arbitrary distributions of (incoherent) radiation, through the correspondence $I\Delta\omega\Delta\Omega = Uv$, to yield:

$$\sigma = M + \mathbf{1}\,\frac{\rho}{c}\int I\,\frac{\partial n}{\partial\rho}\,d\omega d\Omega \tag{73}$$

where M is now given by:

$$M = -\frac{1}{c}\int I\hat{k}\hat{k}n\,d\omega d\Omega\,. \tag{74}$$

Note that the above formulae apply specifically in an inertial frame in which the material is locally at rest. A discussion of the Lorentz transformation properties of these tensors is given elsewhere [33,37]. The Maxwell tensor M belongs to the Minkowski energy-momentum tensor and therefore transforms as the three-tensor component of a pseudo 4-tensor. Radiative stress-tensors are not in general Galilean invariants in the sense that the transformations give rise to terms that are first order in the material velocity and that cannot be neglected if a consistent treatment of radiation pressure work is required. This is in contrast to thermokinetic stress tensors, including the *generalised thermokinetic stress tensor*

$\sigma + \sigma^0 - M$, which are Galilean invariant.

With the aid of (26) and (31), the earlier result (69-72) can be expressed in terms of the local rms fields $\bar{E}, \bar{H}$ etc yielding the following:

$$\sigma = \bar{E}\bar{D} + \bar{H}\bar{B} + \frac{1}{2}\rho\left[\bar{E}\cdot\bar{D}\,\frac{\rho}{\mathrm{Re}\epsilon}\,\frac{\partial}{\partial\rho}\left(\frac{\mathrm{Re}\epsilon}{\rho}\right) + \bar{H}\cdot\bar{B}\,\frac{\rho}{\mathrm{Re}\mu}\,\frac{\partial}{\partial\rho}\left(\frac{\mathrm{Re}\mu}{\rho}\right)\right]\mathbf{1} \quad (75)$$

which again posesses the same form as that which is applicable to quasistatic fields.

Photostriction and the photocaloric effect

The right hand term on the right-hand sides of each of equations (69)-(72) represents the *electrostrictive* ($\mu = 1$) or *magnetostrictive* ($\epsilon = 1$) contribution to the stress. We suggest the term *photostriction* to apply to the general corresponding effect (attributable to the density dependence of the refractive index) due to radiation in a refracting medium. (However the effect is nearly always pure electrostriction.)

The quantity

$$P - P^0 \equiv -\frac{\rho}{c}\int I\,\frac{\partial n}{\partial \rho}\,d\omega d\Omega \quad (76)$$

is the photostrictive contribution to the pressure which is the change in the material thermokinetic pressure induced by the electromagnetic field of the radiation. The corresponding contributions to the energy E and entropy S within a volume $\int dV$ of material follow directly from the above equations. From equation (67) one obtains:

$$S - S^0 = -V\,\frac{\partial\bar{\varphi}}{\partial T} = V\bar{\varphi}\,\frac{1}{n}\,\frac{\partial n}{\partial T}$$

and hence, using (60-64),

$$E - E^0 = T(S - S^0) = VT\bar{\varphi}\,\frac{1}{n}\,\frac{\partial n}{\partial T}$$

which leads to the general result:

$$E - E^0 = T(S - S^0) = \frac{T}{\rho c}\int I\left(\frac{\partial n}{\partial T}\right)d\omega d\Omega dV \quad (77)$$

The effect whereby an amount of heat $E - E^0$ is transferred to the system when the radiation field is established under isothermal, isochoric conditions shall here be known as the photocaloric effect (on account of its relation to the magnetocaloric and electrocaloric effects).

Another way of presenting the above results is through the general variation of the free energy, F:

$$\begin{aligned}
\delta F &= \delta F^0 + \delta(\bar{\varphi}V) \\
&= \delta F^0 + \bar{\varphi}\delta V + \left[\bar{H}\cdot\delta\bar{B} + \bar{E}\cdot\delta\bar{D} + \delta\bar{\varphi}\Big|_{\bar{B},\bar{D}}\right]V \\
&= \delta F^0 + V\mathbf{M}:\mathbf{e} + V\delta\bar{\varphi}\Big|_{\bar{B},\bar{D}}
\end{aligned} \quad (78)$$

where $\delta\bar{\varphi}\big|_{\bar{B},\bar{D}}$ is the variation in the field free-energy density due to changes in the local properties of the medium, and also in the frequency, (the rms fields $\bar{B}$, $\bar{D}$ being held fixed) ie

$$\begin{aligned}
\delta\bar{\varphi}\Big|_{\bar{B},\bar{D}} &= \frac{\partial\bar{\varphi}}{\partial\rho}\delta\rho + \frac{\partial\bar{\varphi}}{\partial T}\delta T + \frac{\partial\bar{\varphi}}{\partial\omega}\delta\omega + \ldots \\
&= (P - P^0)\frac{\delta\rho}{\rho} - (S - S^0)\frac{\delta T}{V} - \zeta\delta\omega + \ldots
\end{aligned}$$

In a non-isotropic medium (for which equations (29-31) remain valid) when the eigenvalues of ϵ and μ are tensors, the generalisation of this is straightforwardly found to be (writing ϵ_{ik} for $\text{Re}\epsilon_{ik}$ etc):

$$\delta\bar{\varphi}\Big|_{\bar{B},\bar{D}} = -\frac{1}{2}\left[\bar{E}_i\bar{E}_k\delta\epsilon_{ik} + \bar{H}_i\bar{H}_k\delta\mu_{ik}\right] ,$$

where

$$\delta\epsilon_{ik} = \frac{\partial\epsilon_{ik}}{\partial\rho}\delta\rho + \frac{\partial\epsilon_{ik}}{\partial T}\delta T + \frac{\partial\epsilon_{ik}}{\partial\omega}\delta\omega + \ldots \quad , \text{ etc.}$$

Equation (78) and those following provide a generalisation of a certain result of Pitaevskii [25] in justification of the conjecture of Washimi and Karpman [32]. Note that we have kept the term involving the Maxwell stress tensor in (78) as we have not yet invoked any decomposition into "field" and "matter". For reasons already indicated above in connection with the "momentum-of-light problem", this decomposition has to be made with care (see below) if one is to obtain correctly the higher order term involving the "slow" time derivatives.

The ponderomotive force

With the canonical decomposition of the energy-momentum tensor made as described above (see [37]), the material equations involve a total thermokinetic stress tensor $\sigma + \sigma^0 - \mathbf{M}$ having the appropriate transformation properties (eg Galilean invariance). This thermokinetic stress tensor can be separated into a photostrictive part $\sigma - \mathbf{M}$ and an ambient (as in the absence of radiation) part σ^0.

The total radiatively induced force, or *ponderomotive force*, acting on the material comprises an external "radiative" part associated with M, and the photostrictive part associated with σ − M. The ponderomotive force is therefore given in terms of σ. Part of this is the force f^{coll} due to scattering and absorption of radiation by the medium. The remaining "non-collisional" ponderomotive force is given by:

$$f^{pmf} = \nabla \cdot {}^T\sigma - \frac{\partial g}{\partial t} - f^{coll} \tag{79}$$

where

$$f^{coll} = \frac{1}{c} \int I n \kappa d\omega d\Omega$$

is the collisional part of the radiative force acting on the material and g is the Abraham momentum density which is given by

$$g = \frac{1}{c^2} \cdot \int I \hat{k} d\omega d\Omega \tag{80}$$

so that the second term on the right hand side of (79) represents the rate of increase of momentum in the "external" electromagnetic field. On the other hand, the radiative momentum transport is governed by the following equation which follows directly from (28)

$$\frac{\partial m}{\partial t} - \nabla \cdot {}^T M - \frac{1}{c} \int I \nabla n \, d\omega d\Omega + f^{coll} = 0 \tag{81}$$

where m is the Minkowski momentum density,

$$m = \frac{1}{c} \int I \frac{n}{v} \hat{k} d\omega d\Omega. \tag{82}$$

Equation (81) is directly and rigorously (in the context of geometric optics) obtainable from the Liouville equation approach to photon dynamics [33,37]. (This approach moreover provides for a more straightforward — and therefore more reliable — treatment of effects due to motion of the medium and local time-dependence of medium properties.) Hence one finds the "non-collisional" ponderomotive force is given by:

$$f^{pmf} = -\nabla(P - P^0) - \frac{1}{c} \int I \nabla n \, d\omega d\Omega + \frac{\partial}{\partial t}(m - g) \tag{83}$$

in which the third term on the right hand side is "the momentum-of-light" contribution, also sometimes called the "Abraham force" [39,49], which represents the rate of change of the mechanical component of the momentum associated with the radiation. We note that, for quasimonochromatic quasiunidirectional radiation in a non-dispersive medium, this result is equivalent, in the quasistatic limit, to the original Landau and Lifshitz result [1].

Generalisation to time-dependent and moving media

The generalisation of the above formalism to a time dependent medium is achieved by replacing κ in (18-19ff) by

$$\kappa' = \kappa + \frac{\omega}{nc}\left[n\frac{\partial^2 n}{\partial\omega\partial t} + \frac{\partial n}{\partial\omega}\frac{\partial n}{\partial t}\right]\hat{k} + \frac{\dot\omega}{nc}\left[\frac{\partial}{\partial\omega}\left(\omega n \frac{\partial n}{\partial\omega}\right)\right]\hat{k} \qquad (84a)$$

$$= \kappa + \frac{\omega}{2nc}\left[n\frac{\partial^2\epsilon'}{\partial\omega\partial t} + \frac{\dot\omega}{\omega}\frac{\partial}{\partial\omega}\left(\omega\frac{\partial\epsilon'}{\partial\omega}\right)\right]\hat{k} \qquad (84b)$$

where $\epsilon' = \mathrm{Re}(\epsilon\mu)$
and where, for freely propagating radiation, we have from the classical Hamilton's equations of motion [33,37,51]:

$$\dot\omega = \left[\frac{\partial\omega}{\partial t}\right]_k = -\frac{\omega\nu}{c}\frac{\partial n}{\partial t}. \qquad (85)$$

Hence, (84) can also be expressed as:-

$$\kappa' = \kappa + \frac{\omega}{c}\left[\frac{\partial^2 n}{\partial\omega\partial t} - \frac{\nu\omega}{c}\frac{\partial n}{\partial t}\frac{\partial^2 n}{\partial\omega^2}\right]\hat{k} \qquad (86)$$

This substitution has the properties (a) that the radiative momentum equation, (28) in its final form – and hence (81), are, when all time derivatives are properly evaluated in accordance with (17) (taking due care over the ambiguous definition of $\partial/\partial t$ remarked upon prior to equation (13)), recovered unchanged; and (b) that the energy equation, which follows straightforwardly from (18-19) in the case of a time-independent medium,

$$\frac{\partial U}{\partial t} + \nabla\cdot F + \kappa\cdot F + \frac{U\nu}{c}\frac{\partial n}{\partial t} = 0 \qquad (87)$$

is similarly recovered unchanged. These are necessary because equations (28) and (87) are known to be generally valid (in the geometrics limit) for time-dependent inhomogeneous media. For example, they are derived for the general case in [33,36] using the Liouville equation approach to photon dynamics. Both (81) and the corresponding integral form of (87) can be shown to be Lorentz covariant [33,37].

We observe that (84) differs from Pitaevskii's result [25] through the presence of an additional term proportional to $\dot\omega$. This is perhaps not unexpected as Pitaevskii restricts his argument to the case when a constant frequency is imposed on the system in the limit of $k = 0$. However this means that his result is insufficiently general to encompass a treatment of freely propagating radiation ($k \neq 0$) that has ceased to interact with its source. We can regard Pitaevskii's formula as being analagous to the isothermal case (in the sense of being at constant frequency, see above), while freely propagating radiation is "adiabatic" in the sense that it is isolated from the influence of its source. Formulae for the Ponderomotive force that are dependent on Pitaevskii's result, such as are to be found in [32,49 etc], are similarly restricted.

Pitaevskii's formula does have the required property that (with $\omega = 0$) it leaves (28) invariant. However the energy equation is then recovered with the fourth term incorrectly replaced by $(U/n)\partial n/\partial t$. (For a start, this destroys the general Lorentz covariance property of this equation). It will however be noticed that the formulae are in agreement for a non-dispersive medium.

We also note that the essential properties of κ' above are retained if an arbitrary transverse component (perpendicular to k) is added to (84) or (86). However, the longitudinal component is found to be uniquely given by (84). This is the second key result of this paper.

Equations (28,81 and 84 etc) apply generally in an inertial frame in which the properties of the medium may be varying slowly in time. However it is generally necessary to relate the medium properties (esp. κ and n) to the properties calculated as if the medium were at rest. This is best done by evaluating the force locally in the inertial frame in which the material is instantaneously at rest. We call this the local co-moving frame. In a non-relativistic situation one can regard the force thus obtained as a Galilean invariant.

The generalisation of the result (83) to cases when the medium is in motion requires further consideration of equation (79). This is discussed in [37]. It is commonly supposed that equation (79) holds exactly in the local co-moving frame. This turns out not to be the case unless the flow is solenoidal (incompressible flow) or $m - g = 0$ (eg. simple plasma for which $v = nc$). We find that, in the in the local co-moving frame, there is a term proportional to $\nabla \cdot u$ (where u is the material velocity field) which arises in connection with the Abraham force and the associated kinetic radiative momentum density $m - g$. Note that the body forces f are defined in the usual manner such that ρf represents the force acting on a fixed mass of material. Thus it is apparent that a given amount of material can acquire an increasing contribution from the radiative kinetic momentum through expansion. As this momentum is carried by the material, the outcome is that there is an additional contribution to the total ponderomotive force equal to $(m - g)\nabla \cdot u$. Therefore, the total ponderomotive force in the local co-moving frame $(u = 0)$ is:—

$$f^{pmf} + f^{coll} = \nabla \cdot T_\sigma - \frac{\partial g}{\partial t} + (m - g)\nabla \cdot u \qquad (88)$$

which, in an arbitrary (inertial) frame, generalises to

$$f^{pmf} + f^{coll} = \nabla \cdot T_\sigma - \frac{\partial g}{\partial t} + (m - g)\nabla \cdot u + u \cdot \nabla(m - g)$$

$$= \nabla \cdot T_\sigma + \rho \frac{D}{Dt}\left[\frac{m - g}{\rho}\right] - \frac{\partial m}{\partial t} \qquad (89)$$

in which we have invoked mass continuity, and where D/Dt denotes the material derivative $\partial/\partial t + u \cdot \nabla$.

We now invoke the property of Galilean invariance of the ponderomotive force f^{pmf} given by (83) of which the three terms on the right hand side can be seen to be separately effectively Galilean

invariant (GI): the photostrictive pressure $P - P^0$ is GI, while, upon neglecting terms $O(\partial u/\partial t \, (m - g)/c)$ — such terms clearly being $O((m_0 - g_0)/\rho c)$ relative to the total force $\rho \partial u/\partial t$, while also being typically $O(u/c)$ relative to (already small) $\nabla u (m_0 - g_0)$ terms — we find that $(\partial/\partial t)(m - g) = (\partial/\partial t)(m_0 - g_0)$. The implied invariance of the remaining gradient-force term requires that
$\int I \nabla n \, d\omega d\Omega = \int I_0 (\nabla n)_0 d\omega_0 d\Omega_0$ (where $_0$ denotes a quantity related to the co-moving frame). But, if the medium is in a state of non-uniform motion (in which case, in the co-moving frame, $n = n_0$ only at the local point of reference), it is evident that $\nabla_0 n_0 \neq (\nabla n)_0$. Application of the Lorentz transformations for ω and k (and hence n) together with the corresponding transformation of the gradient operator in accordance with

$$\nabla \equiv \left[\frac{\partial}{\partial r}\right]_{\omega, \hat{k}} \quad ; \quad \nabla_0 \equiv \left[\frac{\partial}{\partial r}\right]_{\omega_0, \hat{k}_0} ,$$

while retaining terms only as far as $O(u/c)$, yields [33,37]

$$(\nabla n)_0 = \nabla_0 n_0 + \left[1 - \frac{n_0 c}{v_0}\right] \nabla_0 \left[\frac{u \cdot \hat{k}_0}{c}\right]$$

and hence
$(1/c) \int I_0 (\nabla n)_0 d\omega_0 d\Omega_0 =$

$(1/c) \int I_0 \nabla_0 n_0 d\omega_0 d\Omega_0 + (g_0 - m_0) \cdot \nabla u + (g_0 - m_0) \times \text{curl}\, u.$

With these considerations, the generalisation of (83) is found to be:

$$f^{pmf} = f_0 + \rho \frac{D}{Dt}\left[\frac{m_0 - g_0}{\rho}\right] + (m_0 - g_0) \cdot \nabla u + (m_0 - g_0) \times \text{curl } u$$
(90)

where the subscript $_0$ now denotes a quantity that is computed in the local comoving frame as if the medium were *entirely* at rest in that frame, and f denotes the static-medium ponderomotive force (83) *without* the Abraham force term.

Note that the above results, especially (89), indicate that the Abraham force manifests itself through the velocity of the medium acquiring a "drag" component $(m - g)/\rho$ that is solely attributable to the presence of (non-isotropic) radiation. In this connection it must be remarked that, in the above, the velocity u represents the local velocity of the medium *in the presence of the radiation*. ie. if the radiation were to be instantaneously switched off, then the residual velocity would be $\sim u - (m - g)/\rho$.

One should also remember that, in equation (90), we have neglected terms of order u^2/c^2, $(m_0 - g_0)^2/\rho^2 c^2$ and $(m_0 - g_0)(u/\rho c)$. We emphasise that the additional velocity-gradient-dependent terms appearing in (90) arise *because* of the requirement that the force be effectively Galilean invariant and are therefore clearly not in contradiction of this requirement.

The formulae as expressed by (88ff), represent the third key result of this paper.

Conclusions

We have given a general derivation of the ponderomotive force in a slow time-varying moving medium in the geometric optics limit. Our general result, as expressed *inter se* by equations (81-83 & 88-90) is in broad agreement with earlier proposals that express the force in terms of the Helmholtz stress tensor. However, by careful consideration of the momentum-of-light effect in particular, we have obtained new results for the correction terms aring in respect of slow time variation and non-uniform motion of the medium. (In connection with this we find that it is necessary to be careful to include the effect, on freely propagating radiation, of frequency modulation due to time-dependence of the medium refractive index. This is neglected in previous "definitive" accounts of the problem.) Moreover our final result, as expressed by (89-90), gives a simple interpretation of the Abraham force in terms of a drag velocity imparted to the medium.

This paper contains three key results, namely

(i) Equation (59) which is the relationship between the geometrical strain and the resulting change in the time-averaged electromagnetic field. This leads directly to the stress tensor (69ff) through a standard thermodynamic calculation. The greater part of the paper is devoted to an original proof of this result.

(ii) Equations (84-86) which give the effective attenuation coefficient in terms of the time-variation of the medium properties. This is a generalisation of an earlier result of Pitaevskii which was derived with the implicit restriction that $\omega = 0$.

(iii) Equations (88-90) which are expressions for the ponderomotive force in a (slowly-varying) time-dependent medium in an arbitrary state of non-uniform non-relativistic motion. Reference is made to equation (83) which gives the non-collisional ponderomotive force in a static medium.

References

[1] L D Landau and E M Lifshitz, Electrodynamics of Continuous Media (1st edition, Pergamon Press, 1960) ch.IX.
[2] V S Letokhov and V G Minogin, Phys.Rep. **73** (1981) No.1, pp.1-65.
[3] H Hora, Physics of Laser Driven Plasmas (Wiley, NY, 1981).
[4] J A Stamper, in Laser Interaction and Related Plasma Phenomena, vol.**4B**, edited by H J Schwarz and H Hora (Plenum Press, NY and Lond.,1977) p.721.
[5] H Hora, in Laser Interaction and Related Plasma Phenomena, vol.**4B**, edited by H J Schwarz and H Hora (Plenum Press, NY and Lond.,1977) p.841.
[6] G W Kentwell and H Hora, Plasma Physics **22** (1980) 1043.
[7] B J B Crowley, in proceedings of the 1st International Workshop on Atomic Physics for Ion-Driven Fusion, Orsay, France, March 1983, J.de Physique, Colloque C8, **44**, suppl.to no 11, (1983) 25-38.
[8] H Hora, J.Opt.Soc.Am.**65** (1975) 882.
[9] H Hora and E L Kane, Appl.Phys. **9** (1977) 221.
[10] H Hora and E L Kane, in Laser Interaction and Related Plasma Phenomena, vol.**4B**, edited by H J Schwarz and H Hora (Plenum Press, NY and Lond.,1977) pp.913-939
[11] P Kaw, G Schmidt, T Wilcox, Phys.Fluids **16** (1973) 1522-1525.
[12] R W Short, R Bingham, E A Williams, Phys.Fluids **25** (1982) 2302-2305.
[13] M S Sodha, A K Ghatak, V K Tripathi, in Progress in Optics, edited by E.Wolf (North Holland, 1983) pp.451-517.
[14] J N Elgin, Phys.Bull. **33** (1982) 90-92.
[15] N M Kroll, Phys.Quantum Electronics **5** (1978) 115.
[16] N M Kroll, P L Morton and M N Rosenbluth, IEEE J.Quantum Electron. **QE-17** (1981) 1436.
[17] O Willi and P T Rumsby, Opt.Comm. **37** (1981) 45-48.
[18] V S Starunov and I L Fabelinskii, Sov.Phys.Upekhi **12** (1969/70) 463-489.
[19] P K Kaw, W L Kruer, C S Liu and K Nishikawa, in Advances in Plasma Physics, eds. A Simon and W B Thompson (Wiley, NY, 1976) pp.3-269.
[20] R C Davidson, Methods in Nonlinear Plasma Theory (Academic Press, NY, 1972).
[21] P Mulser, A Giulietti and M Vaselli, Phys.Fluids **27** (1984) 2035-2038.
[22] N Bloembergen, Rev.Mod.Phys. **54** (1982) 685-695.
[23] P Mulser, preprint submitted to J.Opt.Soc.Am (1985).
[24] P Penfield and H A Haus, Electrodynamics of Moving Media (Research monograph No.40, MIT Press, Camb., Mass., 1967).
[25] L P Pitaevskii, Sov.Phys. JETP **12** (1961) 1008-1013.
[26] H A H Boot, S A Self and R B R-Shersby-Harvie, J.Electronics and Control **4** (1958) 434-453.
[27] A V Gaponov and M A Miller, Sov.Phys. JETP **7** (1958) 168.
[28] R Z Sagdeev, Izd.Akad.Nauk. SSSR **3** (1958) 346.
[29] V I Veksler and L M Kovrizhnykh, Sov.Phys. JETP **8** (1959) 781.
[30] M Kono, M M Skoric and D ter Haar, J.Plasma Phys. **26** (1981) 123-146.
[31] G Statham and D ter Haar, Plasma Phys. **25** (1983) 681-698.
[32] H Washimi and V I Karpman, Sov.Phys. JETP **44** (1976) 528-531.

[33] B J B Crowley, Classical Radiation Hydrodynamics in Inhomogeneous Refracting Weakly-Dispersive Media, UK Ministry of Defence report ref: AWRE O-20/82 (1982).
[34] C L Inman, Astrophys.J. **142** (1965) 201.
[35] R M More, J.Phys.A:Math.Gen. **9** (1976) 1979-1985.
[36] M Born and E Wolf, Principles of Optics (5th edition, Pergamon Press, 1975).
[37] B J B Crowley, Classical Radiation Dynamics in Inhomogeneous Refracting Dispersive Media, UK Ministry of Defence report ref: AWRE O-12/84 (in preparation).
[38] H Minkowski, Nach.Ges.Wiss.Gottingen, (1908) p.53; Math.Annaln. **68** (1910) 472.
[39] M Abraham, Rc.Circ.Mat.Palermo, **28** (1909) 1; **30** (1910) 33; Theorie der Elektrizitat (3rd edition, Teubner, Leipzig, 1914) §§38-39.
[40] R E Peierls, Proc.R.Soc.Lond.A, **347** (1976) 475-491.
[41] M G Burt and R E Peierls, Proc.R.Soc.Lond.A, **333** (1973) 149-156.
[42] F N H Robinson, Phys.Rep. **16** (1975) 313-354.
[43] I Brevik, Phys.Rep. **52** (1979) 134.
[44] Y S Barash, Sov.Phys. JETP **52** (1980) 1149-1154.
[45] H M Lai, W M Suen and K Young, Phys.Rev.A, **25** (1982) 1755-1763.
[46] L D Landau and E M Lifshitz, Electrodynamics of Continuous Media (1st edition, Pergamon Press, 1960) ch.X.
[47] G C Pomraning, The Equations of Radiation Hydrodynamics (Pergamon Press, 1973).
[48] D G Frood and B K P Scaife, J.Plasma.Phys. **31** (1984) 319-323.
[49] L D Landau, E M Lifshitz and L P Pitaevskii, Electrodynamics of Continuous Media (2nd edition, Pergamon Press, 1984) ch.IX.
[50] L D Landau, E M Lifshitz and L P Pitaevskii, Electrodynamics of Continuous Media (2nd edition, Pergamon Press, 1984) ch.XIII.
[51] H Goldstein, Classical Mechanics (2nd edition, Addison-Wesley, 1980) ch 8.
[52] M W Zemansky, Heat and Thermodynamics (5th edition, McGraw-Hill, 1968) ch.11, problem 11-2(a).

RAYMOND H. FOGLER LIBRARY
DATE DUE

BOOKS ARE SUBJE